25th Offshore Symposium 2020

Houston, Texas, USA
27 February 2020

ISBN: 978-1-7138-0813-8

Printed from e-media with permission by:

Curran Associates, Inc.
57 Morehouse Lane
Red Hook, NY 12571

CURRAN ASSOCIATES INC.
proceedings
.com

Some format issues inherent in the e-media version may also appear in this print version.

Society of Naval Architects and Marine Engineers (SNAME) Texas Section
99 Canal Center Plaza, Suite 310
Alexandria, VA 22314
USA

Phone: +1-703-997-6701

www.sname.org/texassection

Additional copies of this publication are available from:

Curran Associates, Inc.
57 Morehouse Lane
Red Hook, NY 12571 USA
Phone: 845-758-0400
Fax: 845-758-2633
Email: curran@proceedings.com
Web: www.proceedings.com

TABLE OF CONTENTS

FRONTIER TECHNOLOGIES IN OFFSHORE / SUBSEA SYSTEMS

SNAME
THE INTERNATIONAL COMMUNITY
FOR MARITIME AND OCEAN PROFESSIONALS

Propulsion Solutions for Large Arctic Offshore Vessels

Samuli Hanninen
ABB Inc. Marine & Ports, Houston, Tx

Sampo Viherialehto, Pirjo Maattanen, Torsten Heideman
ABB Oy Marine & Ports, Helsinki, Finland

Proceedings of the 25th Offshore Symposium, February 27th 2020, Houston, Texas
Texas Section of the Society of Naval Architects and Marine Engineers

ABSTRACT

With diminishing multi-year ice, the traffic volumes in arctic waters are growing substantially. Completely new fleet of double acting arctic icebreakers and cargo carriers have successfully proven their ability. The size of vessels in the arctic has also grown after introduction of 172k arctic icebreaking LNG carriers. Azipod propulsion is an attractive solution and makes innovative icebreaking ship concepts technically and commercially feasible. ABB Marine has continued the development of the Azipod propulsion concept to meet the demands of ever larger icebreaking vessels with power requirement up to 20 MW units.

Arctic operation capable vessels are the result of decades of systematic propulsion technology development work, from direct shaft lines to the azimuth thrusters. However, it wasn't until the 2006 that the arctic container vessel, the MV Norilskiy Nickel, equipped with a single 13 MW thruster was commissioned. She was the first vessel able to operate independently without icebreaker assistance even in the worst arctic ice conditions.

With growing size of the Arctic vessels there is evident need for propulsion units with power up to 20 MW and with ice classes up to IACS PC2 or even PC1.

Due to this market demand ABB has decided to take the challenge and commence development program aiming to deliver the biggest Azipod units ever built to meet these requirements. This paper will present some of the key design aspects when developing propulsion system for high power icebreaking offshore application.

New technological solutions will help shipowners to access opportunities in the Arctic areas by enabling safe and reliable operation. This paper describes the development of the Arctic vessel propulsion systems from the conventional fixed pitch propellers connected to the diesel engines to the modern electric podded drive systems.

Keywords: Azipod, icebreaker, ice management, electric propulsion, Polaris

ICEBREAKING AND PODDED PROPULSION

Icebreakers in the historic context

Icebreaking provides the nations of the North an important continuous supply of goods and raw material through the cold season[1]. In the early day icebreakers were steam operated. However, as early as 1933 the first diesel electric icebreakers were introduced[2]. The diesel electric propulsion remained rare outside the icebreaking sphere for decades[3]. In the passing years however, the diesel electric propulsion has gained momentum. Diesel electric powertrain offers many advantages over the traditional mechanical diesel shaft line propulsion. To mention few, the power management speed is increased and the dynamic power handling made more reliable.

Science, cargo transport or Line icebreaking?

In icebreaking the general operations are broadly divided between scientific operations, cargo transport and assistance of merchant marine vessels. Scientific operations are subject to the same environmental conditions as assisting but lack the time pressure found in line icebreaking. Indeed, the distinction between icebreaking in science and assistance of vessels can be made with the difference in requirement for efficiency. In line icebreaking the number of vessels assisted in a unit of time is a key term. Different sets of realities encompass the different icebreaking duties. In some countries the number of icebreaking assistance cases annually is less than hundred[4]. In countries which are dependent on icebreaking the assistance is an intense operation yielding to more than 3000 assistance cases annually[5]. Azimuth thrust increases the efficiency of icebreaking assistance[6]. From operating point of view there are however many similarities between line icebreaking and scientific operations. In both the vessels are sometimes required to force consolidated hummocks and ridges. This operational requirement also extends to icebreaking cargo vessels proceeding independently through ice infested waters.

Responsibility over other vessels

In scientific research the timeframe to commence the voyage is relatively long[7]. This is due to the nature how science is performed through analysis and sampling etc. In cargo transport the charter of goods carried is always to work in a set time frame. There is however a difference between these aforementioned vessels and line icebreaking. A vessel only carrying responsibility over its own operations need a different capability as compared to line icebreaker managing ports and vessels, pilotage and VTS functions all at once. In ship design it is also an important feature to distinguish whether the vessel will be primarily engaged in assisting other vessels or trying to complete a voyage independently.

Azimuth thrust

The Handling characteristics of ships stayed static in many respects until the introduction of the Azimuth thruster[8]. The Azimuth trust enables the ship's main engine power to be directed 360 degrees. This freedom of power vector use has had a profound change to icebreaking[9]. The main categories in Azimuth thrust are the mechanical thursters and the electric gearless thrusters. In the mechanical thruster the electric or diesel motor are located inside the ship's hull. In this configuration the power is transmitted via axles and bearings to the propeller. In the case of electric azimuth thrust, the thrust is generated in the pod unit by an electric motor. The electric pod does not require gears to

[1] (Riska, 2020)
[2] (The first diesel-electric icebreaker in the world, 2019)
[3] (Wärtsilä, 2019)
[4] (Dufour, 2019)
[5] (Huoltovarmuusorganisaatio, 2018)
[6] (Keinonen, 2008)
[7] (Australian Antarctic Division: Leading Australia's Antarctic Program, 2019)
[8] (Azipod® gearless propulsion, 2019)
[9] (Keinonen, 2008)

transfer power. These features have made the electric propulsor to become the leading technology choice in icebreaking[10]. It has also allowed the unit size to grow towards the 20MW range.

Line icebreaking principles

Line icebreaking can be roughly divided in four distinctive tasks. Depending on the traffic direction the sequence varies. Assuming a vessel is in-bound to port from sea, the operation starts with variable sea ice conditions and moving ice fields out at sea. Often the vessel is beset in ice requiring first the icebreaker to cut her loose[11]. With higher wind conditions the ice field also has a significant pressure component preventing vessel movements. Secondly the vessel must be assisted to the fairway area. In the fairway assistance is limited by the draft and buoyage. When the convoy reaches port entrance, icebreaker ends assistance and the port icebreaking assumes responsibility[12]. The azimuth propulsion plays a key role in all the above situations by reducing the time needed for ice management[13].

Sea ice conditions challenging icebreaking

Sea ice conditions are greatly varied and present an ever-changing environment. For icebreaking these offer unique challenges. Sea ice for example in the arctic, is often a mix of drift first year ice with multiyear ice inclusions. The ice out at sea is always moving. Ice movement is by experience ranging from 0,1Kts to 1,1kts[14]. When the ice is packed against a land mass or fast ice pack the combination of wind and ice movement creates ice pressure. This pressure again forms different ice features. In a level ice field the pressure begins to crack and bend the ice creating ridges and hummocks[15]. If the ambient air temperature is well below freezing, the hummocks consolidate and are a very difficult to force. At the ice edge where the swell and wind work against the ice edge the result is different. Here the drift ice is ground into shuga and brash ice barriers. If the temperatures are again well below freezing point the brash ice barrier eventually consolidates into floes very difficult to force[16].

Track creation width

Icebreaker with solid axles creates a track as she forces through the ice. The track width is defined by the vessel breadth, ice quality, ice pressure and direction of travel in regards of ice movement. In fast ice with low ice pressure the track may be 1.5 times the breadth of the icebreaker. In heavy ice pressure the track closes in seconds after the icebreaker has passed. For a merchant vessel to be able to follow the icebreaker there must be sufficiently wide track to follow. As the track closes the speed of the vessel in assistance decreases and finally comes to a stop. To increase the track width the breadth of the icebreaker must be increased. This however has practical upper limits. The power requirements and fuel consumption grow with the size of the icebreaker increasing. The azimuth thrust offers a solution for this problem. The azimuth thrust enables the ships thrust vector to be directed not only aft, but also to the sides of the ship at an angle. This creates a power vector moving the icebreaker forwards, but also creates a wider track. In relative terms this makes it feasible to assist larger vessels and operate in difficult ice fields with increased efficiency. It is often critical to achieve increased track width to be able to conclude assistance without stopping. The azimuth thrust enables increased track width and improves assisting efficiency. Channel widening test results for three ice breakers are presented in Figure 1[17].

---

[10] (ABB, 2019)
[11] (Clare, 2017)
[12] (Clare, 2017)
[13] (Keinonen, 2008)
[14] (Vainio, 2002)
[15] (Vainio, 2002)
[16] (Clare, 2017)
[17] (G. H. Taimuri, P. Kujala. Aalto University; T. Leiviska, Aker Arctic; Pirjo Maattanen, ABB, 2018)

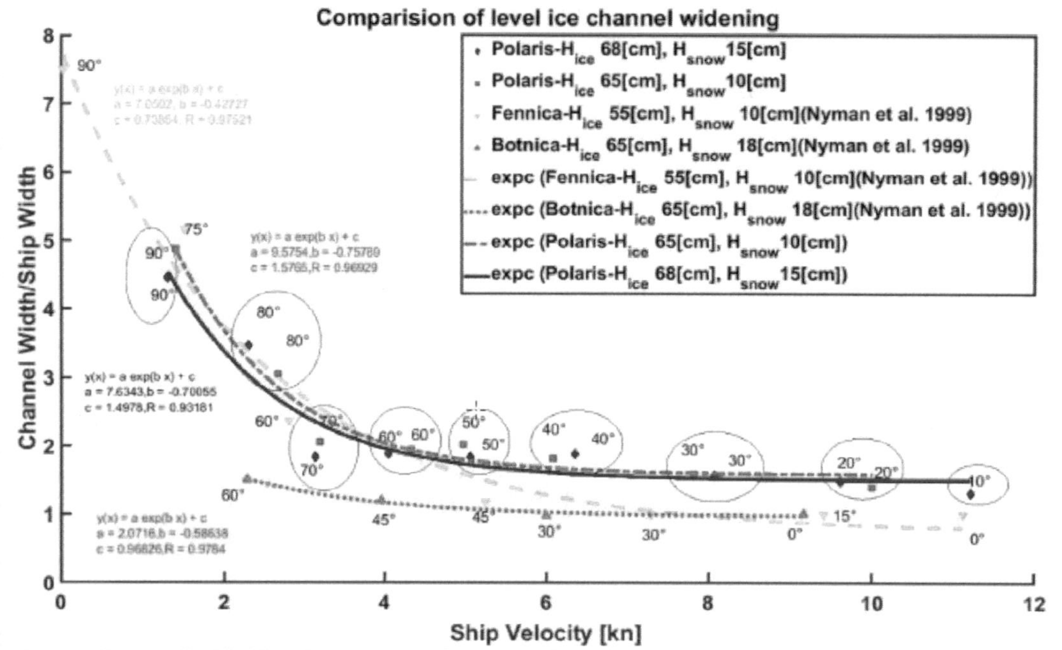

Figure 1. Track creation width in ice trials with vessels equipped with azimuth propulsion at different thrust vector angles (10° to 90°)[18].

A wide channel can also be made by steering the ship in approximately 40 degree angle sideways. This oblique icebreaking is possible with icebreakers equipped with aft and bow mounted azimuth thrusters[19]. Oblique icebreaking results in the ice channel being least twice the width of the ship. Example of wide channel and propulsor steering angles is presented in Figure 2.

Figure 2. Oblique icebreaking test widening the ice channel and thruster orientation manipulation.

Cutting vessels free

Cutting loose a vessel beset in ice requires operations at close quarters. In heavy ice conditions this is done with high power output. Together these contribute to high risk of collision. The Azimuth thrust increases the maneuverability of the icebreaker by a level of magnitude. The turning circles are decreased and control overall is increased[20].

---

[18] (G. H. Taimuri, P. Kujala. Aalto University; T. Leiviska, Aker Arctic; Pirjo Maattanen, ABB, 2018)
[19] (Hovilainen, 2014)
[20] (ABB Inc, 2019)

Ice pressure can create dangerous conditions where the ice floes begin to climb up the side of the ship. In these conditions the icebreaker makes safety circles around the beset vessel to alleviate pressure. In these conditions the increased maneuverability of the azimuth thrust enables making closer and faster safety circles[21].

Between icebreakers with solid axle lines and azimuth thrust the mode of operation is different when cutting ships loose. With solid axles the icebreaker moves along the beset ship at very close quarters and cuts close in front of her bow. At the moment of passing by, the icebreaker instructs the beset vessel to take full ahead. When in front of her bow, the two begin to move as a convoy. The icebreaker leads the way and adjusts the distance between the vessels according to the ice conditions. When the ice pressure increases, or ridges form choke points, the vessel is escorted with very close proximity to the icebreaker. Sometimes only few meters apart. The Azimuth thrust changes these sequences by allowing the icebreaker to pass by and flush the ship free by manipulating the thrust vector towards the beset vessel. Eliminating the need to move in front of the bow, it is also possible to flush loose multiple vessels sequentially. The convoy assistance made with azimuth thrust also differs from the example above. With azimuth thrust the icebreaker can adjust to the prevailing ice pressure and choke points incrementally by spreading the thrust vector. This allows greater distance and added safety between the vessels. This in turn reduces the risk of collision in case of a sudden stop because of unexpected ice feature.

AZIPOD FULL-SCALE ICE LOAD MEASUREMENT

First azimuthing podded propulsors for ice going ships were launched 90's. Due to the novelty of the concept there was no information or guidance about azimuthing propulsor ice loads. Ice loads needed to be measured for verifying dimensioning loads.

First measurements showed, that dimensioning was conservative, as expected. But that does not mean that structural dimensioning methods should be lightened. Ice loads vary greatly depending on ship type and operation area. During the years propulsors in different ice going ships with high ice class – supply vessel, cargo vessel, icebreakers – has been instrumented for ice load measurements. When load calculation and dimensioning methods - including power and propeller torque usage - are verified to cover wide range of propulsors in different ships, it gives solid tools to develop reliable higher power equipment.

Latest ice load measurements were conducted to Finnish icebreaker Polaris propulsors[22,23,24,25]. Ship is equipped with three-pod solution, with 6 MW Azipod unit at the bow and two 6.5 MW units at the stern, Figure 3. It is the first icebreaker with propulsor unit at bow, and excellent possibility to compare ice load distribution between stern and bow propulsors.

[21] (Clare, 2017)
[22] (VTT, VTT-CR-03151-17_Polaris_ice_Trials, 2016)
[23] (VTT, VTT-CR-00587-18_Polaris_season_report_2017, 2017)
[24] (VTT, VTT-CR-04121-18_Polaris_season_report_2018, 2018)
[25] (VTT, VTT-CR-00870-19_Polaris-season_report_2019, 2019)

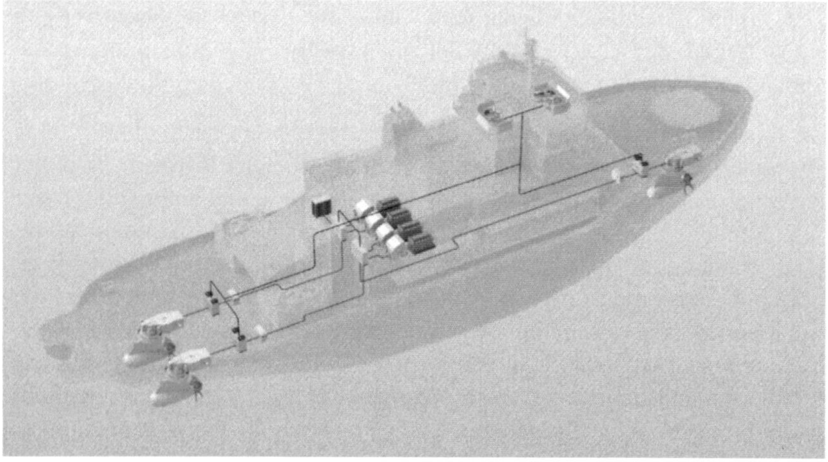

Figure 3. Schematic layout of Azipod arrangement onboard IB Polaris.

Polaris has been in operation at Bay of Bothnia for three winters, now starting the fourth. Operational experience so far indicates that the maximum measured loads for bow and stern propulsors were about 40 % of design loads, Figure 4. It has been also concluded that the maximum loads experienced by bow and stern mounted thrusters are close the same magnitude.

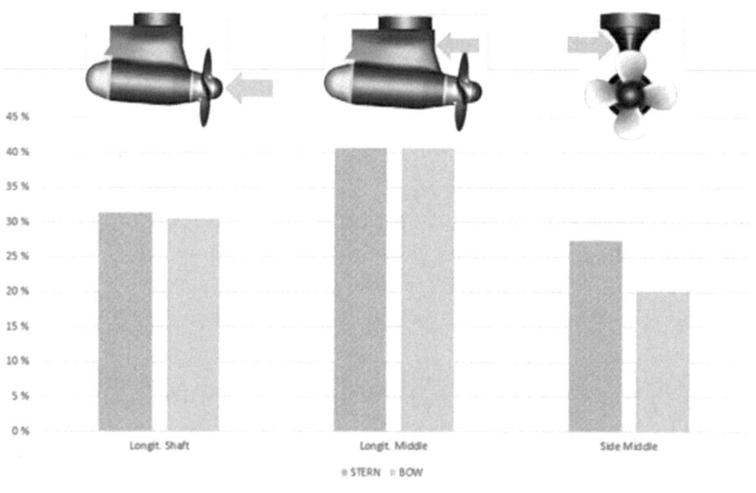

Figure 4. Measured loads compared to design loads

Based on the long-term measurements it can be concluded that Polaris type icebreaker can operate without speed or other operational restrictions in all Baltic Sea ice conditions. This includes ramming operations in ridged ice fields in the Bay of Bothnia.

The development of Polaris was closely followed by other icebreaker operators. After Polaris, Russian giant company Gazprom Neft ordered two icebreakers "Alexander Sannikov" and "Andrey Vilkitsky" based on the similar ship concept. Gazprom Neft operates Noviy Port oil fields on the Yamal peninsula in demanding ice conditions. Supporting the operation close to Arctic loading terminal two highly maneuverable icebreakers were needed and the Polaris concept fitted perfectly. The power was increased to two times 7.5 MW Azipod units in the stern and a single 6.5 MW Azipod unit in the bow. The ice class was also increased to Russian Maritime Register of Shipping Icebreaker 8, which is the second highest RMRS IB class corresponding to IACS PC2. Both vessels were delivered in 2018 and the operational experience have been highly successful.

CONCEPT OF UNIVERSAL ICEBREAKER

The vast operational experience of podded propulsion in icebreakers and ice load measurements as presented above, gives solid background and knowledge to design to build Azipod to the highest PC ice classes.

ABB together with Aker Arctic made a concept study of heavy polar icebreaker concept to take full advantage of a propulsion system with two 17 MW azimuth propulsion units, see Table 1. and Figure 5. This configuration gives significant advantages to vessel design and operational aspects when compared to traditional twin or triple shaft line vessels:

- The podded propulsion units are very compact as opposed to long shaft lines and large propulsion motors inside the vessel. This allows the vessel to be more compact and therefore more economical to build and operate.
- In "all azimuth" propulsion configuration, all machinery spaces are protected by double bottom and sides, improving the watertight subdivision
- Having two propulsion units is advantageous when designing redundant systems.
- Azimuth propulsion allows having the propeller in a hydrodynamically favorable position and designing the aft ship hull form for better efficiency and lower resistance.
- In addition, the polar icebreaker has all the advantages of azimuth propulsion demonstrated, such as superior maneuverability and overall operational capability in both ahead and astern directions.
- The power generation is based on Dynamic AC (DAC) concept with varying frequency for operational efficiency and fuel economy

Figure 5. Universal heavy polar icebreaker concept.

Length, over all	135.3 m (444 ft)
Beam, waterline	25.3 m (84 ft)
Draft, design	10.5 m (34.5 ft)
Depth, to upper deck	16.5 m (54 ft)
Displacement	Abt. 20,000 mt

Deadweight	Abt. 5,000 mt
Propulsion power	2 x 17 MW Azipod
Speed, max	20.7 knots
Icebreaking@3knots	2.44 m (8 ft)
Ice class	PC2 Icebreaker

Table 1. Polar Icebreaker main characteristics.

The concept study confirmed that polar icebreaker with two high power podded propulsion units can operate in first year ice with thickness 2.4 m continuously and is capable penetrate thick Arctic sea ice ridges. As a conclusion, such vessel would be among the most capable polar icebreaker designs ever built. With azimuth propulsion, the maneuverability of the vessel in difficult ice conditions is superior to all existing and proposed polar icebreakers.

To further enhance the Universal Polar Icebreaker Concept ABB and ABS agreed that ABS will perform an Approval in Principal process for the propulsion system and depending on the outcome of the AIP-process ABB will improve and refine the design so that it fulfills the requirements of ABS. This will eventually speed up the delivery schedule of such Azipod units in case they are chosen for an urgent project.

The Universal Polar icebreaker concept presents an exciting opportunity for the industry to take advantage of a variety of technologies and systems that can deliver economic, operational and environmental benefits. It is essential, that these technologies and systems are identified early on and included in the vessel specification. The result will be a state-of-the-art ship that will deliver superior performance well into the future.

NEXT PHASE IN AZIPOD DEVELOPMENT

The changes in ice concentration and ice extent in the Arctic have made possible for the vessel traffic to grow. The effect of climate change is often viewed as the arctic being easy to navigate in terms of ice. This misperception is emphasized by the news portraying new ice minimums around the arctic. In shipping terms, the rather more correct way might be to say that where many operations were not feasible in the past, they now are. This is the result of both the ice conditions lessening in severity and technology moving ahead. The VI series Azipod has been a key factor enabling arctic shipping to grow. A good example is the offshore oil and gas industry in the arctic. To be economically feasible the vessels must be able to operate without continuous icebreaker assistance. The Azipod VI has provided the necessary pollard pull levels, but perhaps more importantly the capability to navigate the vessel stern first through the consolidated ridges and hummocks.

The trend of vessel size and frequency in the arctic is on the growth path. The number of ice classed expedition cruise vessels are on the rise. To meet this increasing demand for high ice class and high tonnage capacity for the Arctic ABB has launched the design of a new Azipod product called "Mörkö". This Finnish word loosely translates to the boogeyman, and not without a reason. Mörkö will be the biggest ever Azipod with power range in excess of 20MW. By Ice class Mörkö will be designed to tackle the harshest ice conditions on the planet. The case for a bigger Azipod unit for icebreaking vessels is also to enable cargo carriage capacity to further expand the Arctic offshore operations as well.

ACKNOWLEDGEMENTS

The authors acknowledge Arctia Oy for the co-operation throughout Icebreaker Polaris measurement campaign and trials enabling long-term measurement and data collection. Warmest thanks to Aker Arctic for Universal icebreaker concept development and ABB team taking part of new product development.

References

ABB. (2019, 12 17). *Azipod® Gearless Propulsion.* Retrieved from Azipod® – State-of-the-art gearless propulsion: https://search-ext.abb.com/library/Download.aspx?DocumentID=3AJM005000005l&LanguageCode=en&DocumentPartId=&Action=Launch

ABB Inc. (2019). Propulsion Technology for Sustainable Polar Shipping. *ASNE Arctic Day.* Washington: American Society of Naval Engineers.

Australian Antarctic Division: Leading Australia's Antarctic Program. (2019, 12 17). Retrieved from Shipping Schedule for 2019/20: https://secure3.aad.gov.au/public/schedules/voyage.cfm

Azipod® gearless propulsion. (2019, 12 17). Retrieved from Benefits of Azipod® propulsion: https://new.abb.com/marine/systems-and-solutions/azipod

Clare, H. (2017). *Icebreaker.* London: Chatto & Windus.

Dufour, S. (2019). Coast Guard Update A/Regional Director, Navigational Programs, CCG, Central and Arctic Region. *Canadian Marine Advisory Council Prairie and Northern Region.* Yellowknife.

G. H. Taimuri, P. Kujala. Aalto University; T. Leiviska, Aker Arctic; Pirjo Maattanen, ABB. (2018). Level ice Clearing in Model and Full Scale Using Azimuthing Propulsion. *Arctic Technology Conference, 2018.*

Hovilainen, M. (2014). Next Generation to Break the Ice -The Oblique Icebreaker. *OTC Arctic Technology Conference.* Houston: Offshore Technology Conference.

Huoltovarmuusorganisaatio. (2018). *Merenkulun huoltovarmuus.* Retrieved from – TOIMINTAYMPÄRISTÖN: https://cdn.huoltovarmuuskeskus.fi/app/uploads/2019/01/14160416/Merenkulun-huoltovarmuus.pdf

Keinonen, A. (2008). Ice Management Experience in the Sakhalin Offshore with Azimuth Icebreakers. *Icetech 2008,* (p. 6). Banff, Canada.

Riska, K. (2020, 1). *UNESCO - Encyclopedia Life Support Systems (UNESCO-EOLSS).* Retrieved from http://www.eolss.net/Sample-Chapters/C05/E6-178-45-00.pdf

The first diesel-electric icebreaker in the world. (2019, 12 17). Retrieved from Scandinavian Shipping Gazette: https://archive.ph/20130809053724/http://www.shipgaz.com/old/magazine/issues/2003/04/oldship_0403.php

Vainio, J. (2002). *Sea Ice Nomenclature.* Helsinki: Merenkulkulaitos.

VTT. (2016). *VTT-CR-03151-17_Polaris_ice_Trials.*

VTT. (2017). *VTT-CR-00587-18_Polaris_season_report_2017.*

VTT. (2018). *VTT-CR-04121-18_Polaris_season_report_2018.*

VTT. (2019). *VTT-CR-00870-19_Polaris-season_report_2019.*

Wärtsilä. (2019, 12 17). *Diesel-electric propulsion.* Retrieved from Wärtsilä: https://www.wartsila.com/encyclopedia/term/diesel-electric-propulsion

THE INTERNATIONAL COMMUNITY
FOR MARITIME AND OCEAN PROFESSIONALS

INTRODCUTION OF AN ECO-FPSO CONCEPT

Jun Zou
DIRECTOR OF NAVAL ARCHITECTURE
HOUSTON OFFSHORE ENGINEERING, LLC
MEMBER OF THE SNC-LAVALIN GROUP
JUN.ZOU@ATKINSGLOBAL.COM

Proceedings of the 25th Offshore Symposium, February 27th 2020, Houston, Texas
Texas Section of the Society of Naval Architects and Marine Engineers

ABSTRACT

A Ship-shaped FPSO is one of most popular hull forms in oil and gas industry. The only alternative hull in operations with minor market shares is the cylindrical FPSO hull. In order to add contractual options and competitions to reduce field development costs, an innovative hull, eco-FPSO, is proposed and presented in this paper.

The proposed hull form shall make it possible to utilize the proven technologies and does not need any step-out applications, technologies and qualifications including equipment limitations, fabrication, transportation, installation, and health, safety and environmental (HSE) considerations.

The development philosophy is to focus on entire life cycle of the FPSO to reduce the CAPEX, OPEX and development risk. The proposed hull form shall meet the following criteria,

- *Simple hull geometry without any moving parts and/or mechanical jacking system*
- *Without scarifying safety with similar ship-shaped FPSO topsides layout and having adequate deck area and separation spaces between topsides modules*
- *No turret (Turretless) system and no thruster system*
- *No need to disconnect and re-connect even in harsh environmental conditions*
- *Low motions and accelerations for*
    - *SCRs*
    - *Sensitive topsides facilities and utilities*
- *Conventional offloading operations*
- *Efficient hull and topsides connection and loading transfer*
- *Good constructability and easy for pre-assembled topsides modules*
- *Good transportability and installability*

Overview of an eco-FPSO hull is proposed and introduced and a case study with 1MM bbls oil storage in harsh environmental condition is described and highlighted. Preliminary analysis results of motions, moorings and risers are presented and discussed. Brief conclusions on the features of the eco-FPSO hull form are drawn.

Keywords: Ship-Shaped FPSO; Cylindrical FPSO; eco-FPSO; Turretless; Low Motions; Low Steel Weight; Low Cost; Constructability, Transportability, Installability

INTRODUCTION

According to a report by Fortune Business Insights 2019, titled, "FPSO Market Size, Share and Global Trend by Storage Capacity (Less than 1 MMBBL, 1 MMBBL - 2 MMBBL, more than 2 MMBBL), by Construction Type (Converted, New-Build, Redeployed), by Ownership (Contractor Owned, Operator Owned), by Water Depth (Shallow Water, Deepwater, Ultra-Deepwater) and by Geography Forecast till 2026, the global FPSO market stead growth is anticipated in the forthcoming years.

Due to long lasting low oil and gas price, the pressure on driving down CAPEX and OPEX remains strong to keep field development economical. FPSOs have some advantages over production semisubmersible platforms, Spar and TLPs, such as, eliminating the need for expensive underwater infrastructure; suitable to work in remote locations; with large storage.

On the global FPSOs market, there are two general types of hull forms as follows,

1) Ship-shaped/barge-shaped hull (to simplify, it is called Ship-shaped in this paper), either with turret mooring (internal or external) or spread mooring
2) Non-ship-shaped/non-barge-shaped hull (to simplify, it is called Non-ship-shaped in this paper) with spread mooring (turretless)

Ship-shaped FPSO hulls are clearly dominated the market and only non-ship-shaped FPSO hull (installed and field proven) is the Cylindrical hull developed by Sevan Marine which occupied the small share of the market. There are many non-ship-shaped FPSO hulls having been proposed and investigated for years as illustrated in Figure 1 as an example (selected, not a complete list).

- SSP hull – developed by SSP Offshore and now owned by Sembcorp Marine (Figure 1a).
- MONOBR – introduced by Masetti et al. (Ref. [17], 2006; Figure 1b). Cylindrical hull is one kind of MPSOs (monocolumn production, storage and offloading system) in principal.
- ULFPSO (Ultra Large FPSO) – introduced by Oliveira, AC and Vilameá, EM (Ref. [18], 2015; Figure 1c).
- LM FPSO hull – developed and owned by INTECSEA (Ref. [14], 2015; Figure 1d).

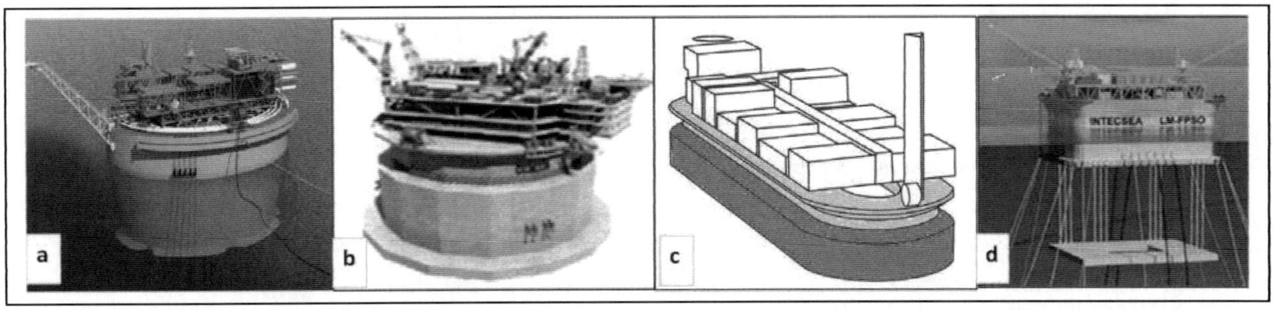

Figure 1: Non-Ship-Shaped Turretless Hulls: a) SSP hull; b) MONOBR; c) ULFPSO and d) LM FPSO

(All images shown in Figure 1 were either copied from corresponding company websites or corresponding papers)

There are large similarities between SSP hull (Figure 1a) and MONOBR (Figure 1b) as well as field proven Cylindrical hull. It is also observed that large similarities of topsides layouts among SSP, MONOBR and Sevan FPSOs which are vastly different from those of conventional ship-shaped FPSOs topsides modularizations for construction and integration philosophy. ULFPSO was aimed to maintain the traditional topsides construction and integration fashion and have good motion characteristics to host SCRs risers in offshore Brazil. LM FPSO hull (Mansour et al, 2015) presented a unique arrangement of hull, a low-hanging solid ballast tank connecting to the bottom of the main hull through 12 tendons and seakeeping of the FPSO was provided by spread moorings. The authors also displayed low motion characteristics of their proposed hull which were suitable for both wet tree and dry tree applications.

There are many features of Ship-shaped FPSOs which are welcomed by industry and holding on domination of the FPSO market for decades. For these merits, we would like to keep them in our innovative eco-FPSO development. As illustrated in Figure 2, the fashion of these pre-assembled topsides modules (PATM) connecting to hull structure is a typical arrangement with some advantages, such as,

- Efficient loading transforms between topsides modules and hull structures
- Constructions of hull and PATMs can be proceeded in parallel to shorten the schedule
- Lifting capability of cranes for modular topsides approach can be lowered to open for more fabrication yards for contractual competitions
- The clearances between topsides modules and hull can provide,
    - Space for piping and equipment
    - Space for accessing hull tank and material handing
    - Space for easing ventilation of mixtures of explosive gases
    - Space for potential green water passage on deck to avoid water impacts on topsides modules,
    - Space as a barrier for accident fire between topsides and hull

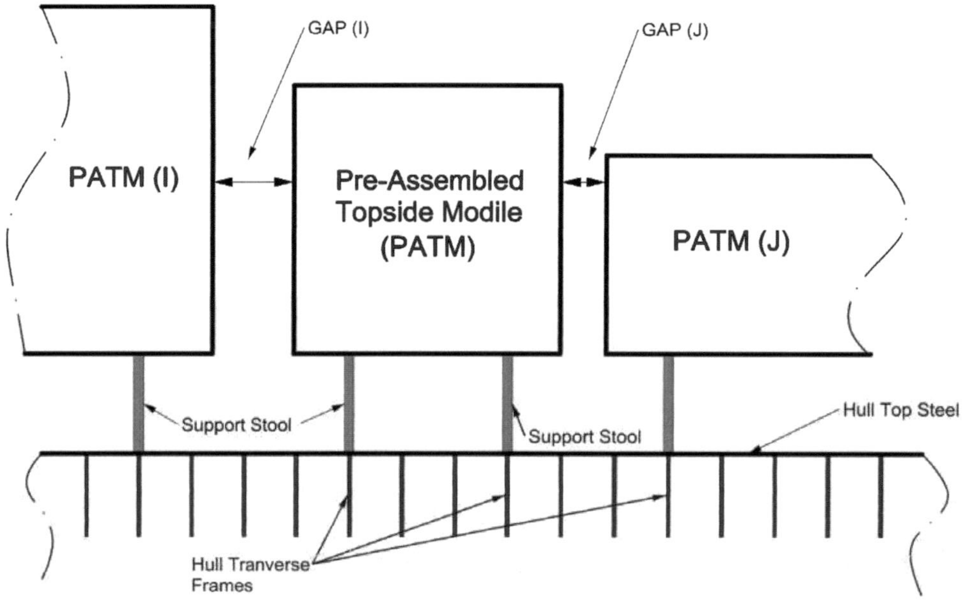

Figure 2: A Typical Schematic Arrangement of Topsides Modules and Hull for a Ship-Shaped FPSO

(For illustration purpose; not to scale)

Since a typical ship-shaped FPSO with length over breath ratios in the range of 4 to 6, the turret mooring is vital to allow the FPSO to weathervane and minimize environmental loads on the vessel and mooring/riser system in harsh metocean conditions. It is known that turret system is very complicated, expensive, long lead item for project schedule and occupying some deck space, as displayed in Figure 3. In order to reduce project cost, complexity and schedule, there are a few turretless options were proposed as displayed in Figure 1.

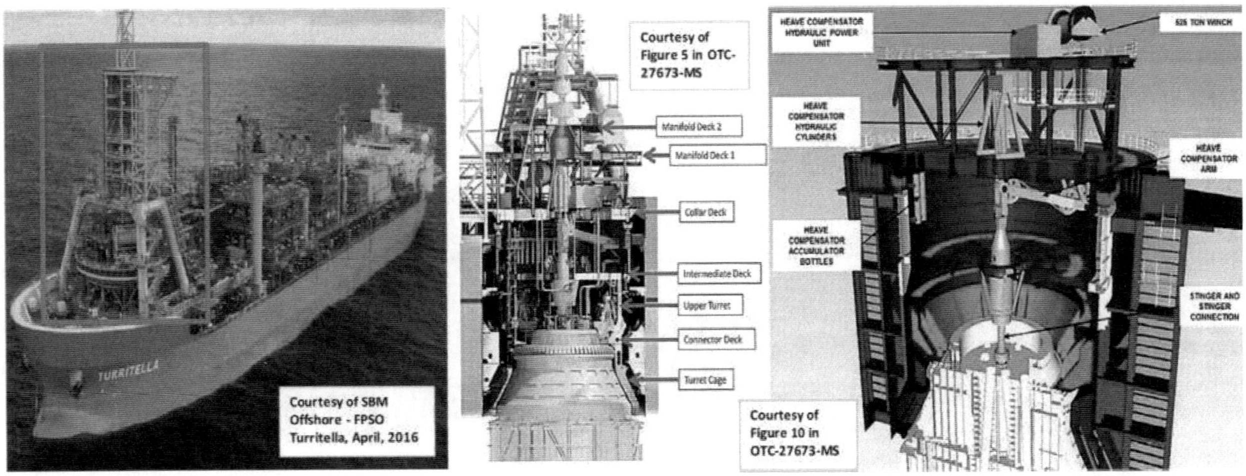

Figure 3: One Example of FPSO Turret System: External Impression – in Red Box of the Left Picture; Internal Impression – Section View in the Middle Picture; Heave Compensation System – Schematic View in the Right Picture

Even SSP (Figure 1a), MONOBR (Figure 1b) and cylindrical Seven hulls are directionless since hull geometry is symmetrical; however, satisfying minimum airgap of these turretless hull forms is a crucial design challenge. The airgap is defined as the vertical distance between the bottom of steel (BOS) of topsides modules to the highest instantaneous water surface (green water, not splash/white water). As indicated in Vidic-Perunovic et al (2017), green water has been observed in wave basin model tests with 100-yrp central GoM hurricane condition (irregular wave with Hs=15.8m). For the ULFPSO hull (Figure 1c), the ratios of length over breath are typical in the range of 3-4 and LMFPSO (Figure 1d), the ratios of length over breath are about 2 or less. The minimum airgap in beam sea heading is anticipated to be governing. As displayed in Mansour et al (2017), the 100-yrp sea state selected for their wave basin model testing was Hs=7.0 m which is less than half of that RPSEA study considered. How to mitigate the design impacts, induced by the pronouncedly enhanced wave upwelling in beam sea heading, for turretless FPSO in harsh environmental conditions, is vital for the proposed concept to be recognized and accepted by Oil and Gas industry. This is one of our ambitious goals set for the innovative hull eco-FPSO to achieve.

Since 2016, HOE started to develop its own version of low cost turretless FPSO hull, named eco-FPSO. The development philosophy is to focus on entire life cycle of the FPSO to reduce the CAPEX, OPEX and development risk. The proposed hull form shall meet the following criteria,

- Simple hull geometry without any moving parts and/or mechanical jacking system
- Without scarifying safety with similar traditional ship-shaped FPSO topsides layout and having adequate deck area and separation spaces between topsides modules
- No turret (Turretless) system and no thruster system
- No need to disconnect and re-connect even in harsh environmental conditions
- Low motions and accelerations for
    - SCRs
    - Sensitive topsides facilities and utilities
- Conventional offloading operations
- Efficient hull and topsides connection and loading transfer
- Good constructability and easy for pre-assembled topsides modules
- Good transportability and installability

The outline of the paper is as follows. First, the brief of an eco-FPSO hull is introduced. Second, descriptions of a case study basis of design are highlighted. Third, the topsides layouts and hull configuration are presented. Fourth, the numerical simulation results, such as motions RAOs, wave upwellings RAOs, extreme motions, extreme accelerations, min airgaps; mooring strength, polyester bottom clearances and chain fatigue analysis; and riser strength and fatigue analysis are presented and discussed; Fifth, offloading plans, offloading rates and shuttle tank capabilities are described and Sixth, constructability, transportability and installability have been addressed. Finally, discussions are given, and main conclusions are drawn.

BRIEF OF AN ECO-FPSO HULL

As displayed in Figure 4, an eco-FPSO hull is presented,

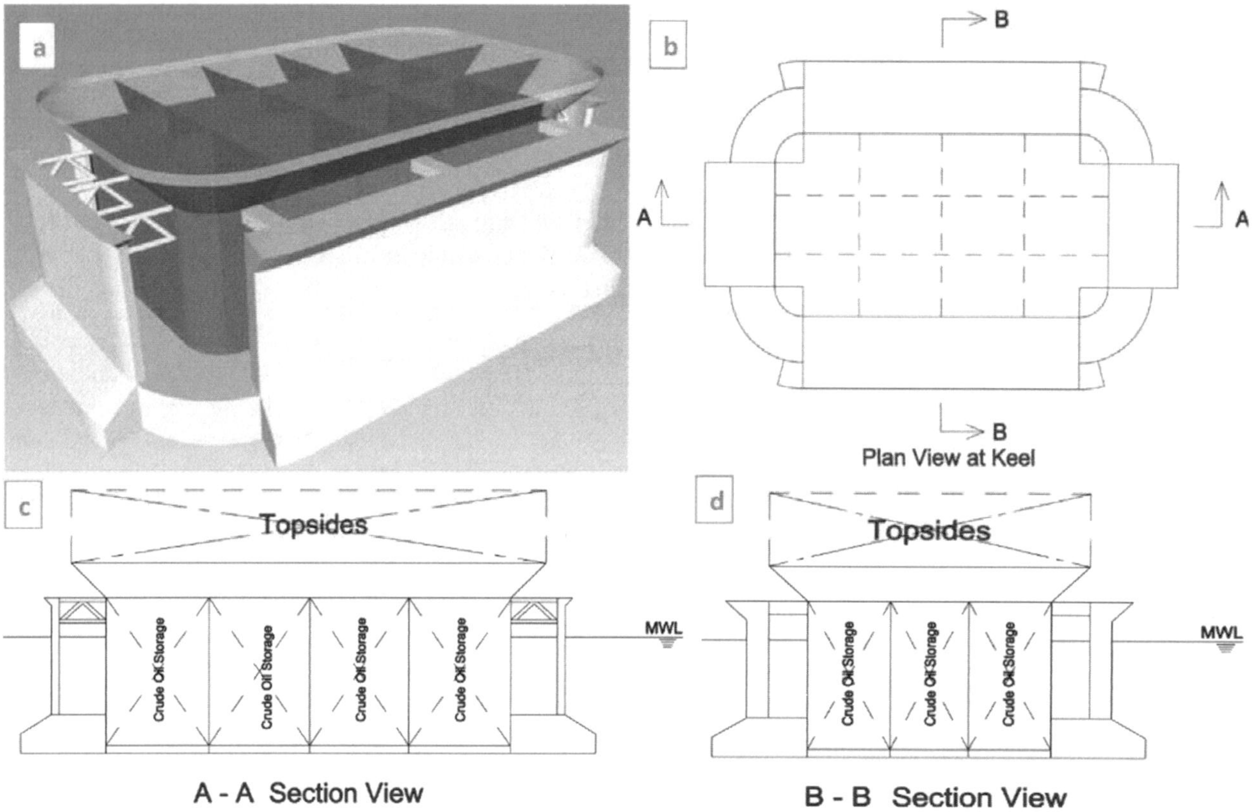

Figure 4: An Eco-FPSO Hull Configuration: a) A Perspective 3D View; b) Plan View at Keel; c) Longitudinal A-A Section View; d) Transverse B-B Section View

In order to get rid of expensive and costly turret system and vessel motion characteristics suitable for SCRs in harsh environmental condition, the proposed hull shall behave like a deep draft semi-submersible platform. Typical natural periods of heave roll and pitch of a shape-shaped FPSO are in the rangers of 9.5 to 15 seconds for heave and pitch motions and 11 to 16 seconds for roll motion. For harsh environmental regions, such as central GoM, North Sea and Offshore Western Australia, those natural periods are right at extreme hurricanes/storms/cyclones wave periods which will result in large and unacceptable motions for SCRs, even SCRs with lazy wave. For the innovative eco-FPSO hull, the preferred heave natural period shall be longer than 22.0 seconds and roll and pitch natural periods are around 30 seconds. In addition to suitable natural periods of heave, roll and pitch, motion response characteristics also shall display similar merits of a deep draft semi-submersible platform.

From Figure 4d, there are two external walls with 3-5m wall depth along longitudinal A-A section view, their functions are highlighted below,

- Holding entrapped water to improve heave natural period

- Increasing GMt effectively since they are farther away from vessel center and with minimum water plane area increment and having quadratic contributions to GMt.
- Effective ballast tanks
- Acting as wave deflectors to mitigate wave upwelling and improving minimum airgap
- Acting as protection walls to avoid direct external collisions with crude oil storage tanks to improve safety

From Figure 4c, there are two external walls with almost no wall depth (just plate with stiffer type structure) along transverse B-B section view, their functions are highlighted below,

- Holding entrapped water to improve heave natural period
- Since nearly no water plane area of these two external walls, thus virtually no increase on GMl for pitch and no water plane area increment on heave
- Acting as wave deflectors to mitigate wave upwelling and improve minimum airgap
- Acting as protection walls to avoid direct external collisions with crude oil storage tanks to improve safety

In order to tune eco-FPSO hull heave, roll and pitch natural periods within the preferred ranges and follow development philosophy, the following steps can be taken per eco-FPSO hull inherited features,

1) Shifting a ship-shaped FPSO hull heave natural period from 9.5-15 seconds to 22.0 seconds or above; massive heave added mass as well as entrapped water are needed with minimum increment on water plane area.
2) The ratio of length over breath of a ship-shaped FPSO hull is in the range of 4-6, pitch stiffness is very large which causes pitch natural period is within 9.5-15 seconds. One effective way is to shorten the length and increase breath to make up the loss of deck area. In addition, tuning hull configuration yields more added pitch moment inertial to the preferred pitch natural period range.
3) The tuning roll motion characteristics of a ship-shaped FPSO are a long-lasting challenging topic. How to invent a FPSO hull's roll motions comparable with a deep draft semi-submersible platform inspires many engineers and researchers to explore. The proposed eco-FPSO hull form offers one solution to be examined and validated by readers. A case study and numerical results will be described and presented in the following sections.

A CASE STUDY

Background

In 2013, RPSEA initiated a project called, "Low Cost Flexible Production System for Remote Ultra-Deepwater Gulf of Mexico Field Development.", Project No. 10121-4404-03. Cylindrical hull developed by Sevan Marine was selected for the study (Vidic-Perunovic et al, 2014). For convenience purpose, major key design parameters summarized in "Sizing Basis" were adopted for current study including central GoM metocean criteria. The suitability of the developed hull form, in general, is not restricted only for GoM application and shall be applicable for all harsh environmental regions, such as North Sea and Offshore Western Australia.

Sizing Basis

For convenient discussions, the key design parameters as well as riser configurations in OTC-25406-MS (Vidic-Perunovic et al, 2014) were reproduced as the following subsections below.

Key Design Parameters

- Topsides production capability: 60,000 bopd
- Storage capability: 1 million bbls
- Design life: 20 years
- Water depth: 2,500 m
- Reservoir pressure: 20,000 psig
- Wellhead shut-in tubing pressure: 13,000 psig
- Reservoir temperature: 240°F
- Metocean Criteria: Central GoM

For global performance, mooring and riser strength analysis, extreme metocean data are based on API-RP-2MET

(2014). Since wave scatter diagrams were not given in API-RP-2MET (2014), the generic GoM wave scatter diagrams were assumed and considered per Zou (2012) for mooring and riser fatigue analysis of this exercise.

Storage Capability

Besides crude oil storage copied from RPSEA study, the requirements of diesel, utility water, potable water and slope were assumed by HOE team. The detailed storages are summarized in Table 1 for references.

Table 1: Fluids Stored in Hull

Items	Volumes	
	(bbls)	(m^3)
Crudel Oil	1,000,000	158,987
Diesel	20,000	3,180
Utility Water	6,500	1,033
Potable Water	6,500	1,033
Slope	8,000	1,272
Total Storage	1,041,000	165,505

Extreme Sea States

Only Selected 100-yrp hurricane wave and wind dominants and 1,000-yrp hurricane wave and wind dominants were reproduced from API-RP-2MET and summarized in Table 2.

Table 2: Summary of 100-yrp and 1,000-yrp Hurricanes

Items		1,000-yr	1,000-yr	100-yr	100-yr
		Hurricane Wave	Hurricane Wind	Hurricane Wave	Hurricane Wind
		Survival	Survival	Design Extreme	Design Extreme
Direction		Omni	Omni	Omni	Omni
Wave		JONSWAP	JONSWAP	JONSWAP	JONSWAP
Significant Wave Height (Hs)	(m)	19.80	18.81	15.80	15.00
Spectral Peak Period	(s)	17.20	17.20	15.40	15.00
Maximum Wave Height (Hmax)	(m)	34.90	33.15	27.90	26.51
Maximum Crest Height (Cmax)	(m)	21.18	20.12	17.04	16.19
Peakedness Parameter	(-)	2.60	2.60	2.60	2.60
Wind		NPD	NPD	NPD	NPD
1 hr @ 10 m Elevation	(m/s)	57.0	60.0	45.6	48.0
Current Profile		Normal		Normal	
		Depth	Velocity	Depth	Velocity
		(m)	(m/s)	(m)	(m/s)
		0	2.40	0	1.92
		-50	1.80	-40	1.44
		-101	0.00	-81	0.00
		-2500	0.00	-2500	0.00

Generic Wave Scatter Diagram

Per Zou (2012), the generic wave scatter diagrams of surface plot and contour plot are copied and denoted as Figure 5 and probabilities of significant wave height Hs and peak period Tp are presented in Figure 6.

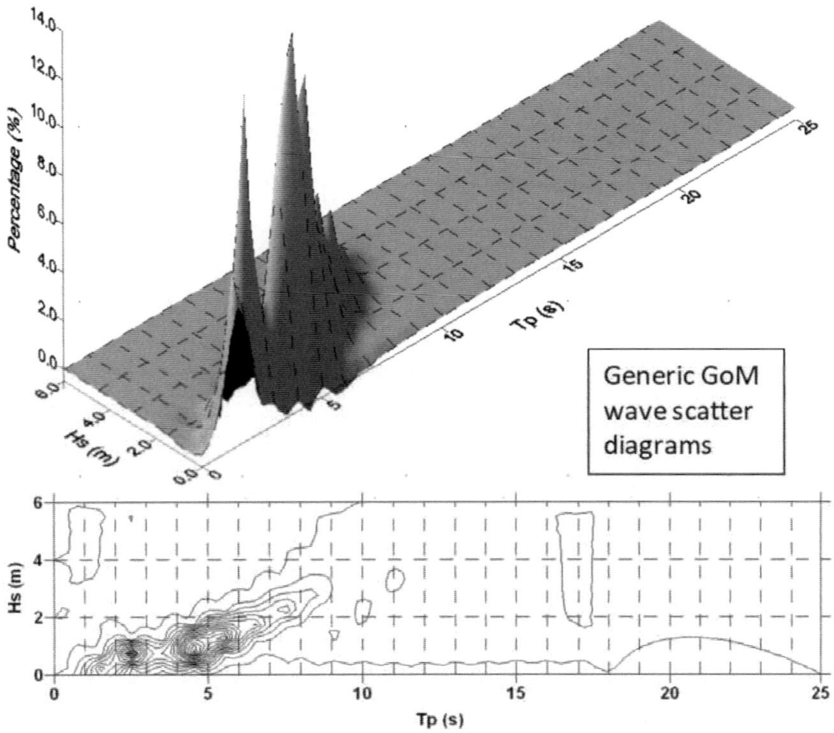

Figure 5: Generic GoM Wave Scatter Diagram: Surface Plot (Top) and Contour Plot (Bottom)

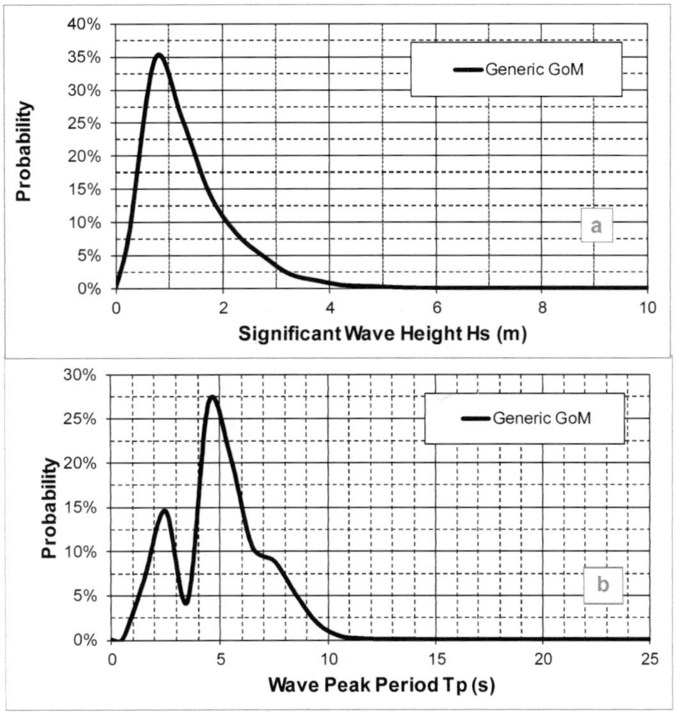

Figure 6: Generic GoM Wave Probability Distributions: a) Significant Wave Height Hs (m); b) Wave Peak Period Tp (sec)

SCR Riser Loads and Field Layout

As an example, production SCR configuration for this exercise is exactly same as the one presented in OTC-25406-MS and summarized in Table 3.

Table 3: Production SCR Configuration Data

Items	Units	Values
Steel Pipe OD	inch/mm	10.75"/273mm
Nominal Wall Thickness	inch/mm	1.614"/41mm
Pipe Material		API X65
Pipe Steel SMYS	ksi	65
Corrosion Allowance	mm	3
Insulation Coating Thickness	inch	3
Insulation Coating Density	lb/ft^3	50
Nominal Hang-Off Angle	deg	10.5

Since there is no data available for water injections, it is our intention to assume the same for a conceptual study in a generic way. The SCR riser horizontal and vertical loads are summarized in Table 4. In Table 4, Positive Fx is towards FPSO forward; Positive Fy is towards FPSO port; Positive Fz points upward. The layout of all SCR risers is displayed in Figure 7.

Table 4: SCR Riser Horizontal and Vertical Loads

Items	Riser Type	Azimuth Angle deg	Departure Angle deg	Horizontal Loads Fx (KN)	Horizontal Loads Fy (KN)	Vertical Loads, Fz (KN)
Prod #1	Production	90	10.5	0	-1099	-5928
Prod #2	Production	270	10.5	0	1099	-5928
Prod #3	Production	285	10.5	-284	1061	-5928
Prod #4	Production	65	10.5	-284	-1061	-5928
WI #1	Water Injection	105	10.5	284	-1061	-5928
WI #2	Water Injection	255	10.5	-284	1061	-5928
		Resultant Loads (KN)		0	0	-35565

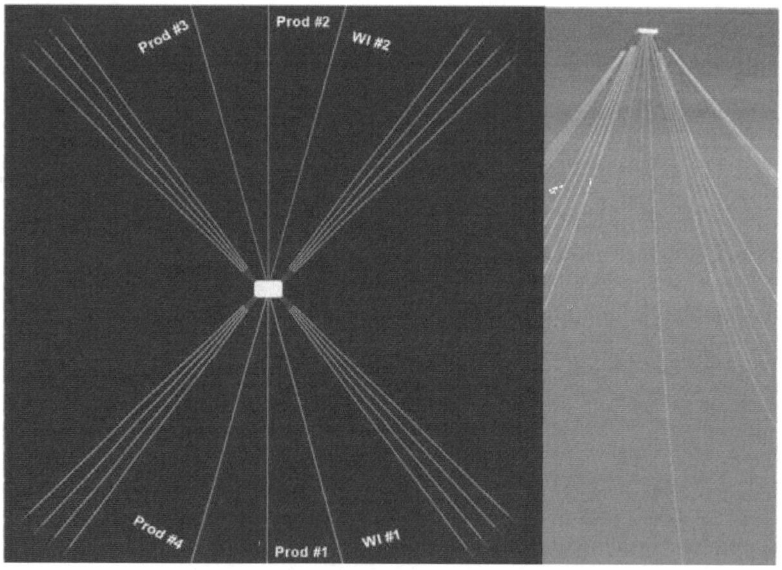

Figure 7: SCR Riser Field Layout and SCR Riser Numbering

TOPSIDES LAYOUT

Topsides Modular Design and Arrangement

As stated before, same as those widely applied in a ship-shape FPSO, topsides modular design and arrangement are adopted for this exercise. Since the breath of our hull is wider than that of similar storage capability ship-shaped FPSO hull, multiple small same functional modules are assembled as one integrated module to maximize the usage of available deck space.

The general philosophy of topsides arrangement is to ensure process and operation safety and efficiency and operation friendly and easy maintenance in service phase and easy construction (modular size and weight) and commissioning in pre-service stage.

Topsides Payloads

Topsides payloads for Base Case in the RPSEA study (Vidic-Perunovic, 2014) are 19,740 mt for dry weights and 25,370 mt for operating weight. For this exercise, topsides weights were assumed to be 20,000 mt for dry weights and 25,500 mt for operating weights.

Offloading Systems

There are two offloading systems considered for this study, the first option is tandem (stern to bow) offloading and the other way is employing SPM. There is one popular offloading way which is called "side by side". Feasibility and suitability will be included for future study. The discussions of offloading systems as well as the suitable operation sea states are presented in one following section of this paper.

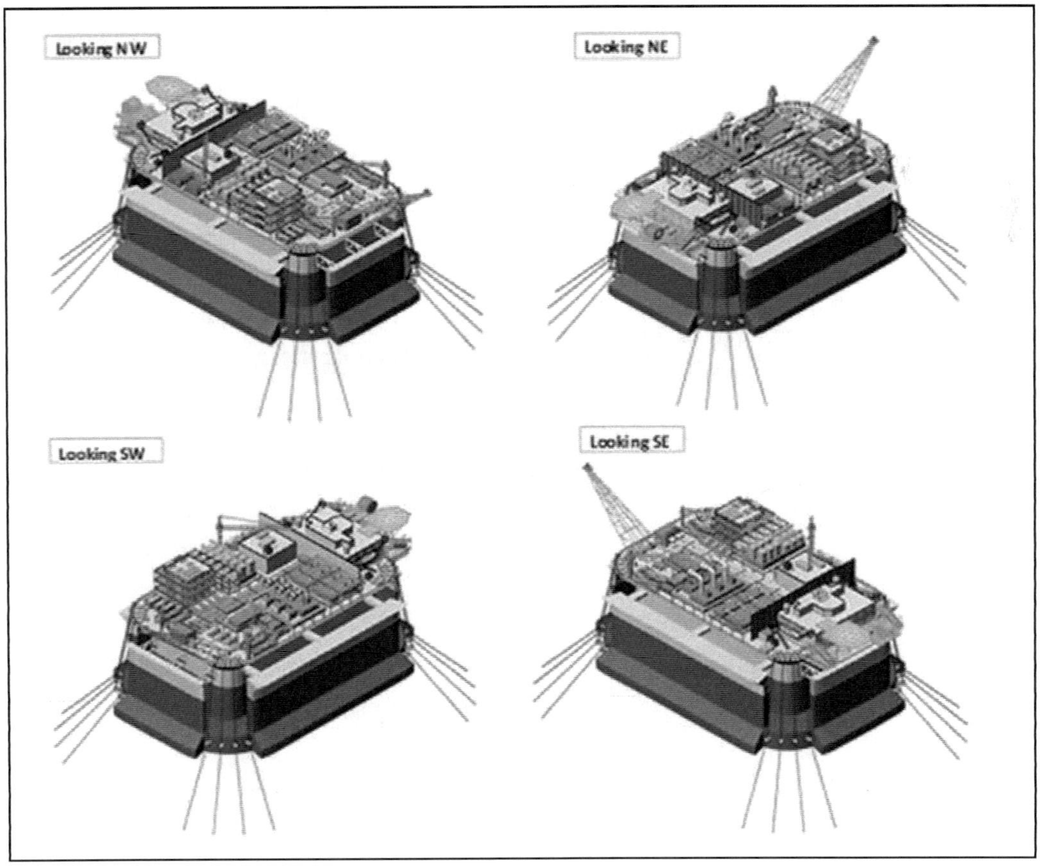

Figure 8: Generic Modular Topsides Layout – Perspective View

Perspective, elevation and top views of modular topsides layouts are illustrated in Figures 8, 9 and 10 respectively.

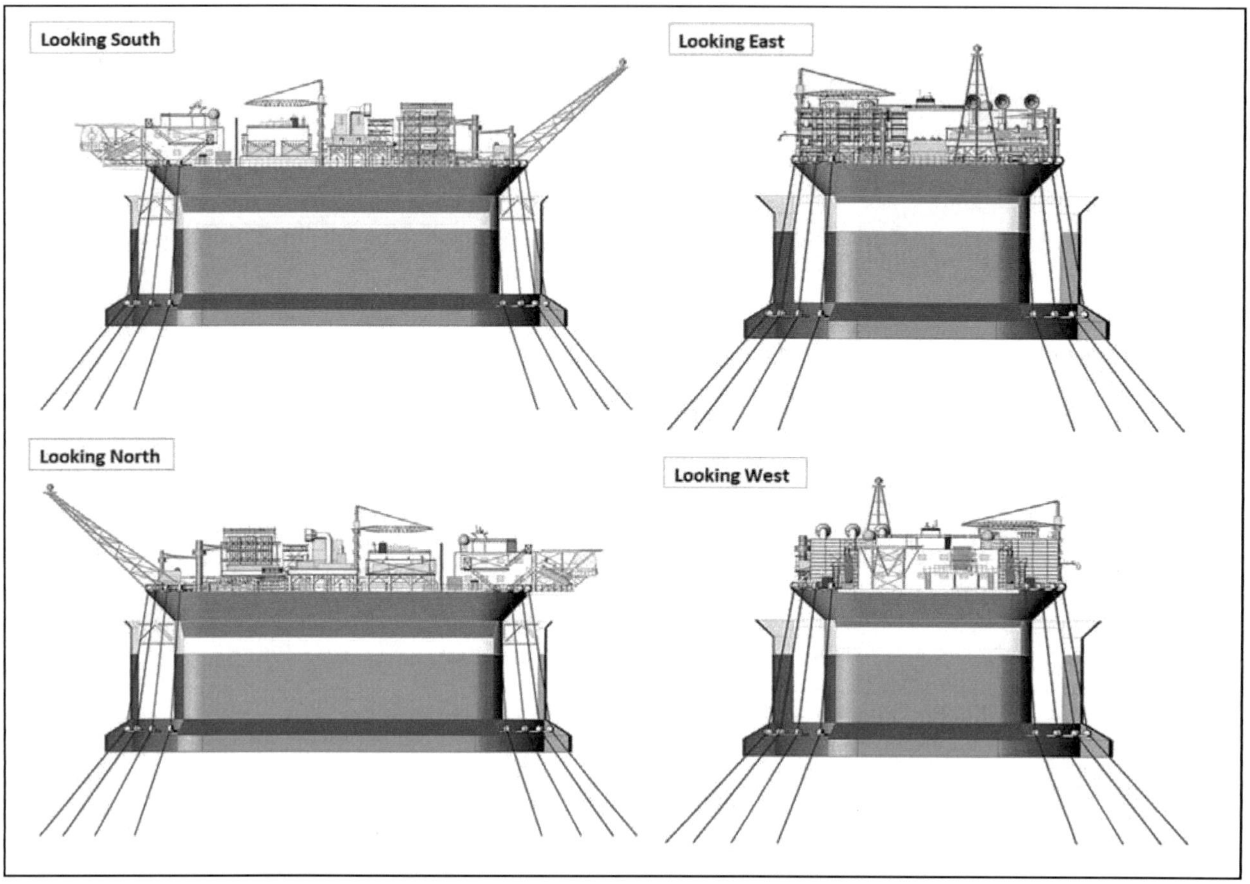

Figure 9: Generic Modular Topsides Layout – Elevation View

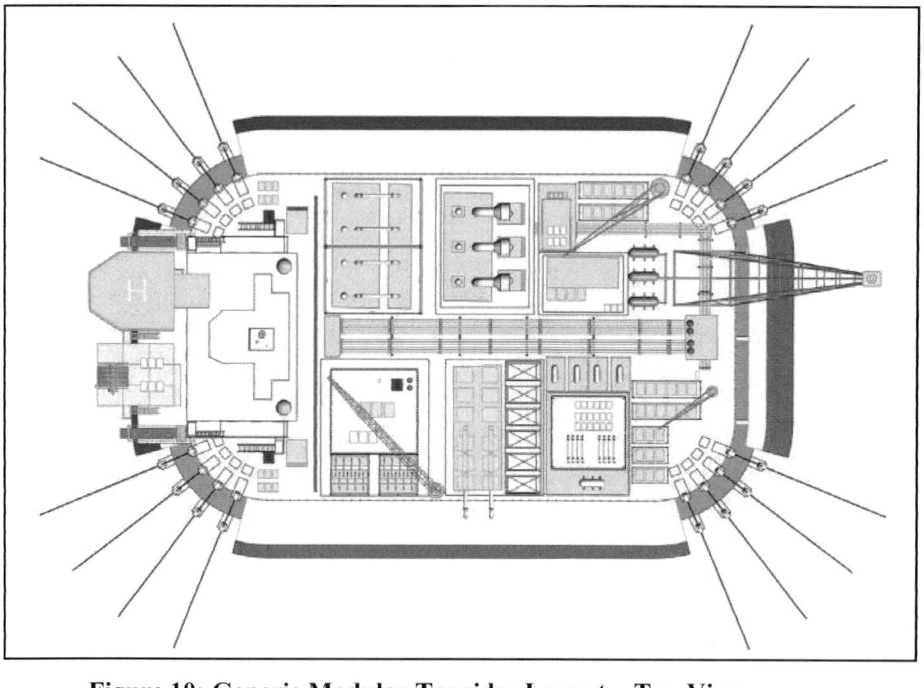

Figure 10: Generic Modular Topsides Layout – Top View

HULL CONFIGURATION

Hull Key Figures

The hull configuration has been illustrated in Figure 4 and key particulars are summarized in Table 5. Since the general philosophy of topsides arrangement is to adopt modular approach, the deck steel is treated as part of hull steel for this study.

Table 5: Hull Key Particulars

Items	Units	Ballast Condition	Fully Loaded Condition
In-place draft	(m)	24.5	30.0
Displacement	(mt)	229,500	267,800
Topsides payload	(mt)	25,500	25,500
Hull structure, fluids storage, marine system and external appurtenance, etc	(mt)	95,900	218,700
Total ballast	(mt)	99,450	15,150
Total vessel weight	(mt)	220,850	259,350
SCRs vertical loads	(mt)	3,650	3,650
Mooring vertical loads	(mt)	5,000	4,800
VCG from keel (mass only)	(m)	17.0	20.3
GMt (with free surface correction & mass only)	(m)	6.8	4.0
GMl (with free surface correction & mass only)	(m)	19.8	15.3
Maximun vessel hull length	(m)	149.0	149.0
Maximun vessel hull width	(m)	100.0	100.0
Nominal deck dimension (LxW)	(m)	125 x 75	125 x 75
Top of bulwark to the mean water line for airgap	(m)	26.5	21.0
Bottom of topsides modules to the mean water line for airgap	(m)	28.5	23.0

In Table 4, SCR riser vertical loads in Table 4 was round-up in Table 5. In addition, there is 4m clearance between the bottom of the topsides module and top of hull deck as illustrated in Figure 2 and bulwark height is assumed 2m tall.

Storage Tanks

Storage tank capabilities of crude oil, diesel, utility water, portable water and slope have been summarized in Table 1. If crude oil storage tank permeability 92% is assumed and the required volume is 1.087 million bbls (172,812 m^3). The permeability 98% has been considered for the other tanks.

Mooring System

There are 16 spread mooring lines in 4 groups and each group has 4 lines. The mooring composition is chain-polyester-chain and taut mooring system. Each line has three main segments, vessel chain, polyester and pile chain. The installation static stiffness (post) and dynamic (storm) of the polyester rope are employed as twelve times (12) and twenty-eight (28) times minimum breaking load, respectively. The mooring line configuration is summarized in Table 6. The line numbering and layout have been illustrated in Figure 11.

Table 6: Mooring Line Configuration

Line Section	Component	Grade	Length	Diameter[1)]	Dry Weight[2)]	Wet Weight[2)]	MBL[3)]	Stiffness
			(m)	(mm)	(kg/m)	(kg/m)	(kN)	(MN)
Upper	Chain	R4S	120	152	462.1	402	20,475	1962
Mid	Polyester	-	3225	266	49.4	12.7	21,575	258.9 (post) 604.1 (storm)
Bottom	Chain	R4S	150	152	462.1	402	20475	1962

Note:

1). The corroded diameter is 144m based on corrosion allowance of 0.4mm/yr.

2). Dry or wet weight is based on the uncorroded diameter.

3). Minimum breaking strength is corresponding to the corroded diameter.

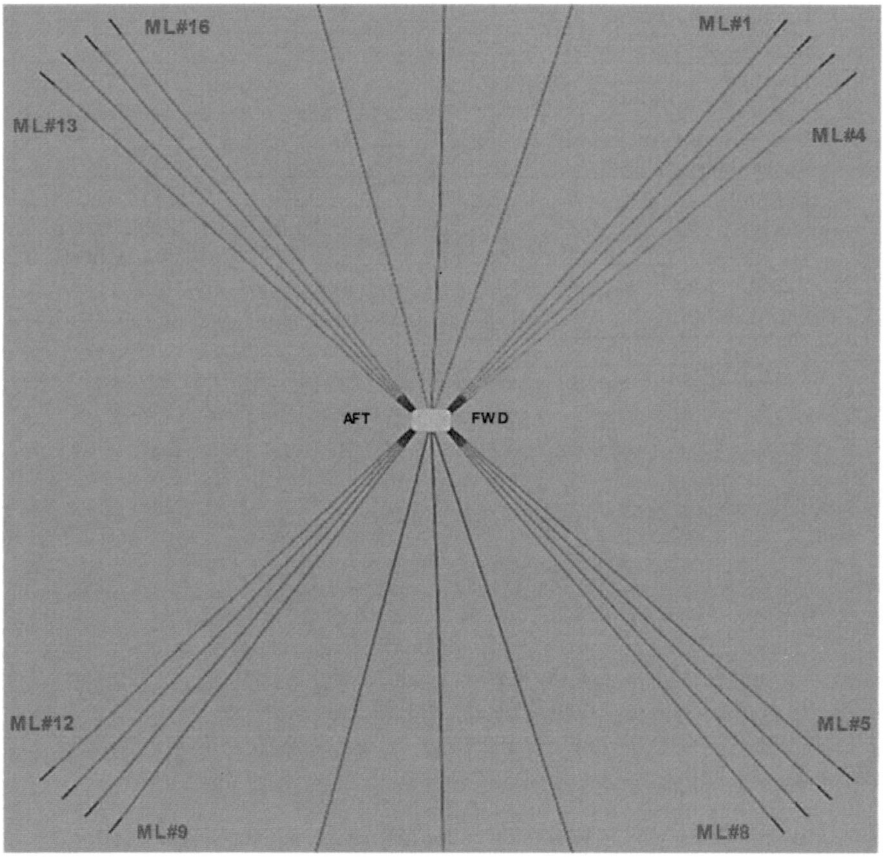

Figure 11: Mooring Numbering and Field Layout

NUMERCIAL RESULTS

Coordinate and Heading Definition

The coordinate and heading definitions are presented in Figure 12 and 13 respectively.

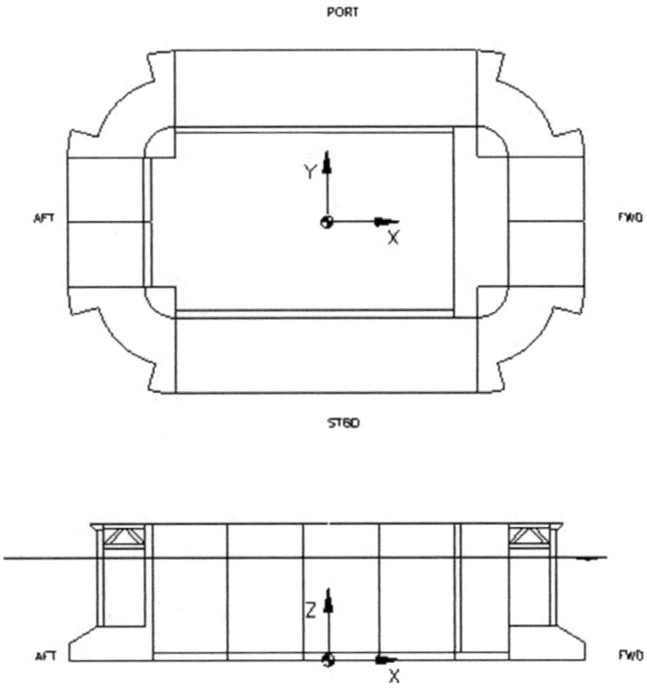

Figure 12: A Schematic Coordinate Definition

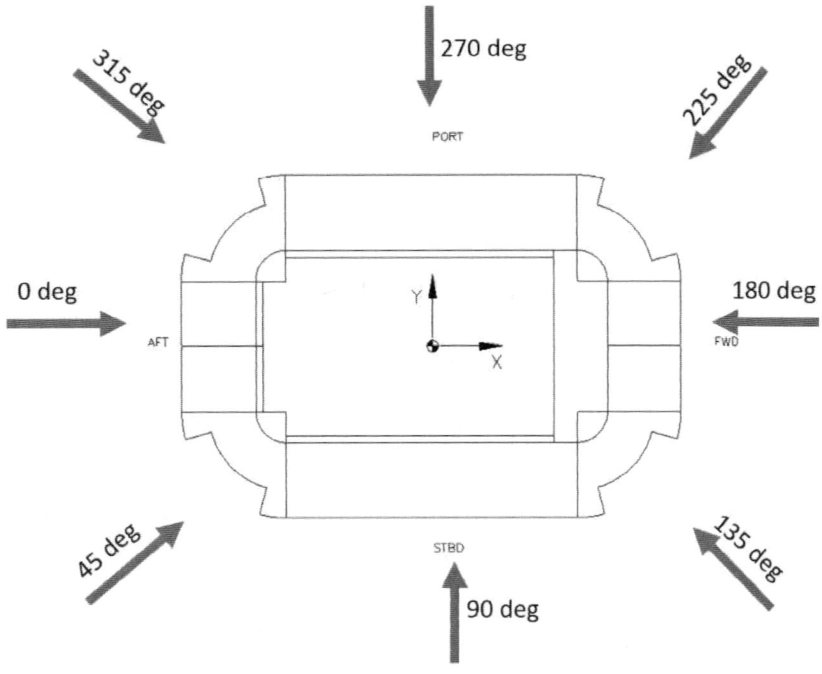

Figure 13: A Schematic Heading Definition

The environmental headings are defined as follows:

- 0-degree heading is defined as incident waves/winds/currents from FPSO aft toward FPSO forward
- 90-degree heading is defined as incident waves/winds/currents from FPSO starboard toward FPSO port
- Other headings shall be determined by rotating with respect to 0 degree heading in counter-clockwise direction as shown in Figure 13.

Diffraction Model

The diffraction model of the proposed eco-FPSO is displayed in Figure 14.

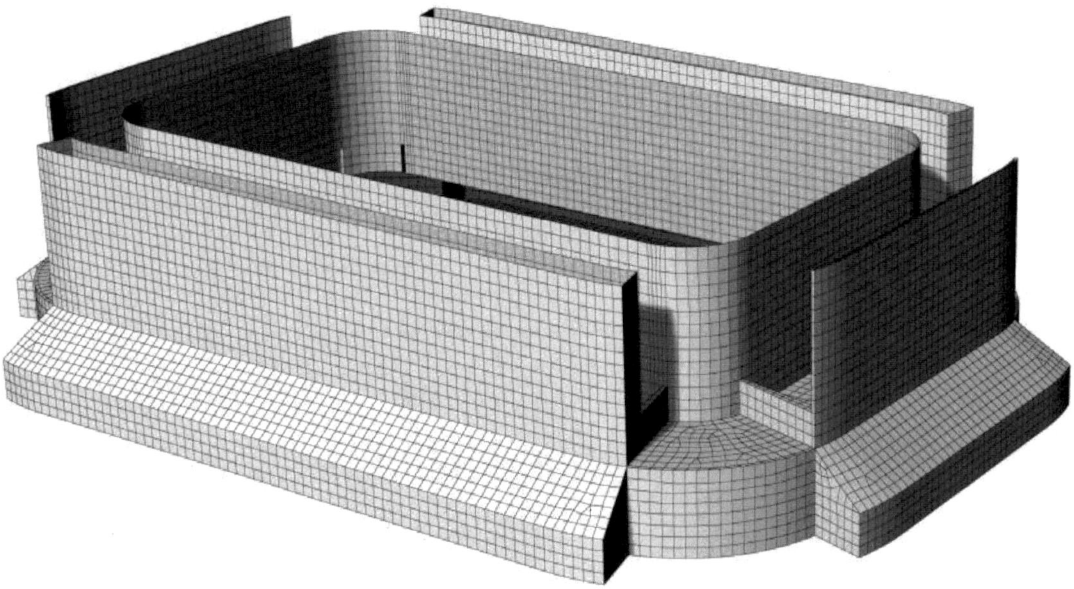

Figure 14: Diffraction Model of the Proposed eco-FPSO

Natural Periods

Natural periods of heave roll and pitch motions of both fully loaded and ballast conditions are summarized in Table 7.

Table 7: Natural Periods of Heave, Roll and Pitch Motions

Items	Units	Ballast Condition	Fully Loaded Condition
Heave	Sec	22.6	22.8
Roll	Sec	34.2	34.8
Pitch	Sec	28.3	28.6

In order to reveal the characteristics of the proposed eco-FPSO hull, comparisons of natural periods of heave, roll and pitch of the conventional ship-shaped FPSO hull in fully loaded and ballast conditions are summarized in Table 8 first. Then comparisons of heave, roll and pitch motion RAOs in fully loaded and ballast conditions are presented and discussed afterwards.

Table 8: Comparisons of Natural Periods of Heave, Roll and Pitch Motions

Items	Units	Ballast Condition		Fully Loaded Condition	
		Ship-Shaped	Eco-FPSO	Ship-Shaped	Eco-FPSO
Heave	Sec	10.0	22.6	12.3	22.8
Roll	Sec	11.3	34.2	14.0	34.8
Pitch	Sec	9.7	28.3	11.0	28.6

Motion RAOs

Heave Motion RAOs

Fully Loaded Condition

Since conventional ship-shaped FPSO with turret system, most of wave headings will be within head sea +/-15 deg and sometimes may be within head sea +/- 30 deg. Therefore, comparisons of heave motion RAOs will be as follow,

a) Head sea: 180 deg
b) Head sea - 15 deg: 165 deg
c) Head sea - 30 deg: 150 deg
d) Beam sea: 90 deg. For the ship-shaped FPSO hull, for illustration purpose; for eco-FPSO hull, it is the most critical heading to display since there is no weather vanning system and it will behavior like production semi-submersible platform and any worst heading is possible.

There are two hurricane sea states considered in RAOs plots as reference spectra acting as indictors for characteristics of vessel dynamic responses, 100-yrp hurricane wave dominant and 1,000-yrp hurricane wave dominant as specified in Table 2, also plotted in Figure 15.

From Figure 15, a few observations are highlighted as follows,

- Even with weather vanning and in most favorable head sea heading (Figure 15a), the ship-shaped FPSO will experience excessively large heave motion in 100-yrp hurricane. This might be one of the reasons why ship-shaped FPSO needs disconnection prior 100-yrp and 1,000-yrp hurricanes.
- For an eco-FPSO, the worst heading for heave motion RAOs is 90 deg as shown in Figure 15d which is equivalently good as a deep draft production Semi-submersible platform. It is obvious that characteristics of an eco-FPSO heave motion RAOs are superior.
- Heave motion RAOs of a ship-shaped FPSO in 90 deg heading are just for reference which will occur only if the turret system is stuck and malfunction. If this situation happens, heave motion will be excessive even in much less severe sea states, such as 10-yrp hurricane or 100-yrp winter storm.

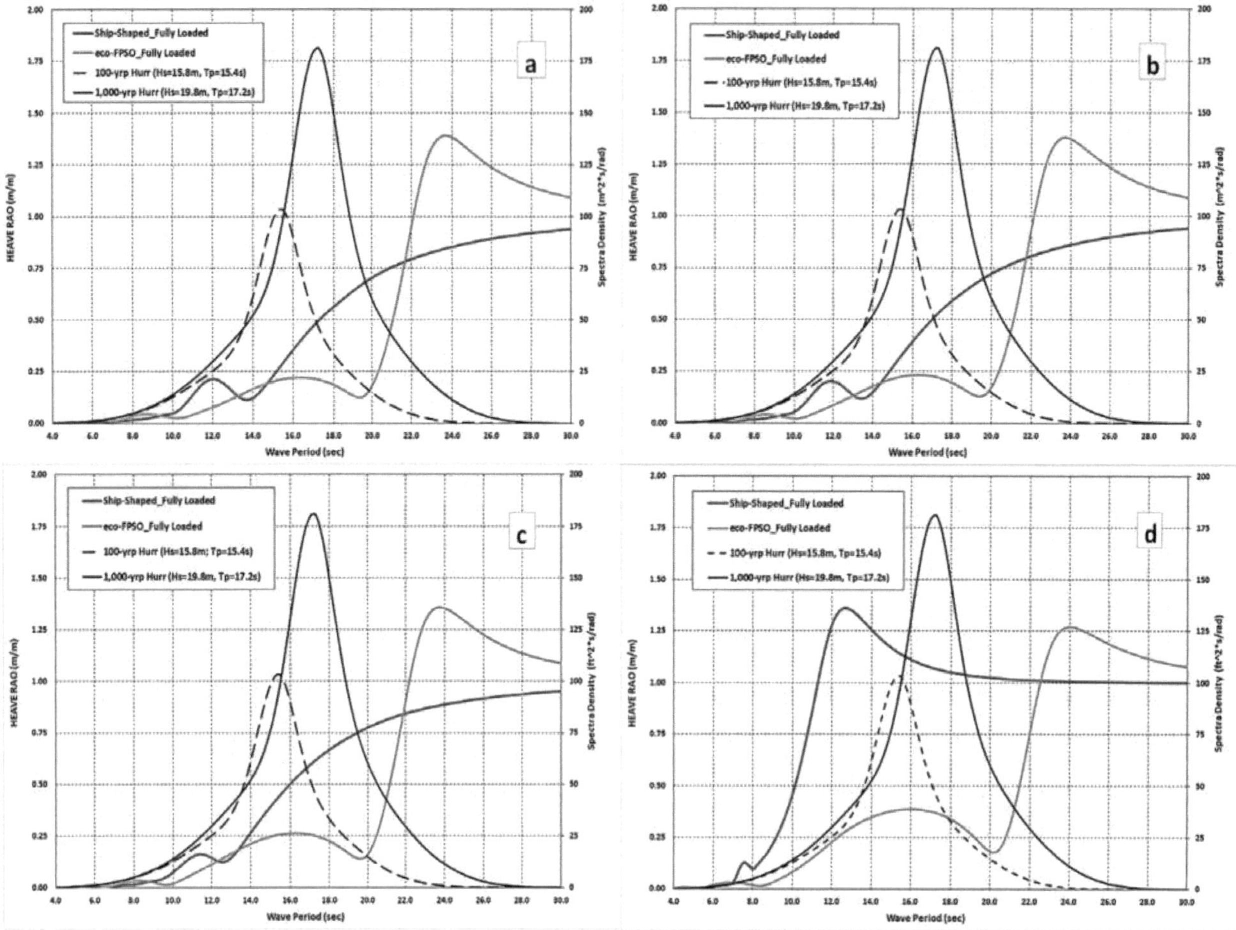

Figure 15: Comparisons of Heave Motion RAOs – Fully Loaded Condition: a) Head Sea-180 deg; b) Wave Heading – 165 deg; c) Wave Heading – 150 deg; d) Beam Sea – 90 deg

Ballast Condition

Like the fully loaded condition, comparisons of heave motion RAOs will be as follow,

i) Head sea: 180 deg
ii) Head sea - 15 deg: 165 deg
iii) Head sea - 30 deg: 150 deg
iv) Beam sea: 90 deg. For the ship-shaped FPSO hull, for illustration purpose; for eco-FPSO hull, it is the most critical heading to present.

Comparison of heave motion RAOs in ballast condition are presented in Figure 16.

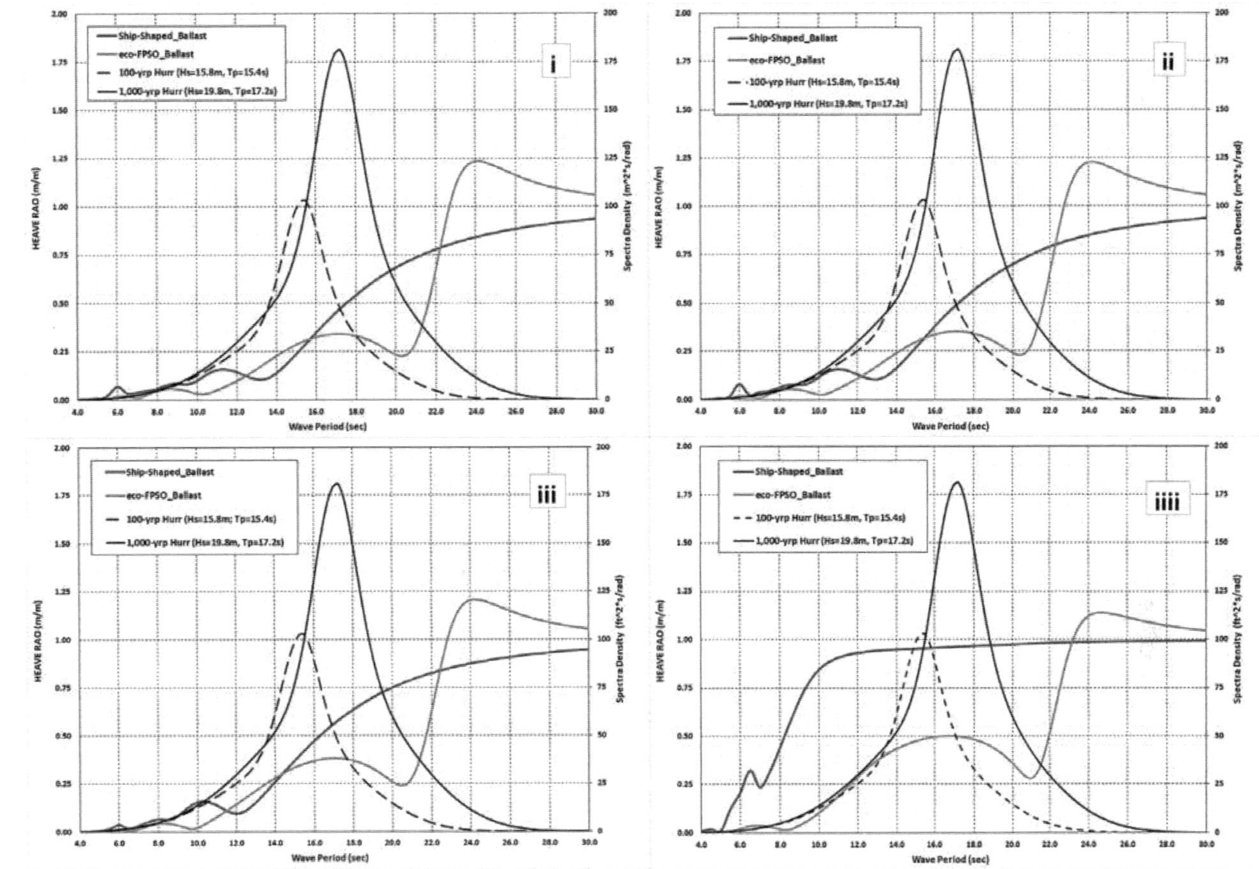

Figure 16: Comparisons of Heave Motion RAOs – Ballast Condition: i) Head Sea-180 deg; ii) Wave Heading – 165 deg; iii) Wave Heading – 150 deg; iiii) Beam Sea – 90 deg

From Figure 16, a few key notes are summarized as below,

- The trends of an eco-FPSO outperform a traditional ship-shaped FPSO in ballast conditions are same as those observed in the fully loaded condition as displayed in Figure 15.
- The heave motion RAOs of an eco-FPSO in ballast condition (Figure 16d) are slightly worse than that of in Fully loaded condition (Figure 15d). Extreme heave motions in various extreme sea states are given in later section.

<u>*Roll Motion RAOs*</u>

Fully Loaded Condition

Theoretically, roll motion RAOs in head sea are zero. Comparisons of roll motion RAOs will be as follow,

a) <u>Head sea - 15 deg</u>: 165 deg
b) <u>Head sea - 30 deg</u>: 150 deg
c) <u>Beam sea</u>: 90 deg. For the ship-shaped FPSO hull, for illustration purpose; for eco-FPSO hull, it is the most critical heading to display

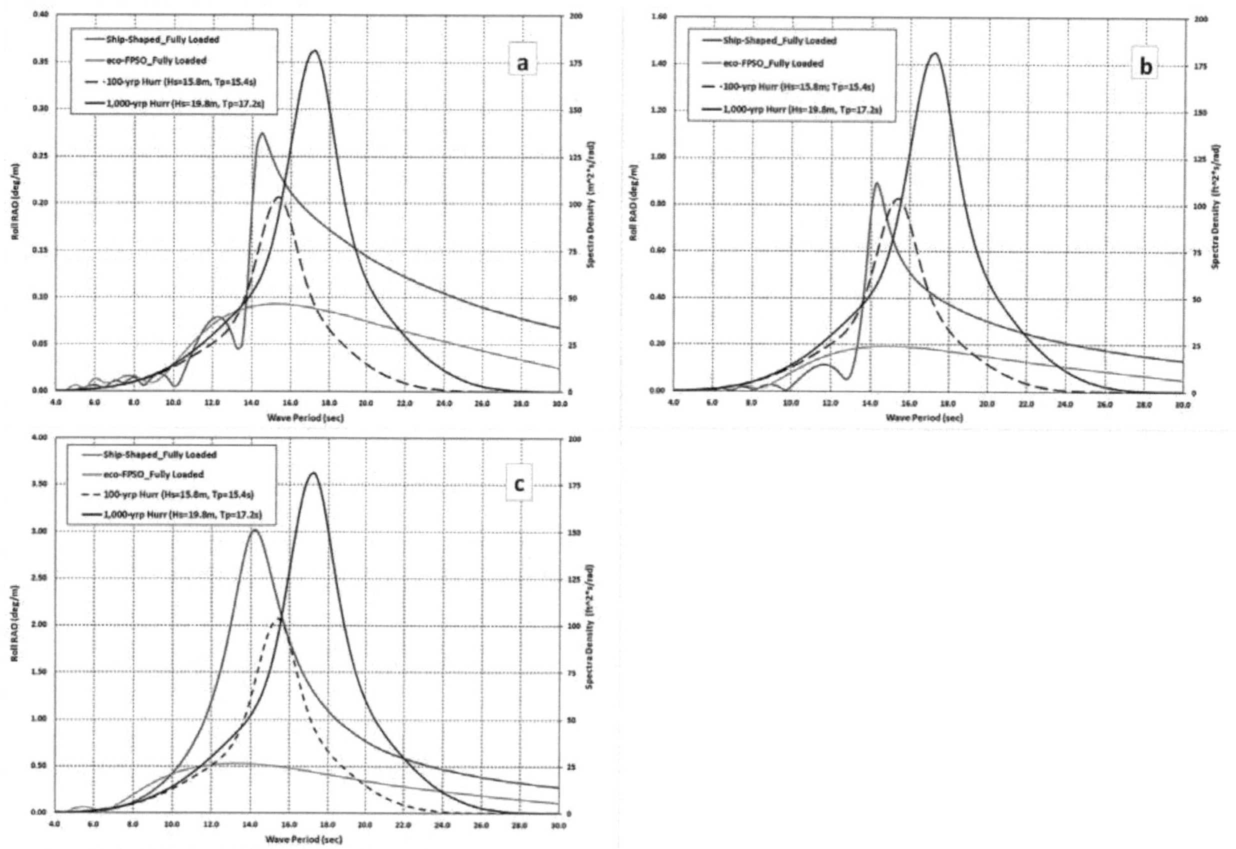

Figure 17: Comparisons of Roll Motion RAOs – Fully Loaded Condition: a) Wave Heading – 165 deg; b) Wave Heading – 150 deg; c) Beam Sea – 90 deg

From Figure 17, a few key observations are presented as follow,

- In theory, there is no roll motion in head sea. However, if there is a small angle, say 15 deg, deviates from head sea as illustrated in Figure 17a, there will be significant roll motions in 100-yrp and 1,000-yrp hurricanes. This will lead to the disconnection requirement even fitted with weather vanning turret system which is the inherited weakness of a ship-shaped hull form.
- If turret system is stuck and wave heading is 90 deg, then the roll motions will be excessive even in a much less serve sea states.
- The worst roll motion RAOs of an eco-FPSO are in 90 deg heading as shown in Figure 17c which are equivalently good as a deep draft semi-submersible platform.

Ballast Condition

Like the fully loaded condition, comparisons of roll motion RAOs will be as follow,

i) Head sea - 15 deg: 165 deg
ii) Head sea - 30 deg: 150 deg
iii) Beam sea: 90 deg. For the ship-shaped FPSO hull, for illustration purpose; for eco-FPSO hull, it is the most critical heading to present.

Comparison of roll motion RAOs in ballast condition are presented in Figure 18.

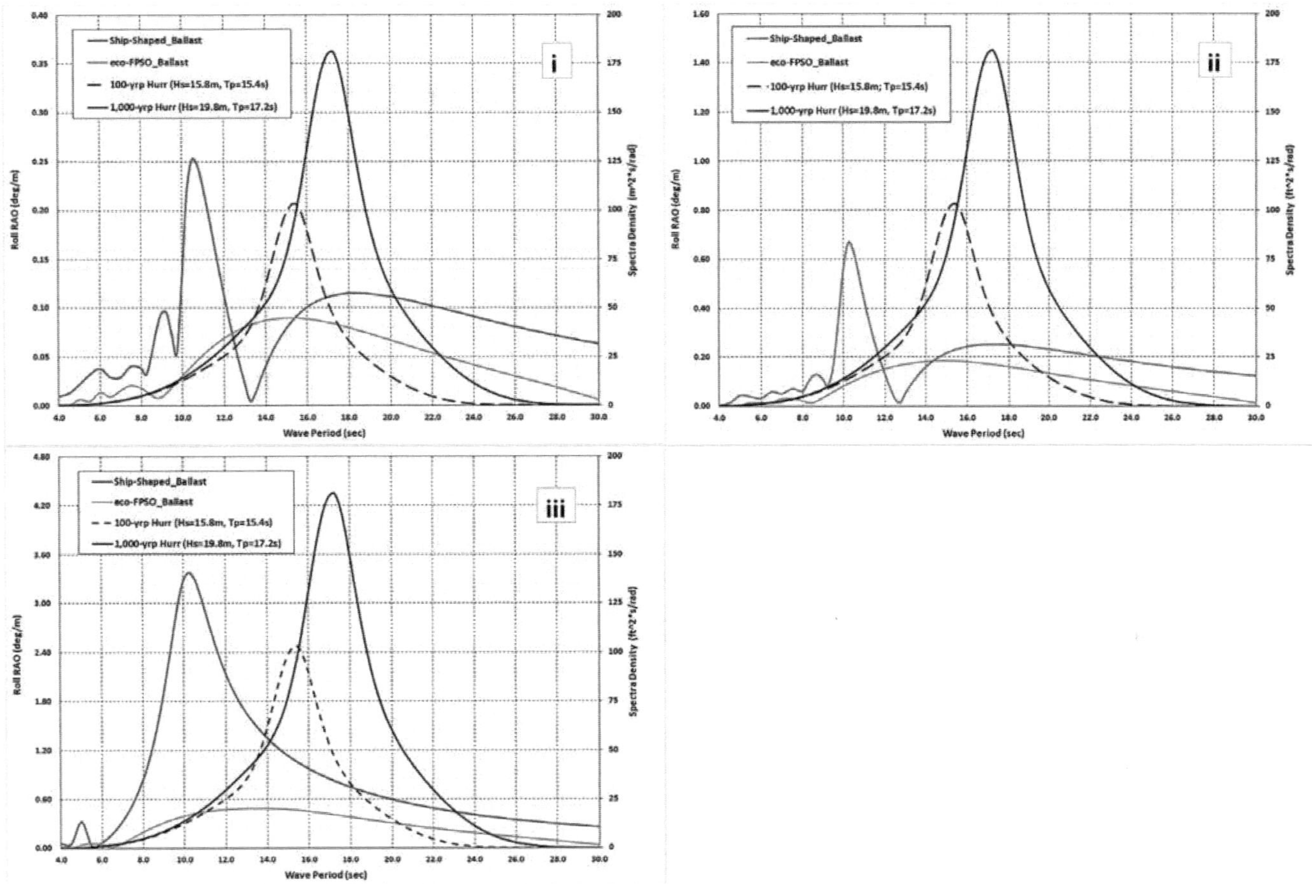

Figure 18: Comparisons of Roll Motion RAOs – Ballast Condition: i) Wave Heading – 165 deg; ii) Wave Heading – 150 deg; iii) Beam Sea – 90 deg

From Figure 18, a few key notes are summarized as below,

- The trends of an eco-FPSO outperform a traditional ship-shaped FPSO in ballast conditions are same as those observed in the fully loaded condition as displayed in Figure 17.
- The roll motion RAOs of an eco-FPSO in ballast condition (Figure 18d) are slightly worse than that of in Fully loaded condition (Figure 17d). Extreme rotation motions in various extreme sea states are given in later section.

<u>*Pitch Motion RAOs*</u>

Fully Loaded Condition

Theoretically, pitch motion RAOs in beam sea are zero. Comparisons of pitch motion RAOs will be as follow,

a) <u>Head sea</u>: 180 deg
b) <u>Head sea - 15 deg</u>: 165 deg
c) <u>Head sea - 30 deg</u>: 150 deg

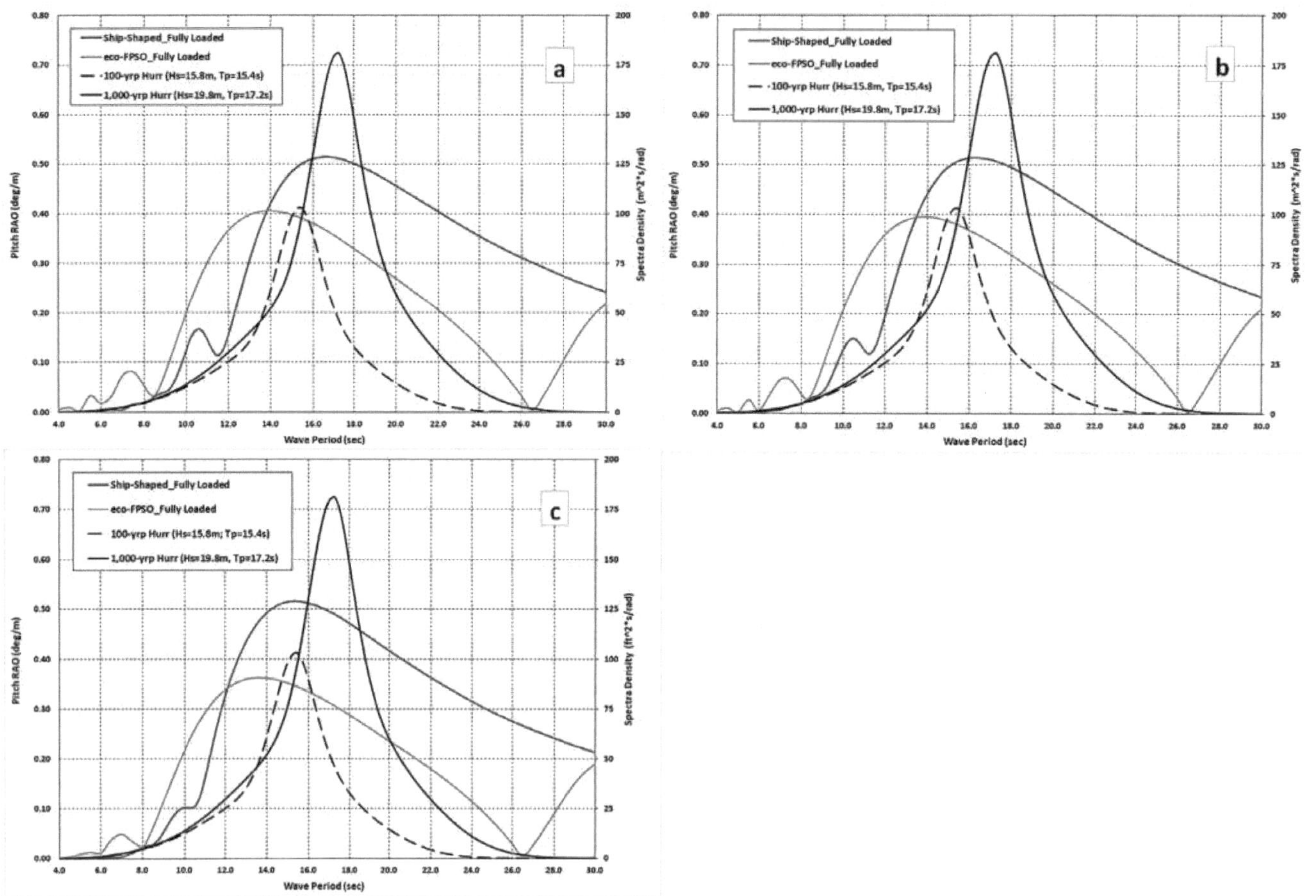

Figure 19: Comparisons of Pitch Motion RAOs – Fully Loaded Condition: a) Head Sea - 180 deg; b) Wave Heading – 165 deg; c) Wave Heading – 150 deg

From Figure 19, a few key notes are presented as below,

- For a ship-shaped hull, the largest peak is about constant as 0.51deg/m in 180 deg, 165 deg and 150 deg headings but the periods corresponding to the peaks are about 17 sec, 16 sec and 15 sec in 180 deg, 165 deg and 150 deg headings respectively
- For an eco-FPSO hull, the largest peak is decreasing from 0.4deg/m in 180 deg, slightly less than 0.4deg/m in 165 deg and 0.36 deg/s in 150 deg headings but the periods corresponding to the peaks are nearly same about 14 sec in 180 deg, 165 deg and 150 deg headings.
- Extreme rotations of an eco-FPSO in various extreme sea states including 1,000-yrp hurricanes are given in the later section which are like those responses of a deep draft production semi-submersible platform.

Ballast Condition

Like the fully loaded condition, comparisons of pitch motion RAOs will be as follow,

i) Head sea: 180 deg
ii) Head sea - 15 deg: 165 deg
iii) Beam sea – 30 deg: 150 deg.

Comparison of pitch motion RAOs in ballast condition are presented in Figure 20.

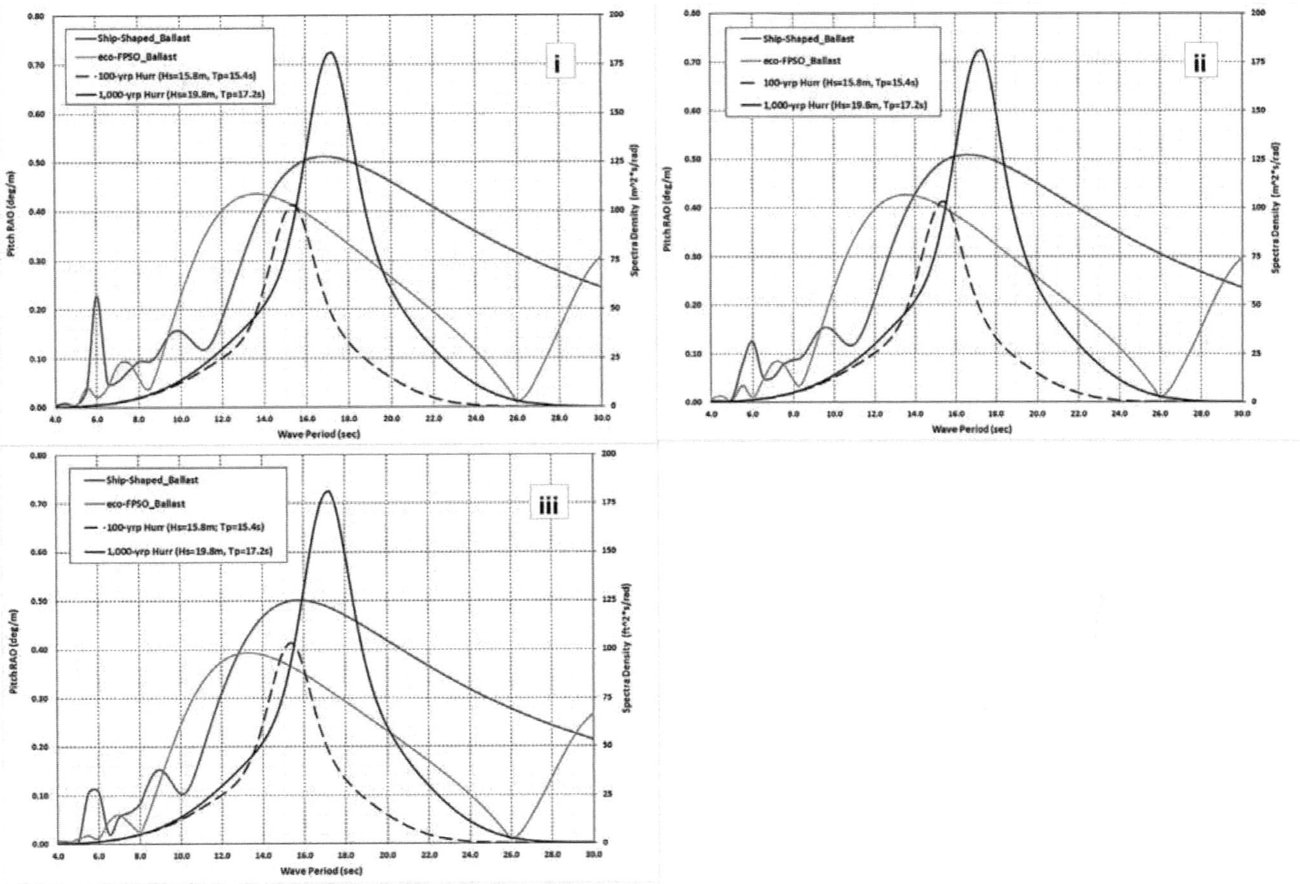

Figure 20: Comparisons of Pitch Motion RAOs – Ballast Condition: i) Head Sea - 180 deg; ii) Wave Heading – 165 deg; iii) Wave Heading – 150 deg

From Figure 20, a few key notes are summarized as below,

- The trends of an eco-FPSO outperform a traditional ship-shaped FPSO in ballast conditions are same as those observed in the fully loaded condition as displayed in Figure 19.
- The pitch motion RAOs of an eco-FPSO in ballast condition (Figure 20d) are slightly worse than that of in Fully loaded condition (Figure 19d). Extreme rotation motions in various extreme sea states including 1,000-yrp hurricane conditions are given in later section.

Wave Upwelling RAOs

As stated in the development philosophy, eco-FPSO hull is turretless concept since turret system is very expensive, very complicated, occupying 10-20% deck space and long lead item. However, if the weather vanning system function is taken away, green water in beam sea is becoming to be an avoidable issue to be solved. In the previous reported turretless FPSO concepts, it hasn't been noticed, to the best our knowledge, there is no any attempting mitigation plan so far. In this study, it is the first time for us to propose an ambitious method to mitigate green water issues for a turretless FPSO hull.

As illustrated in Figure 4, there are 4 external vertical walls. Two are along longitudinal direction while the remaining two are along transversal direction. One of the unique functions of these vertical walls is to act as wave deflectors to mitigate green water issues and reduce still water clearance requirements on minimum airgaps.

There are 18 points distributed in one quadrat as illustrated in Figure 21,

- 9 points in red along longitudinal direction
- 5 points in blue along transversal direction
- 4 points in green along curved region

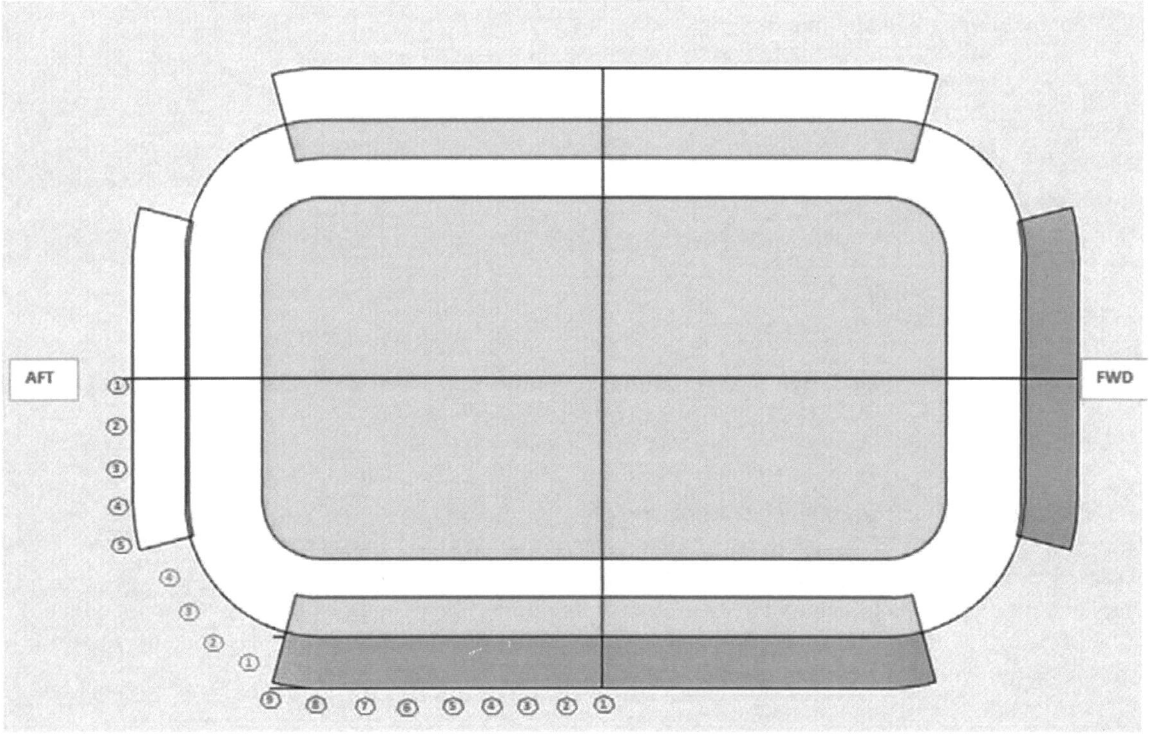

Figure 21: Schematic Arrangement of Point Distributions for Minimum Airgap Analysis

There are 7 headings starting from 0 deg to 90 deg with 15 deg increment having been screened to search for the most critical minimum airgap and corresponding locations. Fully loaded condition is selected for this exercise. It has been found that most critical minimum airgap occurs at point 1 in red in beam sea (90 deg heading). The wave upwelling RAOs at point 1 in red in beam sea are displayed in Figure 22 for discussions.

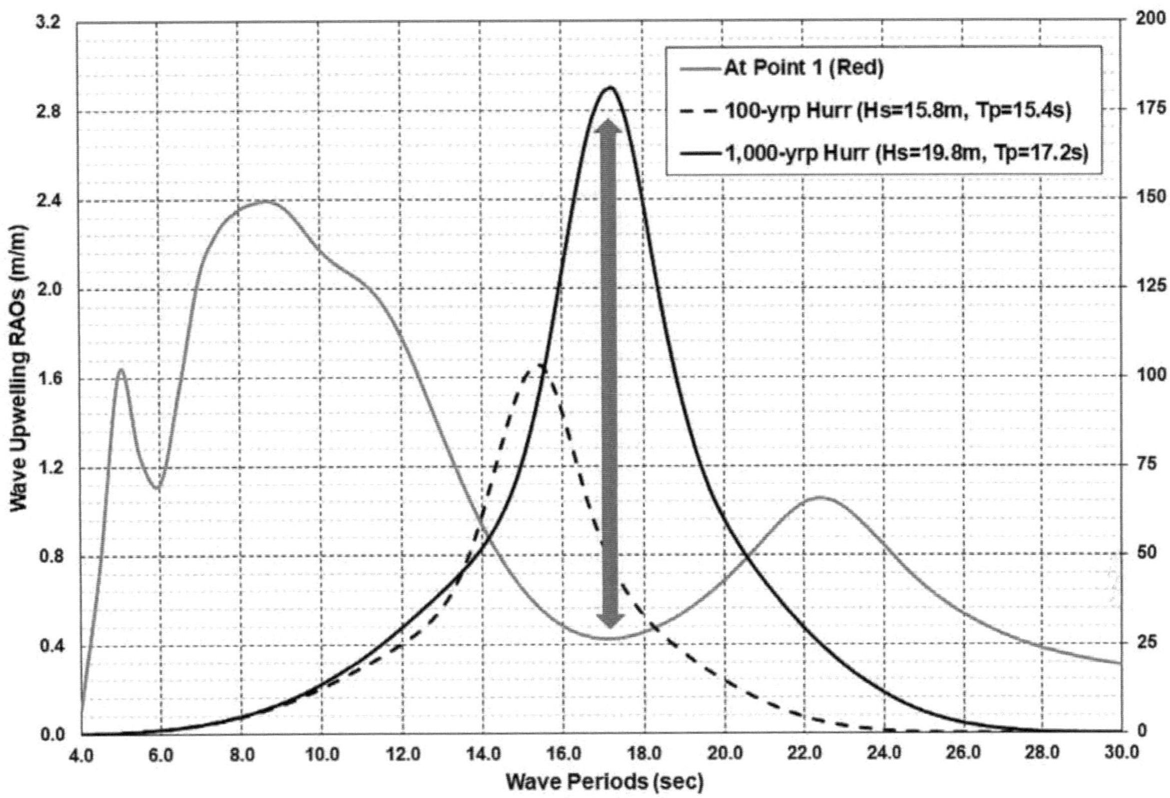

Figure 22: Wave Upwelling RAOs at Point 1 (Red) in Beam Sea - Fully Loaded Condition

From Figure 22, a few key observations are highlighted as follows,

- The current external walls along longitudinal direction have been tuned for an optimized configuration to minimized airgap for 1,000-yrp hurricane. The peak wave energy in 1,000-yrp is corresponding to the trough of wave upwelling RAOs
- Since wave periods of 100-yrp and 1,000-yrp are 15.4 sec and 17.2 sec, best tuned for 1,000-yrp hurricane will not be best for 100-yrp hurricane.
- In addition, minimum airgap criteria for 100-yrp and 1,000-yrp are different,
  - $\geq$1.5 m in 100-yrp hurricane
  - $\geq$0 m in 1,000-yrp hurricane

  It is possible to further tune to yield optimized results for both 100-yrp hurricane and 1,000-yrp hurricanes.
- So far, only eco-FPSO hull has the unique feature to tune the external walls to mitigate green water impacts. Thus, eco-FPSO can stay in field without disconnection/reconnection like a deep draft semi-submersible platform in addition to have large storage capability.

Minimum airgap results in various extreme sea states including 1,000-yrp hurricane have been screened with different headings in fully loaded condition and the most criterial results are presented in a later section.

Load Case Matrix

In order to verify whether the proposed hull meets the goals we set, a comprehensive load case matrix for global performance and mooring strength analysis is given in Table 9 and specially selected load case matrix for riser strength analysis is presented in Table 10.

Table 9: Load Case Matrix for Global Performance and Mooring Strength Analysis

							Design Parameters for Global Performance & Mooring Analysis					
LC#	Vessel Condition	Return Period	Types of Sea States	Headings (deg)	Mooring Condition	Design Category	Extreme Motions	Extreme Accels	Min Airgaps	Line Tensions	Bottom Clearance	Anchor Loads
1	F & B	1	WS	0-90	I	A	X	X	X	X	X	X
2	F & B	10	WS	0-90	I	A	X	X	X	X	X	X
3	F & B	100	HWD	0-90	I	B	X	X	X	X	X	X
4	F & B	100	HWV	0-90	I	B	X	X	X	X	X	X
5	F & B	100	HLC	0-90	I	B	X	X	X	X	X	X
6	F & B	100	MLC	0-90	I	B	X	X	X	X	X	X
7	F & B	1,000	HWD	0-90	I	S	X	X	X	X	X	X
8	F & B	1,000	HWV	0-90	I	S	X	X	X	X	X	X
9	F & B	100	HWD	0-90	D	S	X	X	X	X	X	X
10	F & B	100	HWV	0-90	D	S	X	X	X	X	X	X
11	F & B	100	HLC	0-90	D	S	X	X	X	X	X	X
12	F & B	100	MLC	0-90	D	S	X	X	X	X	X	X

In Table 9, there are a few notes given below,

- Under "vessel condition", "F" stands for "Fully loaded condition" and "B" represents "ballast condition";
- Under "types of sea states",
    - "WS" – winter storm
    - "HWD" – Hurricane wave dominant
    - "HWV" – Hurricane wind dominant
    - "HLC" – Hurricane loop current
    - "MLC" – Maximum loop current
- Under "headings", "0-90" -> 5 headings: 0 deg; 30 deg; 45 deg; 60 deg and 90 deg
- Under "mooring condition", "I" denotes "intact condition" and "D" means "line damaged/broken condition".

Table 10: Load Case Matrix for Riser Strength Analysis

				Inner Pressure					
LC#	Design Stage or Category	Sea State Return Period	Types of Sea States	Value (bar)	Reference Location	Headings (deg)	Vessel Condition	Mooring Condition	Pre-service "X" or In-place "+"
1	Installation	1-yrp	Winter Storm	0	Hang-off	Far-Far	F & B	Intact	X
2						Near-Near			
3	Hydro Test	1-yrp	Winter Storm	1,120	Mudline	Far-Far	F & B	Intact	X
4						Near-Near			
5	Operating	10-yrp	Winter Storm	241	Hang-off	Far-Far	F & B	Intact	+
6						Near-Near			
7	Extreme	10-yrp	Winter Storm	241	Hang-off	Far-Far	B	Intact	+
8						Near-Near			
9		10-ypr	Winter Storm	896	Mudline	Far-Far	B	Intact	+
10						Near-Near			
11		100-yrp	HWD	241	Hang-off	Far-Far	F & B	One line broken	+
12			HWV			Near-Near			
13	Survival	100-yrp	HWD	21	Hang-off	Far-Far	F & B	Intact	+
14			HWV			Near-Near			
15		100-yrp	HWD	896	Mudline	Far-Far	F & B	Intact	+
16			HWV			Near-Near			
17	Robustness	1,000-yrp	HWD	21	Hang-off	Far-Far	F & B	Intact	+
18			HWV			Near-Near			

In Table 10, a few notes are highlighted as follow,

- All load cases specified are based on API-RP-2RD
- Inner pressures assumed were based on Vidic-Perunovic et al (2014) – Ref. [15]. For convenience, they are copied below,
  - Design pressure: 3,500 psi at hang-off
  - Extreme pressure: 13,000 psi at mudline (same as wellhead shut-in tubing pressure)
  - Associated pressure: 300 psi at hang-off
  - Hydro test pressure: 16,250 psi at mudline

Fatigues of moorings and risers due to waves (scatter diagram presented in Figures 5 and 6) in both fully loaded condition and ballast condition have also been undertaken.

Extreme Motions

Detailed global performance analysis in time domain per load cases specified in Table 9 were completed in 2018 and numerical results have been summarized in HOE internal global performance analysis report (Ref. [3]). Only governing results are extracted and presented in Tables 11 to 15 respectively.

Extreme Offsets

Table 11: Summary of Extreme Offsets

Design Category	Mooring Condition	Sea States	Headings (deg)	Extreme Values	Required Criteria	Meet Criteria?	Vessel Conditions
A	Intact	10-yrp WS	90	32m (1.3% of WD)	<5% of WD	Y	Ballast
B	Intact	100-yrp HWV	90	114m (4.5% of WD)	<5% of WD	Y	Ballast
S	Intact	1,000-yrp HWV	60	171m (6.8% of WD)	<7% of WD	Y	Ballast
S	One line broken	100-yrp HWV	60	142m (5.7% of WD)	<7% of WD	Y	Ballast

Extreme offsets have been analyzed per Table 9 and they are governed by wind dominant sea states in ballast condition. Loop current conditions were analyzed in fully loaded condition and found the extreme offsets didn't govern.

Extreme Downward Motions

Table 12: Summary of Extreme Downward Motions

Design Category	Mooring Condition	Sea States	Headings (deg)	Extreme Values	Required Criteria	Meet Criteria?	Vessel Conditions
A	Intact	10-yrp WS	90	-1.2m	-	-	Ballast
B	Intact	100-yrp HWD	90	-7.1m	-	-	Ballast
S	Intact	1,000-yrp HWD	90	-11.5m	-	-	Ballast
S	One line broken	100-yrp HWD	90	-7.5m	-	-	Ballast

Extreme downward motions have been analyzed per Table 9 and they are governed by wave dominant sea states in ballast condition and beam sea headings.

Extreme Rotations

Table 13: Summary of Extreme Rotations

Design Category	Mooring Condition	Sea States	Headings (deg)	Extreme Values	Required Criteria	Meet Criteria?	Vessel Conditions
A	Intact	10-yrp WS	90	3.4 deg	<10 deg	Y	Fully Loaded
B	Intact	100-yrp HWD	90	9.2 deg	<10 deg	Y	Ballast
S	Intact	1,000-yrp HWD	90	14.4 deg	<15 deg	Y	Ballast
S	One line broken	100-yrp HWD	90	9.8 deg	<10 deg	Y	Ballast

Extreme rotations have been analyzed per Table 9 and they are governed by wave dominant sea states in ballast condition except operating condition in the fully loaded condition.

Extreme Accelerations

Extreme horizontal and vertical accelerations at the center of the topsides including gravity effects are given in Table 14.

Table 14: Summary of Extreme Horizontal and Vertical Accelerations

Design Category	Mooring Condition	Sea States	Headings (deg)	Horz. Accel. (g)	Vert. Accel. (g)	Meet Criteria?	Vessel Conditions
A	Intact	10-yrp WS	90	0.07	-	-	Fully Loaded
A	Intact	10-yrp WS	90	-	1.03	-	Ballast
B	Intact	100-yrp HWD	90	0.18	1.12	-	Ballast
S	Intact	1,000-yrp HWD	90	0.20	1.17	-	Ballast
S	One line broken	100-yrp HWD	90	0.19	1.12	-	Ballast

Minimum Airgaps

Minimum airgaps are more critical in fully loaded condition than those in ballast condition since there is 5.5m more clearance in ballast condition as presented in Table 5. After screening all headings as shown in Table 9 and all distributed locations as illustrated in Figure 21, the governing minimum airgaps are summarized in Table 15.

Table 15: Summary of minimum Airgaps

Design Category	Mooring Condition	Sea States	Headings (deg)	Extreme Values	Required Criteria	Meet Criteria?	Vessel Conditions
B	Intact	100-yrp HWD	90	3.4	>1.5m	Y	Fully Loaded
S	Intact	1,000-yrp HWD	90	1.1	>0.0m	Y	Fully Loaded

These analytical results shall be verified by wave basin model tests in near future.

Extreme Mooring Tensions

Detailed mooring strength analysis in time domain per load cases specified in Table 9 were undertaken in 2018 and numerical results have been summarized in HOE internal mooring analysis report (Ref. [4]). Only governing results are extracted and presented in Tables 16 and 17 for reference.

Table 16: Summary of Factor of Safety for Chain and Polyester

No.	Environ. Cond.	Load Condition	Mooring Cond.	Factor of Safety (FOS)		Required FOS		Meet Criteria?
				Chain	Poly	Chain	Poly	
1	Operating	Ballast	Intact	3.44	3.84	1.67	1.82	Pass
2		Full Load	Intact	3.85	4.33	1.67	1.82	Pass
3	Extreme	Ballast	Intact	**1.68**	**1.82**	1.67	1.82	Pass
4		Full Load	Intact	1.81	1.96	1.67	1.82	Pass
5	Extreme	Ballast	One line Broken	1.42	1.52	1.25	1.43	Pass
6		Full Load	One line Broken	1.53	1.65	1.25	1.43	Pass
7	Survival	Ballast	Intact	1.24	1.33	1.10	1.10	Pass
8		Full Load	Intact	1.39	1.49	1.10	1.10	Pass

Table 17: Summary of Polyester Seabed Clearances

No.	Environ. Cond.	Load Condition	Mooring Cond.	Poly Clearance	Required Min. Clearance	Meet Criteria?
1	Operating	Ballast	Intact	81.0	10.0	Pass
2		Full Load	Intact	80.6	10.0	Pass
3	Extreme	Ballast	Intact	25.5	10.0	Pass
4		Full Load	Intact	25.8	10.0	Pass
5	Extreme	Ballast	One line Broken	5.1	0.0	Pass
6		Full Load	One line Broken	8.0	0.0	Pass
7	Survival	Ballast	Intact	**-8.9**	0.0	Failed
8		Full Load	Intact	-3.1	0.0	Failed

From Table 17, the worst polyester bottom clearance case is case 7 in 1,000-yrp hurricane ballast condition. In order to reveal more details, the time series of the polyester bottom clearances with the touchdowns as shown in Case 7 in the Table 17 are plotted in Figure 23. It has been seen there are three times bottom of polyester touched sea bed in 3-hr time series. The first time is about 1/3 of the magnitude of the second time (worst time, -8.9m) and the third time just barely touched sea bed. At this conceptual stage, the bottom of polyester only touched three times in 3-hr duration in 1,000-yrp hurricane, it is believed it is acceptable and can meet the criteria if some minor “tweaks” are carried out in next phase study.

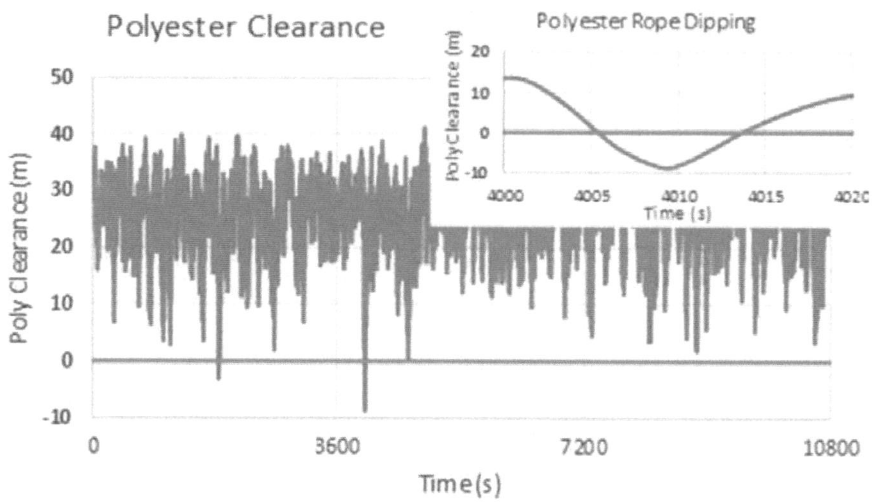

Figure 23: Local View of Polyester Touching Seabed - Ballast Condition

Extreme Riser Tensions

Detailed riser strength analysis in time domain per load cases specified in Table 20 were carried out in 2018 and simulation results have been documented in HOE internal riser analysis report (Ref. [5]). Only governing results are reproduced in Figures 24 and 25 for discussions.

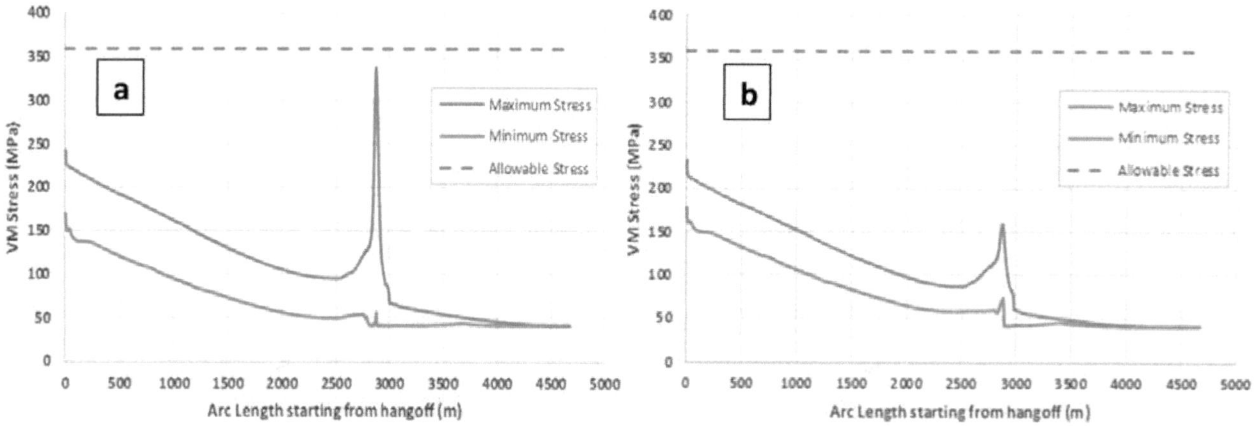

Figure 24: Stress Distributions Along 10" Production SCRs – 100-yrp Hurricane Wave Dominant, Near-Near a) in Ballast Condition and b) in Fully Loaded Condition

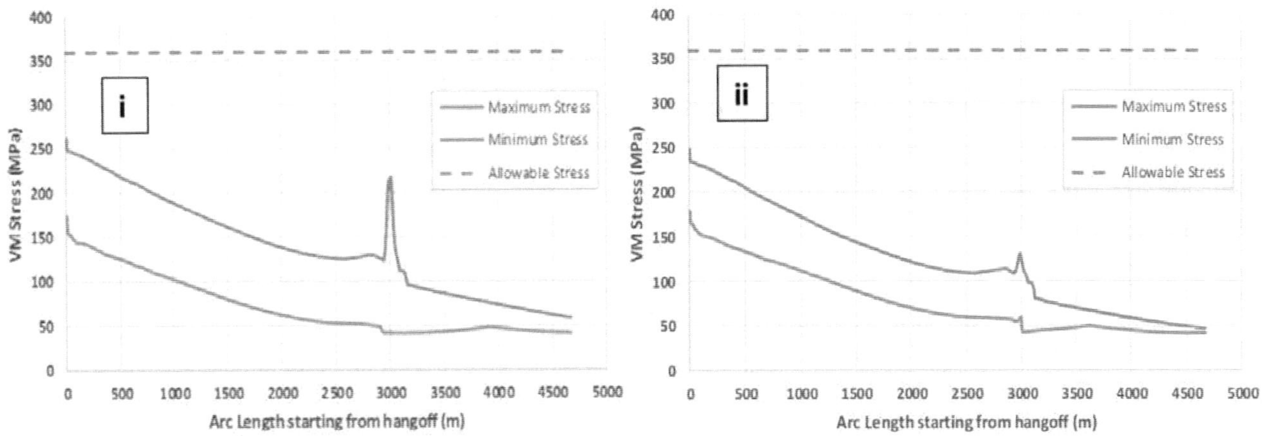

Figure 25: Stress Distributions Along 10" Production SCRs – 100-yrp Hurricane Wave Dominant, Far-Far i) in Ballast Condition and ii) in Fully Loaded Condition

From Figures 24 and 25, some observations on stress distributions for near-near and far-far in bot fully loaded and ballast conditions are highlighted below,

- Maximum stress in near-near ballast condition near touchdown zone is most critical
- Maximum stress in far-far ballast condition near touchdown zone is the second most critical which is about 65% of that in near-near case
- Maximum stress in far-far ballast condition at hang-off location is the highest in all cases and conditions.

Mooring Fatigue

Generic wave scatter diagrams as illustrated in Figures 5 and 6 were employed for mooring fatigue analysis. The design fatigue factor (DFF) of the mooring line is given as:

$$\text{Design Fatigue Factor (DFF)} = \frac{Minimum\ Fatigue\ Life}{Design\ Service\ Life}$$

Per ABS Position Mooring (Ref [1]), the mooring DFFs are given in **Table**18.

Table18: Minimum Required DFFs

Area	Design Fatigue Factor (DFF)
Inspectable areas	3
Non-inspectable[1] and Critical Areas[2]	10

Note:

1. Non-inspectable means that the detailed physical conditions of the component, such as changes of geometry etc., cannot obtained reliably. It may depend on the locations of the component and the inspection techniques.

2. The failure of the component could immediately and directly affect the position keeping capability of the mooring system, such as mooring line link.

The mooring system is also considered as a redundant system. Therefore, the DFF of 3 shall be acceptable for most components. However, for this conceptual study and only wave induced fatigue was considered, DFF of 10 is the minimum required value regardless.

In addition, the half-corroded diameter and ORQ grade of mooring chain are to be applied in the fatigue analysis. T-N curves for assessing mooring fatigue based on API RP 2SK (Ref. [3]) are copied in **Figure** 26 for convenience.

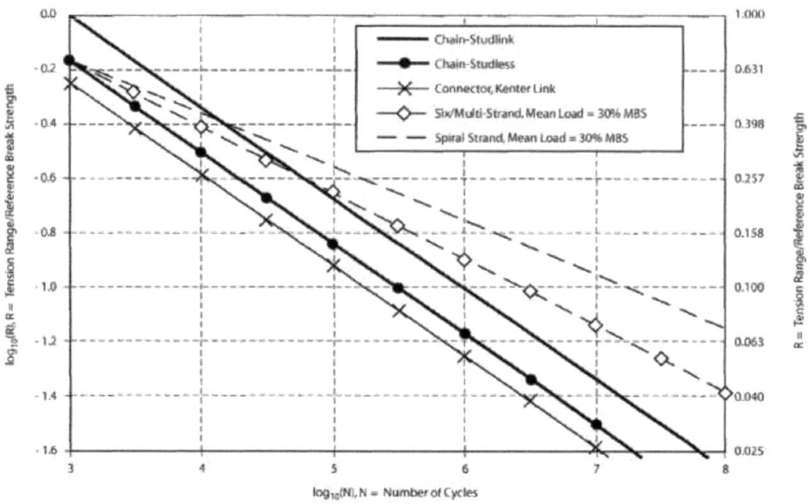

Figure 26: Mooring Fatigue Analysis Curves (Courtesy of API- RP-2SK)

The T-N curve can be expressed by the following equation,

$NT^M = K$

For the studless chain, the values of M and K are 3.0 and 316, respectively.

Wave fatigue damage per year and fatigue life for each of the load groups are cumulated to estimate the total damage per year. The fatigue results of all mooring lines are reported for both ballast and fully loaded conditions in Table19 and Table 20 respectively. A comparison of fatigue damage in ballast and fully loaded conditions is plotted in Figure 27. In this study, it is assumed that the ballast or fully loaded condition each contributes 50% of total fatigue damage. Hence, combined fatigue damage and life for each mooring line are summarized in Table21.

Table 19: Mooring Line Fatigue Damage and Life – Ballast Condition

Mooring Line #	Vessel Condition	Damage (/year)	Fatigue Life (years)	Estimated DFF(/20)	Required Min DFF	Meet Criteria?
ML#1	Ballast	4.01E-03	249	12.5	10.0	Y
ML#2	Ballast	3.63E-03	275	13.8	10.0	Y
ML#3	Ballast	3.29E-03	304	15.2	10.0	Y
ML#4	Ballast	2.99E-03	334	16.7	10.0	Y
ML#5	Ballast	3.50E-03	286	14.3	10.0	Y
ML#6	Ballast	3.88E-03	258	12.9	10.0	Y
ML#7	Ballast	4.30E-03	233	11.6	10.0	Y
ML#8	Ballast	4.76E-03	210	10.5	10.0	Y
ML#9	Ballast	4.70E-03	213	10.6	10.0	Y
ML#10	Ballast	4.24E-03	236	11.8	10.0	Y
ML#11	Ballast	3.82E-03	262	13.1	10.0	Y
ML#12	Ballast	3.44E-03	291	14.5	10.0	Y
ML#13	Ballast	3.12E-03	321	16.0	10.0	Y
ML#14	Ballast	3.41E-03	293	14.7	10.0	Y
ML#15	Ballast	3.75E-03	267	13.3	10.0	Y
ML#16	Ballast	4.12E-03	243	12.1	10.0	Y

Table 20: Mooring Line Fatigue Damage and Life – Fully Loaded Condition

Mooring Line #	Vessel Condition	Damage (/year)	Fatigue Life (years)	Estimated DFF(/20)	Required Min DFF	Meet Criteria?
ML#1	Fully Loaded	2.11E-03	474	23.7	10.0	Y
ML#2	Fully Loaded	1.90E-03	526	26.3	10.0	Y
ML#3	Fully Loaded	1.71E-03	585	29.2	10.0	Y
ML#4	Fully Loaded	1.54E-03	649	32.5	10.0	Y
ML#5	Fully Loaded	1.88E-03	532	26.6	10.0	Y
ML#6	Fully Loaded	2.11E-03	474	23.7	10.0	Y
ML#7	Fully Loaded	2.37E-03	422	21.1	10.0	Y
ML#8	Fully Loaded	2.65E-03	377	18.9	10.0	Y
ML#9	Fully Loaded	2.65E-03	377	18.9	10.0	Y
ML#10	Fully Loaded	2.37E-03	422	21.1	10.0	Y
ML#11	Fully Loaded	2.11E-03	474	23.7	10.0	Y
ML#12	Fully Loaded	1.88E-03	532	26.6	10.0	Y
ML#13	Fully Loaded	1.59E-03	629	31.4	10.0	Y
ML#14	Fully Loaded	1.76E-03	568	28.4	10.0	Y
ML#15	Fully Loaded	1.95E-03	513	25.6	10.0	Y
ML#16	Fully Loaded	2.15E-03	465	23.3	10.0	Y

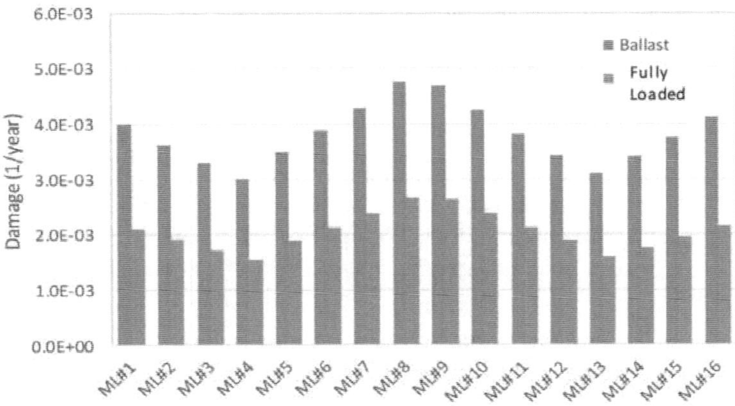

Figure 27: Comparison of Wave Induced Fatigue Damages – in Ballast and Fully Loaded Conditions

Table 21: Combined Mooring Line Fatigue Damage and Life

Mooring Line #	Vessel Condition	Damage (/year)	Fatigue Life (years)	Estimated DFF(/20)	Required Min DFF	Meet Criteria?
ML#1	B and F	3.06E-03	327	16.3	10.0	Y
ML#2	B and F	2.77E-03	362	18.1	10.0	Y
ML#3	B and F	2.50E-03	400	20.0	10.0	Y
ML#4	B and F	2.27E-03	442	22.1	10.0	Y
ML#5	B and F	2.69E-03	372	18.6	10.0	Y
ML#6	B and F	3.00E-03	334	16.7	10.0	Y
ML#7	B and F	3.34E-03	300	15.0	10.0	Y
ML#8	B and F	3.71E-03	270	13.5	10.0	Y
ML#9	B and F	3.68E-03	272	13.6	10.0	Y
ML#10	B and F	3.31E-03	303	15.1	10.0	Y
ML#11	B and F	2.97E-03	337	16.9	10.0	Y
ML#12	B and F	2.66E-03	376	18.8	10.0	Y
ML#13	B and F	2.36E-03	425	21.2	10.0	Y
ML#14	B and F	2.59E-03	387	19.3	10.0	Y
ML#15	B and F	2.85E-03	351	17.5	10.0	Y
ML#16	B and F	3.14E-03	319	15.9	10.0	Y

A few observations from Tables 19 to 21 and Figure 27 are highlighted as follows:

- Fatigue damage and life is governed by Ballast condition. In ballast condition, the mooring line had less fatigue life, which is about 55% of the fatigue life estimated from fully loaded condition.
- Minimum combined fatigue life is 270 years and the minimum DFF is 13.5 which are higher than those minimum required criteria. Thus, it believed that the proposed mooring configuration meets the design criteria.
- To improve mooring fatigue life, it is crucial to optimize the motion characteristics at ballast condition.

SCR Riser Fatigue

The fatigue damage at pipe girth weld location is expected to govern the fatigue life. The stress-cycle (S-N) method is employed and the S-N curve is per DNV RP C203 (Ref [6]). In this study, two-slope S-N curves in seawater with CP are applied for fatigue analysis. The parameters of two-slope S-N curves are copied and reproduced in Table for convenience.

Table 22: Parameters of Two-Slope S-N Curves

Design S-N Curve[1]	m_1	$\log a_1$	m_2	$\log a_2$
DNV Two-slope E Curve	3	11.61	5	15.35
DNV Two-slope D Curve	3	11.76	5	15.61

Note: 1. Both DNV design curves are applied to weld cap (OD) and weld root (ID).

An SCF of 1.25 is assumed for the pipe to pipe welds for steel risers for this study. Safety factor of wave induced fatigue damage, $SF_{wave} = 10.0$

The global dynamic analysis for wave fatigue motion is performed in time domain and the simulation of each fatigue bin is 1-hr duration excluding the ramping time 600 second. Same as mooring analysis, the rain-flow counting method is adopted to calculate fatigue damage. There are eight points uniformly distributed around the riser pipe outer circumference of each location.

The unfactored wave fatigue lives for the 10-inch production riser for both ballast and fully loaded conditions are summarized in Table . The fatigue life along the riser pipe starting from hang-off position to TDZ as well as local enlarged views around hang-off and TDZ areas are displayed in Figure and Figure **29**29, respectively.

Table 23: Unfactored Fatigue Lives for Ballast and Full Load Conditions

Design S-N Curve	Locations	Unfactored Fatigue Life (yrs)			
		Ballast Condition		Fully Loaded Condition	
		Hangoff[1]	TDZ	Hangoff[1]	TDZ
DNV E Curve	Weld Cap	410	469	390	640
	Weld Root	1396	2967	1534	4334
DNV D Curve	Weld Cap	645	840	611	1153
	Weld Root	2262	5356	2504	7827

Note: 1. Assume the first weld located at 2.8m from hang-off.

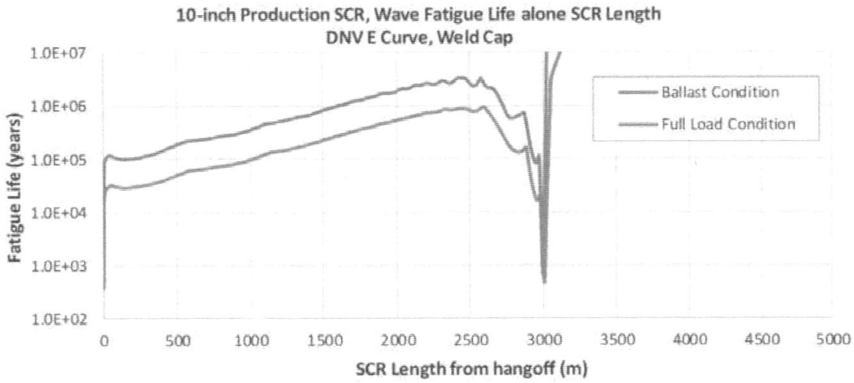

Figure 28: Wave Induced Fatigue Distribution Along Rise Pipe per DNV E Curve

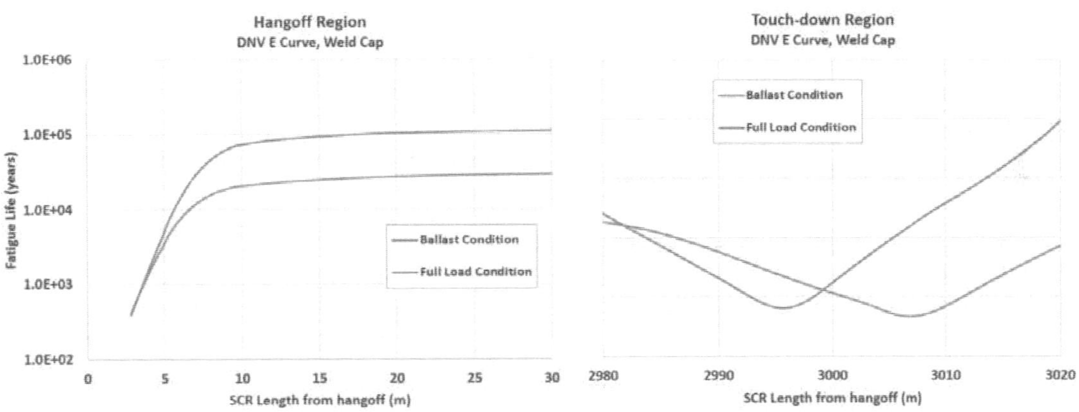

Figure 29: Local Views of Wave Induced Fatigue around Hang-off and TDZ Regions

Same as mooring fatigue analysis, it is assumed that the ballast or full load condition each contributes 50% of total fatigue damage. Hence, the combined wave induced fatigue lives are given in Table .

Table 24: Total Wave Induced Fatigue Lives

Design S-N Curve	Location	Unfactored Fatigue Life (yrs)	
		Hang-off	TDZ
DNV E Curve	Weld Cap	400	911
	Weld Root	1462	5709
DNV D Curve	Weld Cap	628	1631
	Weld Root	2377	10307

A key summary of fatigue results for 10-inch production riser is listed as below,

- Minimum fatigue lives at the rise hang-off and touch-down-zone (TDZ) are 400 years and 911 years respectively per E curve in Table 22 and 628 years and 1631 years respectively per D curve in Table 22.
- Both fatigue damages estimated in ballast and fully loaded conditions contribute equally to the total fatigue damage. The fully loaded condition provides the similar minimum fatigue life at hang-off compared to the ballast condition. However, minimum fatigue life at the TDZ in fully loaded condition (In Table 23) is about 1.5 times of that of the fatigue life in ballast condition.
- Minimum fatigue life along the riser length is observed at the hang-off along the weld cap (riser OD). As

expected, the longer fatigue life at weld cap has been estimated using DNV D curve instead of DNV E curve.

- Since only wave induced fatigue was considered, other contributions, such as VIM and VIV induced fatigue haven't been accounted for. Roughly at this conceptual study, by assuming 50% of the fatigue damage contributing from other sources, the estimated minimum fatigue life is about 200 years which is about 10 times of service life and is acceptable at this early stage. In future study, it will be closely monitored and can be improved through a few mitigation plans if needed.

OFFLOADING PLANS

Offloading Rates and Shuttle Tanker Features

Offloading hose: 20"
Offloading rate: 45,000 bbls/hr (7,155 m^3/hr)
For storage capability 1,000,000 bbls,

- 300,000 bbls DP Shuttle Tanker:
    - Offloading frequency every 4 days
    - Fill-up time: ~7 hrs
- 500,000 bbls DP Shuttle Tanker:
    - Offloading frequency every 6 days
    - Fill-up time: ~11 hrs

Offloading Plans

Since the eco-FPSO is a concept without turret system, advantages with weather-vanning features for offloading operation don't exist. Two preliminary offloading plans having been considered for this study are highlighted as follows:

- eco-FPSO + DP Shuttle Tanker (DPST, Tandem Offloading) as displayed in Figure 30
- eco-FPSO + SPM Buoy + Shuttle Tanker as illustrated in Figure 31

Due to unique features of eco-FPSO hull with two longitudinal external walls, a removable hard-link side-to-side offloading plan is being investigated and is not available for now.

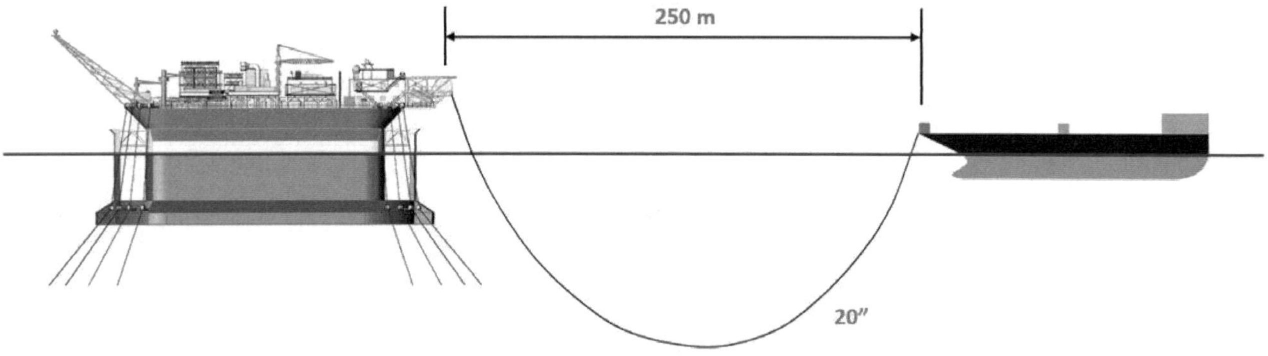

Figure 30: Tandem Offloading Plan – Eco-FPSO + DP Shuttle Tanker

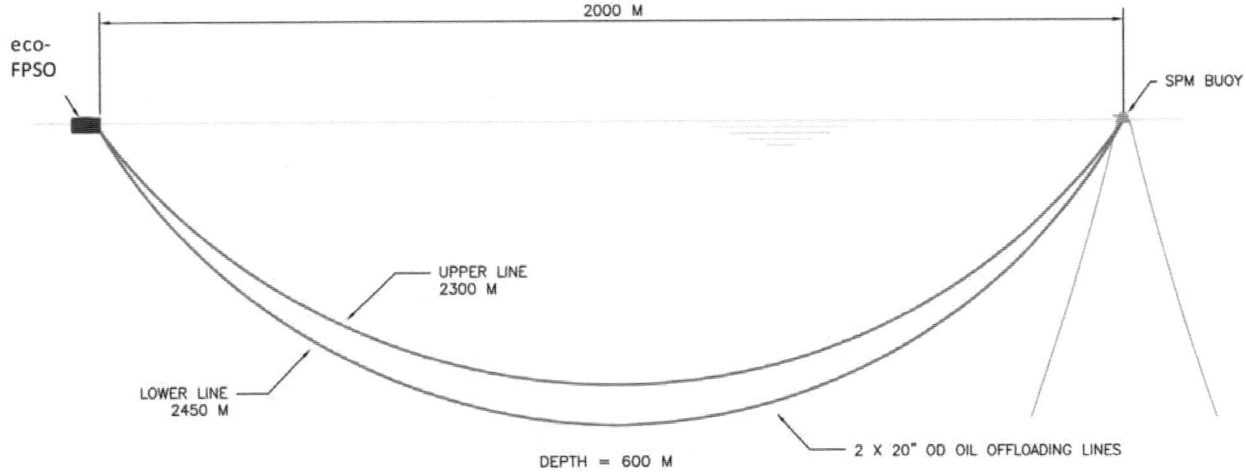

Figure 31: SPM Buoy Offloading Plan – Eco-FPSO + SPM Buoy+ Shuttle Tanker

For Tandem offloading plan, the allowable sea states, significant wave heights Hs are 4.0m to 5.0m and connecting/disconnecting offloading horse times are about 2-3 hours. The zones for normal operation, OESD-I and OESD-II are presented in Figure 32 for illustration and dimensions shown are not to scale.

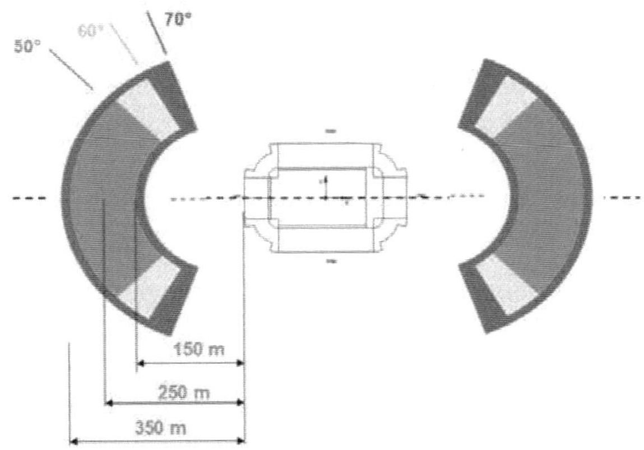

Figure 32: Schematic Guidance for Safe Zone, OESD-I and OESD-II

In Figure 32, Green Zone stands for safe zone; Yellow Zone represents OESD-I and Red Zone means OESD-II.
OESD-I: initiate the first stage of the offloading emergency shutdown procedure
OESD-II: Actuate the emergency disconnection of the loading system offloading horse line and disconnect it

In the SPM buoy operation, shutter tank doesn't need equipped DP system and we may need two 20" offloading lines as shown in Figure 31.

CONSTRUCTABILITY

Hull Constructability

The hull structure as illustrated in Figure 4 is comprised of traditional stiffened flat plate panels for simplified fabrication. The flat panels are especially beneficial for fabrication yards with existing panel-line assembly equipment that can efficiently assemble stiffeners to plate panels prior to subsequent assembly steps. The hull plate thickness, stiffener sizes and girder sizes were conventionally standard hull materials. Stiffener sizes are also selected from commonly available shapes. Girders are fabricated "T" sections, and the associated spans are suitable for the arrangement of the structural panels. The construction sequence and method for the hull is very common and typical.

In general, the block sizes, material usage, plate and shape sizes and overall structural arrangements are well suited to fabrication at many worldwide locations. Any fabrication yards with experience assembling medium to large hulls for floating production facilities would be capable of building the proposed hull configuration. The maximum length of hull is 149m and width is 100m at keel level. A typical candidate list (not complete) is presented in Table 25.

Table 25: Potential Construction Yards and Facility Dimensions

No.	Yard and Country	Dry Dock Dimensions
1	DSME (Korea)	530m x 131m x 14.5m
2	SHI (Korea)	157m x 131m (floating)
3	HHI (Korea)	490m x 115m x 13.5m
4	Sungdong Shipbuilding (Korea)	545m x 126m x 16m
5	Qingdao Beihai Shipbuilding (China)	530m x 121m x 14m
6	Yanti CIMC Raffles (China)	380m x 120m x 12m
7	STX Offshore & Shipbuilding (China)	430m x 135m x 14m
8	COSCO Qidong (China)	170m x 120m x 13m

Topsides Constructability

As described in previous section, general philosophy of topsides layout is to have modular design which maintains competitions between different yards. The weight and dimension of each modular was limited which may offer opportunity to local fabrication yards to meet local content requirement. Project execution plan may vary; however, quayside integration is the preferred option for this study.

TRANSPORTABILITY

Project execution plan may be different, such as hull and topsides are constructed at same fabrication yard or hull and topsides are constructed at different locations. The current hull sizing has adequate feasibility and is suitable for either execution plan.

Dry Transport

Eco-FPSO Hull with Deck Only

As illustrated in Figure 33, hull with deck only (no topsides facilities/utilities modules) can be dry transported on a heavy lift vessel. In this example, BOKA (former Dockwise) Vanguard is considered. Main characteristic data can be found in BOKA website and a few selected parameters are reproduced in Table 26 for convenient discussion.

Table 26: Main Characteristic Data of BOKA Vanguard

Parameters	Dimensions or Weights
Overall length	275m
Breadth molded/max	70m/78.75m
Depth	15.5m
Max submerged draft at FPP/App	31.5m
Max draft	10.99m
Dead weight	117,000 mt

Since eco-FPSO width is 100m and there will be about 15m overhang on both sides. The available length in

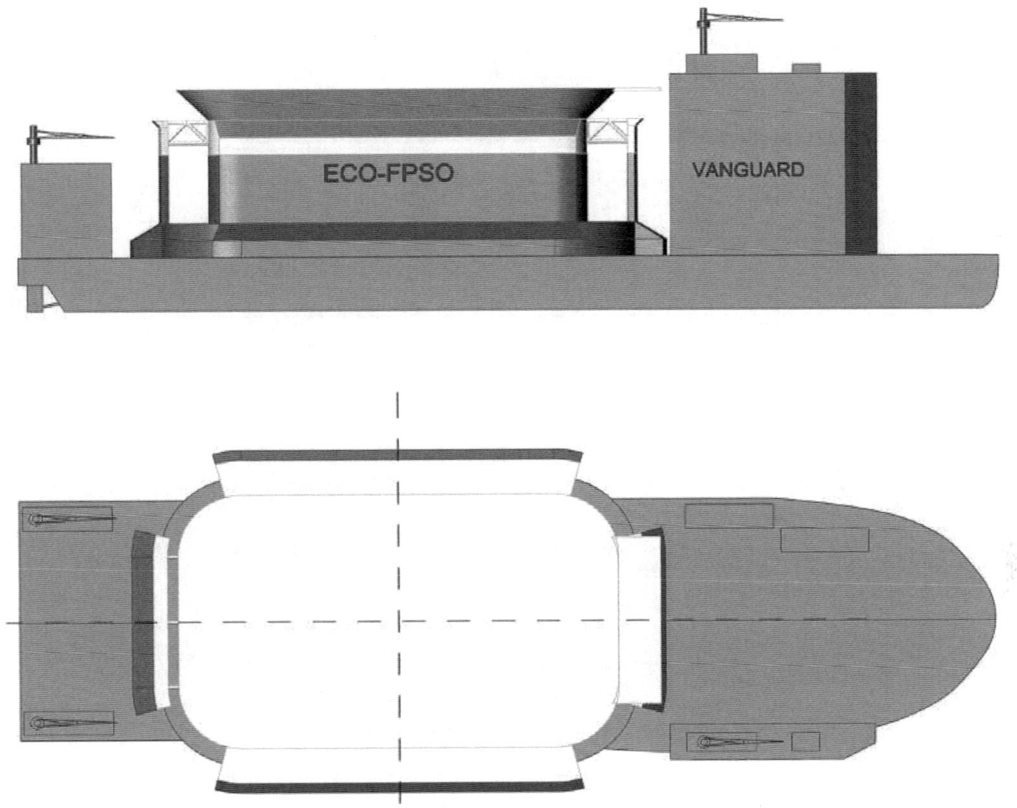

longitudinal direction of Vanguard is 155m on starboard side.

Figure 33: Schematic Arrangement of Eco-FPSO Hull on BOKA Vanguard

Weights and required eco-FPSO draft with deck only are summarized in Table 27.

Table 27: Weights of eco-FPSO Hull with Deck only

Parameters	Dimensions or Weights
Weights	73,452 mt
Eco-FPSO draft	5.45 m
Maximum hull edge to edge width	100 m
Hull overhang on Vanguard	15 m

From Tables 26 and 27, it is obvious that weights of hull with deck only is well within the limit of Vanguard and overhang on both sides of HLV is typical and acceptable.

ECO-FPSO Hull with Integrated Topsides Modules

As presented in Figure 34, eco-FPSO hull with integrated topsides modules can be dry transported on a heavy lift vessel, BOKA (former Dockwise) Vanguard. Weights and required platform draft of hull with integrated topsides are summarized in Table 28.

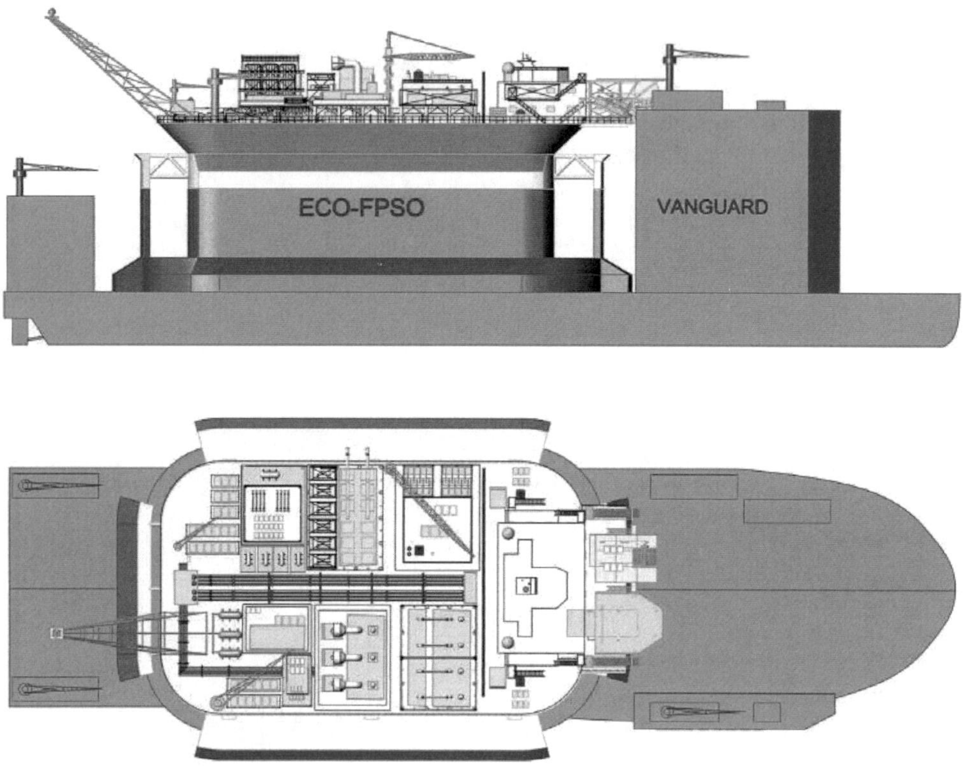

Figure 34: Schematic Arrangement of Eco-FPSO on BOKA Vanguard

Table 28: Weights of eco-FPSO Hull with Integrated Topsides Modules

Parameters	Dimensions or Weights
Weights	93,452 mt
Eco-FPSO draft	7.0 m
Maximum hull edge to edge width	100 m
Hull overhang on Vanguard	15 m

From Tables 26 and 28, it is noted that weights of eco-FPSO hull with integrated topsides modules is within the limit of Vanguard and overhang on both sides of HLV is acceptable.

Wet Tow

Since BOKA Vanguard is the only HLV which can be employed for dry transport of this eco-FPSO, its availability is a risk for project. Feasibility of wet tow of eco-FPSO either without topsides modules or with topsides modules is vital for a robust project execution plan.

Eco-FPSO Hull with Deck Only

If no topsides modules, the wet tow draft can be 5.6m and weight, VCG, GMTc and GMLc of wet tow without topsides modules are summarized in Table 29.

Table 29: Weights of eco-FPSO Hull for Wet Tow

Parameters	Dimensions or Weights
Weights	75,500 mt
Eco-FPSO draft	5.6 m
VCG from Keel	18.8 m
GMTc (Free surface corr, Mass VCG only)	69.8 m
GMLc (Free surface corr, Mass VCG only)	177.2 m

Eco-FPSO Hull with Integrated Topsides Modules

If topsides modules are included, the wet tow draft can be 7.5m and weight, VCG, GMTc and GMLc of wet tow without topsides modules are summarized in Table 30.

Table 30: Weights of eco-FPSO for Wet Tow

Parameters	Dimensions or Weights
Weights	99,500 mt
Eco-FPSO draft	7.5 m
VCG from Keel	20.5 m
GMTc (Free surface corr, Mass VCG only)	62.0 m
GMLc (Free surface corr, Mass VCG only)	163.0 m

INSTALLABILITY

The integrated eco-FPSO can be dry transported to the nearby sheltered area and offloaded; then it is wet towed to the installation site or wet towed the integrated eco-FPSO from the quayside to the installation site. The mooring system is traditional chain-polyester-chain and can be pre-laid and eco-FPSO is a typical vessel with adequate stability. Obviously, there is no "show stopper" to hoop-up hull and mooring. Similarly, it is believed there is "no concern" to pull-in risers.

CONCLUSIONS

An alternative FPSO hull, without turret system and with traditional modular topsides, eco-FPSO, is proposed and introduced and its features and feasibility have been demonstrated through

- Extensive global performance analysis
- Mooring strength and fatigue analysis
- Riser strength and fatigue analysis
- Offloading operations
- Topsides and hull constructability
- Transportability of hull only and with integrated topsides modules
- Installability

Advantages of an eco-FPSO vs. a ship-shaped FPSO are highlighted as follows,

- Simple hull geometry and no moving parts and/or mechanically jack-up system. Low cost hull
- Much less sagging and hogging moments and much larger mid-section modulus stiffness which result in less stress and light hull.
- Considerably less complexity and capital expenditure with spread-mooring vs. internal or external turret system for a ship-shaped FPSO
- Much better heave and roll/pitch motions which are suitable for SCRs application in harsh environmental conditions

- Much better heave motion characteristics which leads to better offloading operations and offers large offloading window and less requirement on offloading arms.
- Much better roll motion which results in better performances of motion sensitive topsides facilities and much less sloshing in storage tanks.
- No thrusters reduce complexity of hull arrangement and cost of operations

Thus, the development philosophy set at the beginning has been completely fulfilled and it is believed that the proposed eco-FPSO alternative is a safe, low motion and low-cost candidate for the field development in future.

This study represents one cycle exercise and it has been found there are many opportunities to optimize hull, mooring and riser during courses of this study which will be captured and implemented in the next phase study. A conceptual wave basin model test is deemed to be necessary to validate the numerical analysis results and provide good indications for further development and maturation.

ACKNOWLEDGEMENTS

The author would like to acknowledge the management of Houston Offshore Engineering, LLC, member of the SNC-LAVALIN Group for their supports of the research on innovative low cost turretless FPSO development and for allowing the publication of the study results. The author is also grateful for valuable contributions from my colleagues during multi-year development programs. Special thanks go to Dr. Vimal Vinayan for his outstanding contributions on hydrodynamic computations; Mr. Sid Siddharth (now he is an independent contractor) for his excellent contributions on global weight management and stability analysis; Mr. Jianfeng Zhang for his significant contributions on mooring and riser strength and fatigue analysis and Mr. Dongxing Jia for his important contributions on global loading and various challenging aspects. Special thanks also go to Mr. Chuck Brown for topsides layouts and Mr. Joey Lopez for tank compartmentation drawings.

REFERENCES

1. ABS Guidance, “Position Mooring Systems”, May 2018.
2. API-RP-2MET, “Derivation of Metocean Design and Operating Conditions”, 1st Edition, November 2014.
3. API RP 2SK, "Recommended Practice for Design and Analysis of Stationkeeping Systems for Floating Structures",3rd Edition, October 2005.k
4. API RP 2SM, "Recommended Practice for Design, Manufacture, Installation, and Maintenance of Synthetic Fiber Ropes for Offshore Mooring",1st Edition, March 2001.
5. Campos, FCR; Cueva, M; Nishimoto, K; and Costa, APS, “Stability Criteria and Analysis of a Monocolumn Concept”. 23rd International Conference on Offshore Mechanics and Arctic Engineering. OMAE2004-51531.
6. DNVGL RP C203,"Fatigue Design of Offshore Steel Structures", April 2016.
7. DNVGL RP F204,"Riser Fatigue", October 2017.
8. “Eco-FPSO Global Performance Analysis Report”, P100056291-NA-REP-006-A1 (Confidential), September 24, 2018
9. “Eco-FPSO Mooring Strength and Fatigue Analysis Report”, P100056291-NA-REP-007-A1 (Confidential), September 13, 2018
10. “Eco-FPSO Riser Strength and Fatigue Analysis Report”, P100056291-NA-REP-008-A1 (Confidential), August 28, 2018
11. Gourdet, G; “Connection hull-topsides: principles, designs and returns of experience”, Bureau Veritas Technical Paper, 2008.

12. Krekel, MH and Kaminski, ML; "FPSOs: Design considerations for the structural Interface hull and topsides", Offshore Technology Conference, OTC 13996, 2002, Houston, Texas.
13. Leverette, S and Carrico, T; "Design of the Disconnectable Buoy Turret Mooring for the Turritella FPSO", OTC-27672-MS; May 2017; Houston, Texas.
14. Mansour, AM; Wu, CF; Zuccolo, R and Azevedo, J; "The low motion FPSO (LM-FPSO): a novel SCR and TTR friendly floater", Proceedings of the Deep Offshore Technology International 2015 (DOT-2015), Houston, USA, October 2015.
15. Mansour, AM; Wu, CF; Peng, C; Zuccolo, R; Bagaria, A; Bhat, S; Yu, J and Hong, GS; "A Low Motion FPSO Design Hosting SCR and TTR in Harsh Environment", OTC-27893-MS, 2017, Houston, Texas.
16. Marshall, RW and Smith, SL, "UKOOA FPSO Design Guidance Notes for UKCS Service", UKOOA, 2002.
17. Masetti, IQ; Costa, APS; Sphaier, SH; Nishimoto; Kazuo and Machado, G; "Development of a Concept of Mono-Column Platform: MONOBR", International Journal of Advanced Manufacturing Technology, 2006.
18. Oliveira, AC and Vilameá, EM; "The Development of the ULFPSO Concept Design", Offshore Technology Conference – Brazil (OTC Brazil) 2015, OTC-26153-MS, Rio de Janeiro, Brazil.
19. SBM Offshore; "FPSO Turritella – Floating Production Storage and Offloading", Key Facts and Figures; April, 2016.
20. Vidic-Perunovic, J; Guo, XS; Wang, L; Hopen, F and Head, WJ; "Steel Catenary Riser Design for Cylindrical FPSO Application in Ultra-Deep GoM", OTC-25406-MS, 2014, Houston, Texas.
21. Vidic-Perunovic, J; "Low Cost Flexible Production System for Remote Ultra Deepwater Gulf of Mexico Field Development", UDW Technology Conference 2014; September 3-4, Norris Conference Center, Houston, Texas.
22. Vidic-Perunovic, J; Lee, L; Glomnes, EB; Mercier, RS and Head, WJ; "Ultra-Deepwater Production System: Model Test Study of Cylindrical FPSO in GoM Hurricane Conditions", OTC-27844-MS, 2017, Houston, Texas.
23. Duhamel, N., "LNG Tandem Offloading Solution: Ready for Development", Offshore Technology Conference 2016, Houston, Texas, USA. OTC 27166.
24. Odeskaug, L., "The Cylindrical Hull Concept for FLNG Application", Offshore Technology Conference 2015, Houston, Texas, USA. OTC 25703.
25. Terashima, T., Shaimada, K., Ichimaru, Y., and Orimo, Y., "Methodology to Define Design Motion Criteria for Performance of Floating LNG Process Facilities", Offshore Technology Conference 2011, Houston, Texas, USA. OTC 21292.
26. Zou, J., "Semisubmersible platforms with Steel Catenary Risers for Western Australia and Gulf of Mexico", International Journal on Ocean System Engineering, June, 2012, vol. (2), pp 99-113.

SNAME
THE INTERNATIONAL COMMUNITY
FOR MARITIME AND OCEAN PROFESSIONALS

Coupled Analysis for Dry Transportation of Large Floating Production Systems

Jay (Huajie) Wen
EXMAR Offshore Company

Aimin Wang
EXMAR Offshore Company

Otto DaSilva
EXMAR Offshore Company

Proceedings of the 25th Offshore Symposium, February 27th 2020, Houston, Texas
Texas Section of the Society of Naval Architects and Marine Engineers

ABSTRACT

It is becoming increasingly common to dry transport floating production systems (FPS), especially when there is a long distance between the shipyard and deployment site. The FPS, heavy lift vessel, and interfacial components (the cribbing and seafastening) need to be properly evaluated and designed, usually through Finite Element Analysis (FEA). Common practice of the analysis on dry transportation is decoupling the FPS and heavy lift vessels. A decoupled analysis is unable to directly consider the interacting effects from other components and simulate the actual loading scenarios. This paper presents a coupled analysis framework, in which the FPS, the vessel and the interfacial components are explicitly included in the FE analysis. The shaped cribbing and seafastening can therefore be designed and evaluated at a more highly optimized level. The interaction between the FPS and transport vessel is also successfully simulated. The approach is able to provide more realistic and accurate response of the system. The methodology and procedure of the coupled analysis used for dry transportation are presented in the paper, followed by sample calculations and results from a generic dry transportation analysis.

Keywords: Floating Production System, Dry Transportation, Coupled Analysis

Introduction

Dry transportation of floating production systems (FPS) is becoming common since the shipyards are more frequently to be distant from deployment sites. The FPS is required to be able to resist the marine transport induced loads, which may significant differ with loads conditions during production of the FPS. Also, the transport vessel needs to have sufficient strength to support the cargo in possible metocean conditions during transportation.

Environmental loads with a reasonable possibility to be exposed are expected to be firstly determined in the early stage of the project, based on the planned transportation route. A group of load conditions derived from environmental loads are then used in the studies on FPS and vessel capacity, usually through Finite Element Analysis (FEA). It is essential to determine the load conditions, and how these loads are applied to dry tow structures. An approach that underestimates the response could lead to unsafe design, while excessively conservative approach will result in unnecessary high cost and possible project delays.

The common practice of dry tow structural analysis is decoupling the cargo and vessel, and separately applying loads derived from either motions study/model tests, or more conservative default motion criteria as recommended by regulatory agencies. This type of decoupling approach neglects the stiffness contributions from vessel and cargo, which could miscalculate some interface forces and vessel/cargo response, and subsequently lowered the reliability level of analysis. In the present paper, a coupled analysis on dry tow transport, based on generic FPS and vessel, was conducted. The coupled analysis simulated both FPS and vessel as well as the interactions between them in one FE model. A traditional decoupled analysis was also conducted for the purpose of comparison. Results indicates that the coupled analysis is able to reflect the variances of seafastening forces at difference location of vessel, while through decupled approach only averaged seafastening forces can be estimated, which implies possible unsafe design in some seafastening locations. The coupled analysis also produces more accurate cribbing pressure distribution compared its decoupled equivalents, which again suggests possible non-conservative design for wood cribbings. Besides, the stress levels at critical locations in coupled analysis turned out to be smaller than their counterparts in decoupled analysis.

METHODOLOGY

In decoupled analysis, only FPS is included in the structural FE model, while the vessel is not covered. The loads are either adopted from the Default Motion Criteria in guidelines for marine transportation (DNVGL-ST-N001), or derived from Motions study or model tests. The interactions between FPS and vessel are either simulated as forces or boundary conditions, based on engineers' knowledge and preference, as shown in Figure 1. There is significant amount of assumptions in this type of approach, in order to substitute the influence from missing vessel, and the accuracy of the analysis will largely depend on the reasonableness of these assumptions. After all, the entire systems, especially the interfacial components, are in motion and in a non-inertial reference frame.

In coupled approach, motion analysis or its equivalence is essentially required. The FPS, vessel and the interfacial components are all included in the FE model. The hydrodynamic forces and motion are derived from motion study, and then mapped as pressure to the wetted part of the vessel, as shown in Figure 2. The assumption in this kind of approach is limited, since all major components have been realistically simulated, and the loads also reflects the real transportation scenarios.

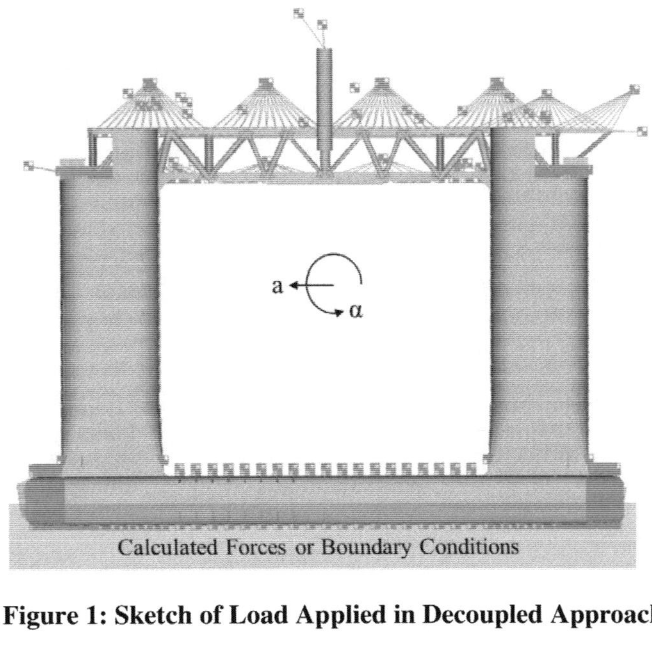

Figure 1: Sketch of Load Applied in Decoupled Approach

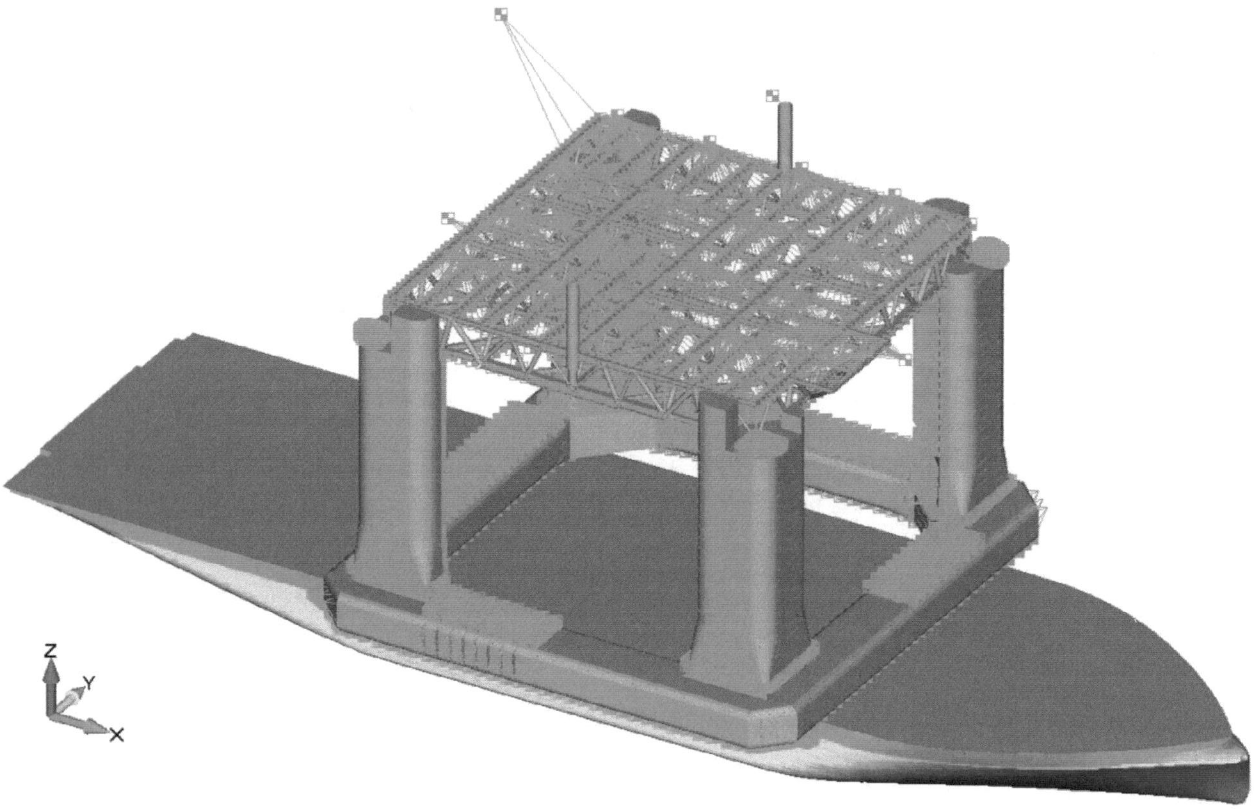

Figure 2: Contour of Pressure Applied in Coupled Approach

Loads based on motions study are less conservative but more accurate compared to the criteria from standards. Also since it is also required by coupled approach, the decoupled analysis in the present adopted loads from motion analysis. For the dry tow transportation, a metocean design threshold of during one out of every 100 voyages or the peak monthly 10 year extreme is recommended. The range of wave spectral peak period corresponding to the significant wave height is compiled using the 3 hour enroute statistics. Short wave periods, particularly for those close to the nature period of the dry tow system, should be evaluated because of the resonance. The motion responses and hydrodynamic forces have been calculated using WAMIT, and a panel model was developed using Multi-Surf for the transport vessel.

Dynamic pressure on vessel hull produced by waves will produce motion of the dry transportation system in all translational and rotational degrees of freedom, and consequently inertial forces are generated in term of accelerations. The accelerations and motions have been calculated by WAMIT. The decoupled analysis will directly adopted the maximum responses in each headings, while for coupled analysis the most critical load cases that cause the most significant response of the FPS and vessel needs to be determined. The decoupled analysis considers the critical load combinations in each heading, from 0 to 360 degrees with intervals of 15 degrees. For each of these headings, 24 wave frequencies ranging from 0.30 to 1.8 radians per second are calculated. The following critical load cases are considered and checked in the coupled analysis.

- Vertical bending moment
- Vertical shear force
- Horizontal shear force
- Horizontal bending moment
- Torsional Moment

All the above load cases are accessed in both positive and negative directions along both longitudinal and transverse axis. For example, the vertical bending moment is evaluated for both hogging and sagging conditions. So there are total 20 load cases included in the coupled analysis. The maximum expected response is determined in a stochastic manner, and a 90% percent percentile in the extreme value distribution for short term response is used.

An "equivalent" regular sinusoidal wave which produce the same maximum expected responses, is selected based on first principle and derived from WAMIT motion analysis results. This wave with selected wave height and period, noted as "Design Wave", is used in the structural analysis of the dry tow transport system. Once the design wave is calculated, hydrodynamic pressures calculated from the hydrodynamic panel are transferred to the structural finite element model through linear interpolation. The mapped values are imposed as loads on the wetted area below the waterline of the finite element model. Two load components, real and imaginary (in phase and out of phase) are considered. Hydrodynamic pressure and corresponding inertia loads are applied to the coupled FE model of the dry tow system.

In order to meet the minimum seafastening forces requirement by DNVGL-ST-N001, 2.5 % of the cargo weight for longitudinal seafastening, and 5% of the cargo weight for transverse longitudinal seafastening, frictions in the interface between vessel and cargo may only be partially used.

FINITE ELEMENT MODEL OF DRY TOW TRANSPORTATION SYSTEM

A generic FPS and vessel are employed in the study, and the properties of the FPS and vessel are listed in Table 1. Shaped cribbing was used in order to achieve uniform pressures on all cribbings under Stillwater load conditions. Seafastening in longitudinal direction, termed as "adapter", and in transverse direction, termed as "strong box", are designed to only resist compression forces. The vessel are usually to expect to have enough width to structurally support the FPS, since the load condition with only narrow support solely on pontoon is so different from the designed operating load conditions, and possible strengthening on FPS column-pontoon connections maybe needed during this scenario. However, wider vessel may lead to higher cost and bigger roll accelerations. On balance, sometimes when the vessel is not wide enough, sponson or outriggers are required to be added on the side shell of the vessel, in order to provide additional structural support to the cargo. The width of current generic vessel meets the requirements, and there is no sponsons or outriggers.

Beam elements are used to model the deck truss members. Mass elements are applied to the deck truss, lower hull to simulate the equipment weight, self-weight. Plate and beam elements are utilized to model the hull structural details. Mass elements are used to simulate the equipment weight, self-weight, and ballast in the vessel. Plate and beam elements are utilized to model vessel structural details. In the coupled analysis, Gap elements are used to simulate the wood cribbings and seafastening, with corresponding shaped cribbing properties. The FE model of the coupled vessel and FPS system is shown in Figure 3, and in decoupled analysis only the FPS, which is the same as its counterpart in Figure 3, is included.

Table 1: FPS and Vessel Properties

	L(m)	W or B (m)	H (m)	Weight (Mt)	Draft (m)
FPS	89	89	52	20000	-
Vessel	210	59	13	40000	9

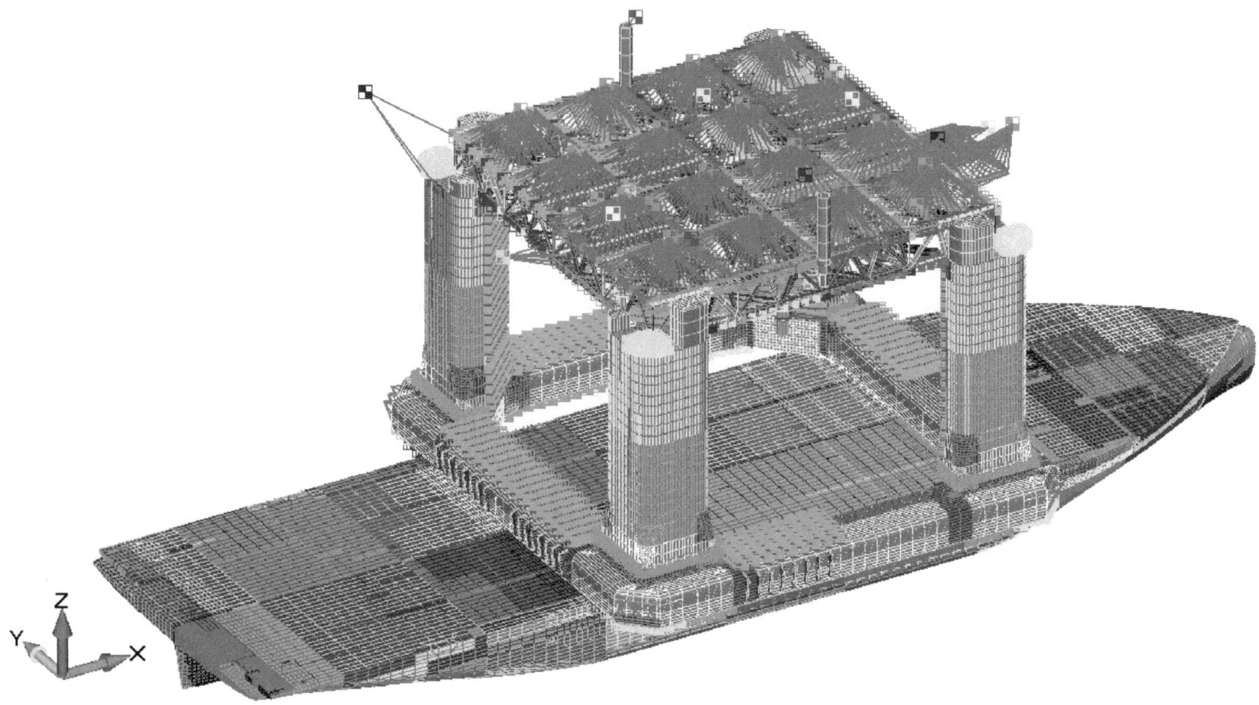

Figure 3: Coupled FE Model of Dry Tow System

RESULTS COMPARISONS BETWEEN COUPLED AND DECOUPLED DRY TOW ANALYSIS

The dry tow structural analysis is evaluated through the following aspects: the critical stress levels at FPS, the seafastening forces and the cribbing pressures.

The critical stresses are located at the pontoon column connections, in 90/270 degree headings, or under the load case of Torsional Moment along Transverse Axis (design wave is also in 90 degree heading). The utilization factor contour for FPS column, pontoon and node in critical load case of both coupled and decoupled analysis are shown in Figure 4. In critical locations, the utilization factor is 0.97 for coupled analysis, and 1.05 for decoupled analysis, which is above allowable. Therefore, regarding on the cargo strength, the decoupled approach is overly conservative, which may result in unnecessary cargo strengthening or more stringent vessel and route selection, and any of which may increase the cost and possibility of schedule delay. It is worth to note that in the present decoupled analysis, the more accurate loads from motions study, instead of more conservative default loads in standards, are adopted. It is anticipated that higher unitization factors can be reached if more conservative loads are applied in decoupled analysis.

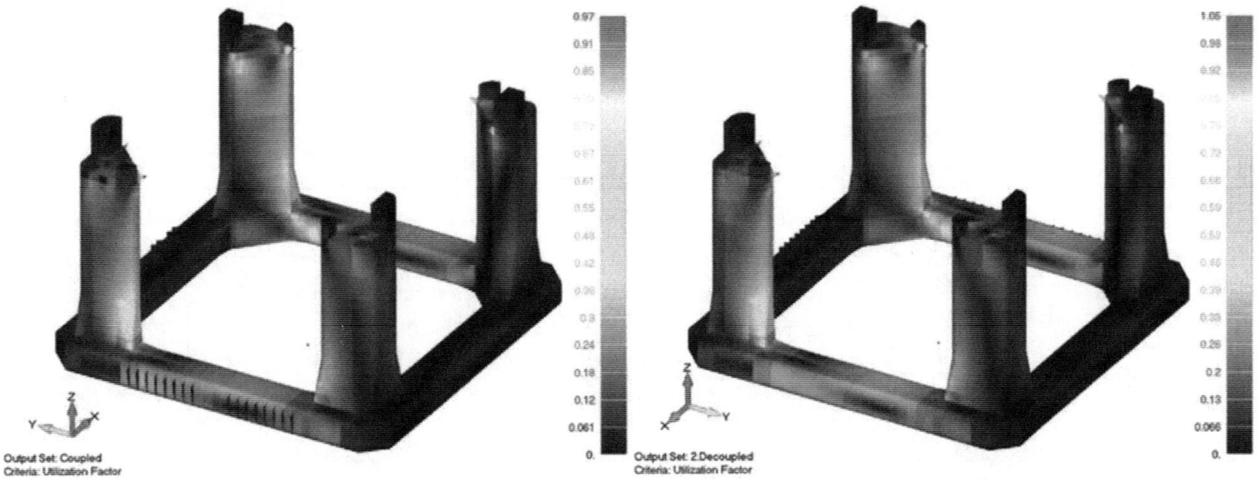

Figure 4: Utilization Factor Contour for Coupled and Decoupled Analysis in Critical Load Cases

In the seafastening forces calculations for decoupled analysis, it is usually presumed that the all seafastening equally contributes. While in the coupled analysis, the forces were derived from the FE model. Seafastening forces in critical load cases in both coupled and decoupled analysis are listed in Table 2. As shown, in the critical case for longitudinal seafastening, considering friction, the maximum seafastening forces in coupled analysis can be up to 5 times bigger than the averaged forces in decoupled analysis considering friction, and even slightly bigger than the case without considering friction. The reason partly lies in the fact that the pontoon deformed as shown in Figure 5, which leads to the adapters on the side resist most of the loads, while adapters in the middle actually have a negative impact. Similarly, in the transverse direction, especially under the yaw condition, the forces in strongbox may be not in the same direction. In the critical load condition, the maximum transverse seafastening force can be up to 3 times bigger than the maximum forces in decoupled analysis. Seafastening design based on the averaged forces could be non-conservative and unsafe.

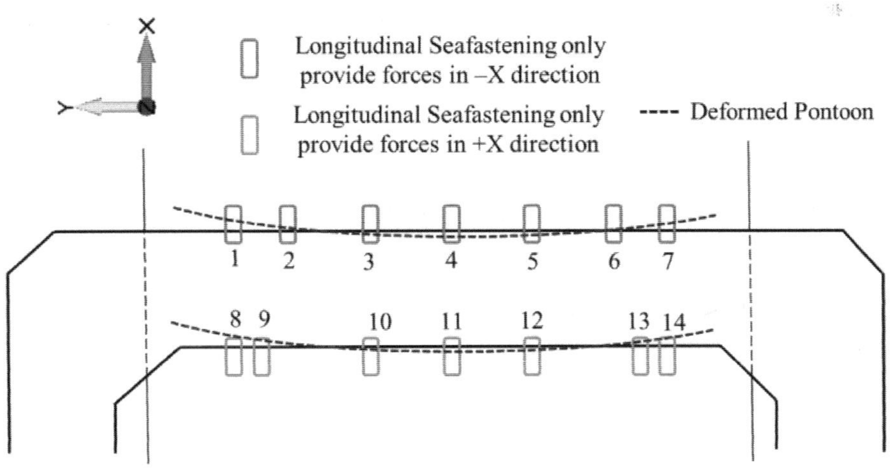

Figure 5: Sketch of Longitudinal Seafastening Layout

Table 2: Forces in Longitudinal Seafastening in Critical Load Case of Coupled and Decoupled Analysis

Coupled Analysis	Adapter #	1	2	3	4	5	6	7
	Seafastening Force (‰ Cargo Weight)	12.44	1.62	0	0	0	2.66	21.03
	Adapter #	8	9	10	11	12	13	14
	Seafastening Force (‰ Cargo Weight)	0	0	-2.89	-6.25	-3.60	0	0
Decoupled Analysis with Friction	Adapter #	1	2	3	4	5	6	7
	Seafastening Force (‰ Cargo Weight)	3.57	3.57	3.57	3.57	3.57	3.57	3.57
	Adapter #	8	9	10	11	12	13	14
	Seafastening Force (‰ Cargo Weight)	0	0	0	0	0	0	0
Decoupled Analysis w/o Friction	Adapter #	1	2	3	4	5	6	7
	Seafastening Force (‰ Cargo Weight)	17.96	17.96	17.96	17.96	17.96	17.96	17.96
	Adapter #	8	9	10	11	12	13	14
	Seafastening Force (‰ Cargo Weight)	0	0	0	0	0	0	0

The maximum cribbing pressure determines the selection of wood type and dimensions for cribbings. The calculated maximum cribbing pressure in the decoupled analysis is 1.78 Mpa, and based on which softwood could be selected since the maximum pressure is under 2 Mpa. However, as shown in Figure 6, the coupled analysis indicates the maximum localized pressure reaches 3.4 Mpa, and hardwood should be used instead of softwood. The reason partly lies in the stiffness contributions from the vessel, because stronger points naturally resist more forces, while the decoupled analysis is unable to take account this effects, and only more averaged pressure can be obtained in decoupled analysis. Also, more accurate forces on cribbing are produced in coupled analysis. The difference between the two critical pressure values suggests there could be unsafe design, which may cause unexpected consequences, such as the crushing of wood cribbing in some locations. Besides, more accurate cribbing pressures will produce more accurate response calculation for both FPS and vessel.

CONCLUSION

In the FPS design, besides the regular load conditions during production, the dry transport load condition is also need to be considered if long transport is expected, and transport load conditions is usually very different from load conditions during production. It is vital to determine the loads and the way of load applying during the transportation. The ideal balanced approach should neither underestimate the load and response, which could results in unexpected consequences, nor overestimate structural behavior, which usually leads to additional cost on unnecessary strengthening on FPS or vessel as well as vessel or route selection.

The traditional decoupled approach may not only provide overly conservative response in FPS structures, but also underestimate the actions in the interface between cargo and vessel. Seafastening and wood cribbing design based on non-conservative forces will be non-conservative and unsafe, which may lead undesirable consequences during dry transportation.

The coupled analysis on dry tow transportation system is able to realistically simulate the load conditions, and provide more accurate results for design. Coupled analysis is able to the reduce cost and risk of project schedule delays.

Figure 6: Pressure on Part of Wood Cribbings

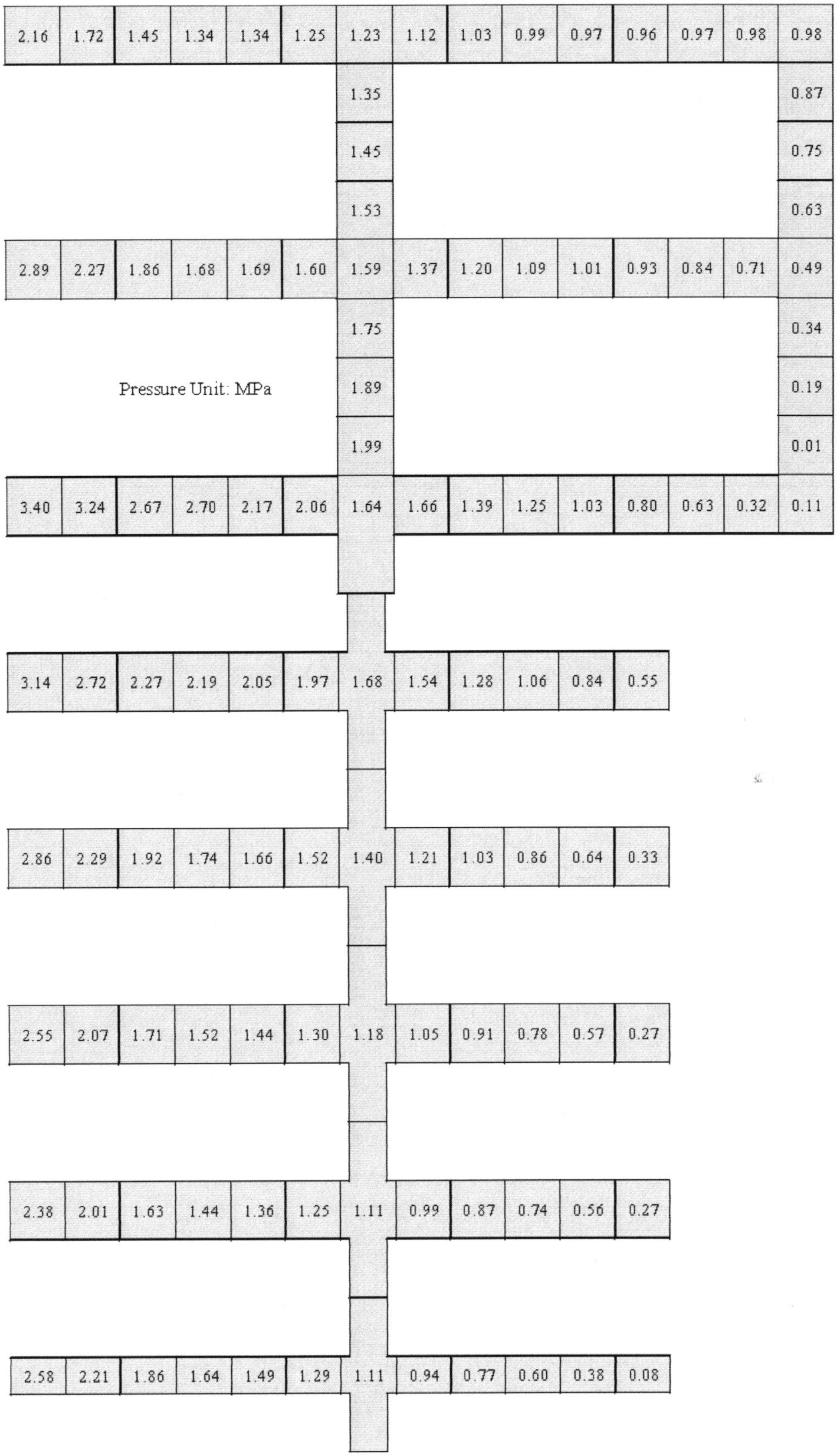

REFERENCES

API, Recommended Practice for Planning, Designing and Constructing Fixed Offshore Platforms – Working Stress Design, 21st Ed

Det Norske Veritas, "Environmental Conditions and Environmental Loads", DNV RP C205, 2010.

Det Norske Veritas, "Marine operations and marine warranty", DNVGL-ST-N001.

Guidelines for Marine Transportations, Technical Standards Committee, GL.

SNAME
THE INTERNATIONAL COMMUNITY
FOR MARITIME AND OCEAN PROFESSIONALS

A Mooring Designer's Approach to a Mooring Monitoring System

Joerik Minnebo
SOFEC, Inc.

Amir Izadparast
SOFEC, Inc.

Arun Duggal
SOFEC, Inc.

Proceedings of the 25th Offshore Symposium, February 27th 2020, Houston, Texas
Texas Section of the Society of Naval Architects and Marine Engineers

ABSTRACT

In recent years, the floating production industry has paid a lot of attention to the integrity of their mooring systems, and live monitoring of the stationkeeping performance and the condition of the anchor legs (with a primary focus on line-failure detection) has taken a prominent role. The requirement for having a mooring monitoring system is now reflected in most regulatory codes, standards, classification rules, and project specifications, but clear guidance on the functional requirements or proper implementation is missing.

This paper introduces a Mooring Monitoring System (MMS) approach that is designed to evaluate both the stationkeeping performance of the mooring system as well as the individual anchor legs, while ensuring the monitoring system is both robust, reliable and in-situ maintainable. Further, it does not only focus on complete anchor leg failure, but also on deterioration of individual anchor leg performance, such as polyester rope elongation, or uneven load sharing within clustered anchor legs. It relies on a thorough understanding of mooring system's stationkeeping characteristics, and on combining data from the DGPS with feedback data from the anchor leg profile under different conditions. Digital advancements can be applied (such as machine learning algorithms), but the principles can also be used deterministically based on design or as-installed data.

Keywords: SNAME, Offshore Symposium, Mooring Monitoring, Anchor Leg, ALMS, DGNSS, DGPS, Virtual Sensor, Machine Learning, Single Point Mooring, Turret, Integrity, Inclinometer, Depth Sensor, Cluster Comparison, Acoustic Communication, Mooring Line, Tension.

Introduction

Until recently, most classification guidelines or regulatory requirements called out load monitoring of the anchor leg system as the means for monitoring the mooring system. This has now more commonly been replaced with the requirement of having a means to monitor the anchor legs, with an emphasis on line failure detection, for example in ABS [1], DNVGL [2] and Oil&Gas UK [3].

Many project specifications however still typically view the direct tension measurement methods as a preferred solution for monitoring the mooring system. In theory, monitoring the tension in each of the anchor legs provides the most intuitive measure to evaluate the status of the mooring system, detailed information on the tension in the anchor leg, and a means for detecting a line failure. Unfortunately, application of direct tension measurement techniques to monitor the mooring system have not been successful in practice. Most of these systems that have been implemented within the industry lack the necessary robustness to provide reliable data for a long-term application, and in most cases servicing or replacing the sensors (typically load cells or strain gauges), power or signal cables is not an option and therefore a failure of a single component jeopardizes the usefulness of the entire system. The case against direct tension monitoring has been made previously in Minnebo et. al. [4], where the main arguments against this method are made in terms of poor industry track record (lack of robustness and therefore poor reliability though the intended service life), accessibility limiting serviceability (placed in the load path, or in hard-to-reach locations), and lack of resolution compared to the range they need to span.

Instead, SOFEC has been using DGPS based monitoring systems to assess the overall stationkeeping performance, as well as Anchor Leg Monitoring Systems (ALMS) assessing the anchor leg condition, based on either inclinometers, or depth measurements, both indicators of the anchor leg profile. The key to either of those systems is that they are robust and reliable, while also maintainable with the anchor leg in service. The ALMS is based on a thorough understanding of how the anchor leg profile changes as the vessel moves around, between near and far offsets. By placing an inclinometer or a depth sensor in a location of large profile variation provides a large range to define the anchor leg profile, with sufficient resolution to detect issues.

When combining the data from the DGPS and the ALMS (and other systems), detection of failure, and anomalous behavior, can be finetuned. SOFEC approaches this by using the DGPS data as the primary method to estimate the anchor leg profile. From this the depth, angle, position and tension at any point in the line can be calculated. By looking at the same location as where the actual ALMS sensor is located, a so called "virtual sensor" is introduced, and the two can be compared.

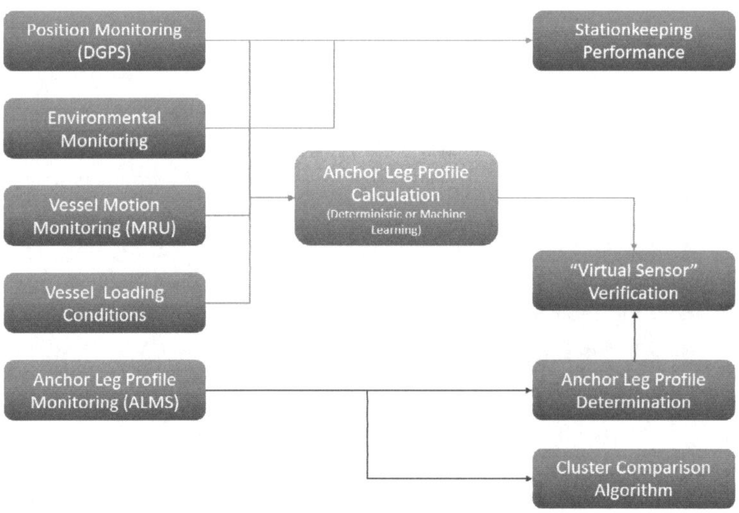

Figure 1: Mooring Monitoring System Setup

Other input data (such as data from environmental monitoring, MRU, draft sensors, etc.) can be utilized to create confidence bounds around the initial estimate, to account for draft changes, vessel wave frequency motions, etc. The actual data from the ALMS is then used to verify the accuracy of the predication. When the calculated anchor leg profile is verified, it can be used for other purposes as well, such as top tension calculation.

Additionally, the ALMS system has a Cluster Comparison Algorithm that monitors for uneven load sharing within groups of clustered anchor legs.

MONITORING THE STATIONKEEPING PERFORMANCE USING DGNSS

Stationkeeping Performance is best monitored using a Differential Global Satellite Navigation System (DGNSS), often referred to as Differential GPS or DGPS (after the American Global Position System), although other commercial services are available as well. This system is used to capture the Low-Frequency Horizontal Motions of the turret centerline (for turret moored vessels) or vessel center (for spread moored vessels), reported in UTM Easting and Northing. Minnebo et. al. [4] provides a detailed overview of the practical application of DGPS for mooring monitoring purposes. The below section is a follow up with some real-life feedback.

Continuously monitoring the vessel position is one of the easiest and most direct means of evaluating the stationkeeping performance of the mooring system, as the main function of any mooring system is to keep a facility stationary in place, to within a certain allowable offset. The actual instantaneous position will continuously be compared to the project defined center position, and referred as an instantaneous offset. A watch circle approach can be used to determine if any design or operational thresholds are exceeded.

Apart from this direct, live feedback to operators, the DGPS collects a wealth of data to assess the overall, long-term performance of the mooring system. In the below examples, long term operational statistical data has been compared to the same parameters that were estimated during the design stage (from 1000+ design fatigue sea states with associated probabilities).

The below heading histogram shows that from design, it was already expected that this single point mooring system would be experiencing very directional conditions (blue bars), and the operational data (orange bars) reflects this as well. But it does also show that the peak of this heading distribution is approximately 30 degrees off from expected. For this specific unit, this means that the FPSO is experiencing significantly more beam-on swell conditions than per the original design.

Another insightful comparison is the turret offset from equilibrium. Similar to the heading comparison, the absolute offset and offset bearings are compared. The overall trend is clearly visible where the mooring system's station keeping performance closely matches with anticipated performance from design. Both the absolute offset curve shows that day-to-day offsets are well predicted, as are the directions in which the offset occurs. If environmental conditions are monitored as well, headings and offsets can be correlated to the input environments as well, further validation of the design accuracy.

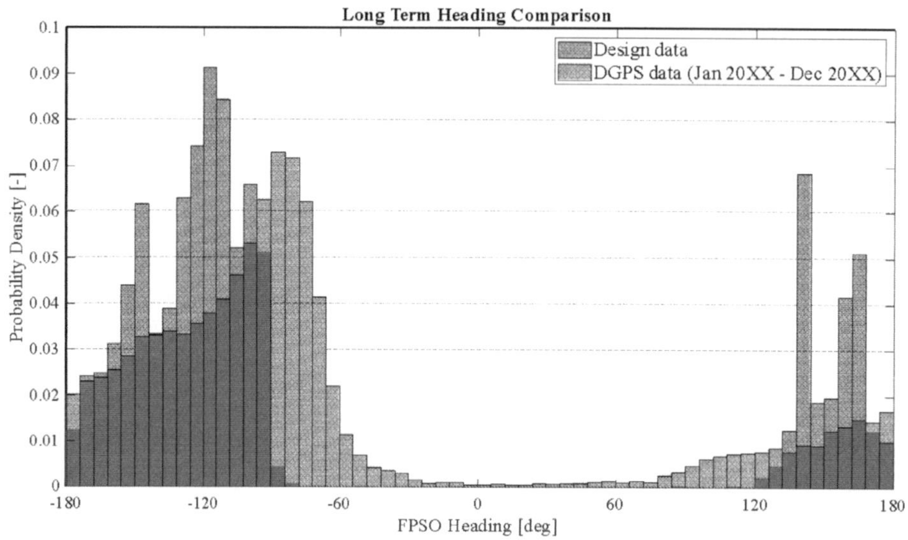

Figure 2: Long Term Heading Comparison using DGPS data

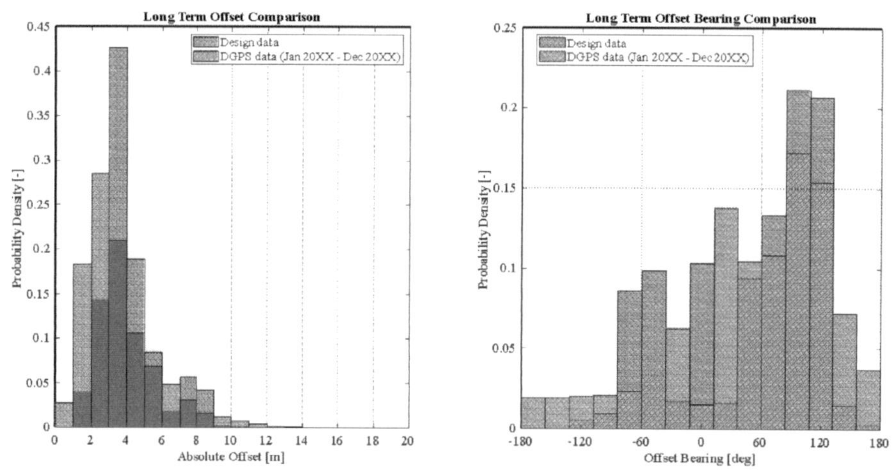

Figure 3: Long Term Offset Comparison using DGPS data

DIGITAL DEVELOPMENTS IN MOORING MONITORING APPLICATIONS

Artificial Intelligence (AI) has become very popular as a tool for detecting anomalous behavior, and multiple applications of AI to mooring monitoring have been published in recent years. Many of them focus on detecting anchor leg failure and estimating anchor leg tension, by using the DGPS information and applying this to a trained Artificial Neural Network.

Prislin et. al. [5] , Sidarta et al. [6] Jaiswal et. al. [7] and Gumley, et. al. [8] for instance all provide methods where an Artificial Neural Network (ANN) is trained based on simulations of vessel position. Some of them enhance the capabilities with additional inputs from measured 6DOF vessel motions, or Metocean data. Prislin et. al. [5] aims to detect a shift in equilibrium position following the loss of an anchor leg's restoring capability, while Sidarta et al. [6] aims to detect the change in a facilities' natural frequency following the change in stiffness from a failed anchor leg.

In both cases, the ANN was provided with a large set of simulations (for many different environmental conditions) for both intact and damaged anchor leg mooring systems, and was thereby conditioned to detect the differences. While still theoretical, the trained algorithms can then be fed live DGPS data in order detect anomalous behavior in the mooring system performance – thereby hopefully detecting an anchor leg failure. Many publications are available on training of ANN, but Sidarta et al. [9] and Kyoung et al. [10] provide good applications of this to mooring systems on Floating Production Facilities. LifeLine JIP is an ongoing Joint Industry Project which also aims to detect anchor leg failure through the use of DGPS data only [11]. Their methodology is based on a combination of a short-term event-based algorithm, as well as long term status-based algorithm, applying advanced data processing to recognize anomalous behavior.

So far, all of these tools have been trained based on numerical simulations, since there are simply not enough actual failures on any asset to provide sufficient "failed" cases to train an ANN. Further, these algorithms can only assume a known condition and configuration of the anchor leg mooring system. Typically the system is trained based on the design condition of the anchor leg mooring system, or sometimes the training is repeated for the as-installed condition, based on feedback from the as-built survey. With subsequent inspections only taking place every 2.5 to 5 years, any intermittent changes to the anchor legs may impact the reliability of the algorithm. Lastly, these algorithms may not be sensitive enough to detect small changes, such as polyester rope elongation or uneven load-sharing between anchor legs within clusters.

In order to capture those subtle changes in a timely manner, some feedback from the anchor legs themselves will be necessary. SOFEC therefore uses an integrated monitoring system, where data from the DGPS is combined with monitoring data from individual anchor legs, based on their profile, so to provide feedback on the accuracy of the DGPS predication. Similar approaches have been presented, for instance by Hageman et. al. [12] which make use of inclinometers on the anchor legs to provide feedback on the predicted response. An application of this principle related to riser system monitoring is presented by Sundararaman et al. [13], which describes this use as a localized response digital twin, or "virtual sensors". In both of these approaches, position and motion data from the floating facility is used to predict the response of the risers or anchor legs. On locations where a response can actually be monitored, virtual (predicted) and real data is then compared in order to validate the prediction accuracy.

While analytic software (such as OrcaFlex and others) makes quick work of transforming vessel motions into dynamic anchor leg responses, it is important to look closely at how the anchor leg profiles changes in order to determine where real and virtual sensors should be applied, and what responses can be observed.

ANCHOR LEG PROFILE DESCRIPTION

The anchor leg profile is mathematically and physically described by a combination of catenary equations and axial stiffness characteristics of the different sections and components of the anchor leg. At each location along an anchor leg, the position (X,Y,Z), inclination and tension are prescribed when the following parameters are fixed: horizontal and vertical end-point separation, combined component and sectional lengths, and individual component and sectional distributed weights and axial stiffnesses, all assuming gravity is constant of course. (While this is not meant to be an exhaustive explanation of catenary equations, the authors do wish to credit the groundbreaking work of Christian Huygens, who first provided proof of these equations in the 17th century, as described in [14])

When in-service, the only variations that typically occur are variations in vertical end-point separation, for instance by changes in water depth (tidal elevation, storm surge) and vessel loading conditions, and in horizontal end-point separation (the vessel drifting under the influence of wind, waves and current). All of these parameters can be carefully monitored: tidal elevations and storm surges are typically monitored by a weather service, vessel draft is typically monitored by the on-board ballasting system, and vessel position is already monitored using the DGPS system described earlier.

It is noted that there could be some long-term variations, such as a change in lengths and axial stiffness (i.e. polyester rope stretching), and distributed weight (i.e. chain corrosion), but those can be assessed periodically when properly accounted for during underwater inspections.

Of the described day-to-day variations, typically the horizontal excursions are most significant to the changes in anchor leg profile. This variation is used to show the influence on the anchor leg profile in the examples below. In both examples, the anchor leg is given a near displacement (offset towards the anchor) and a far displacement (offset away from the anchor), both relative to the neutral equilibrium position.

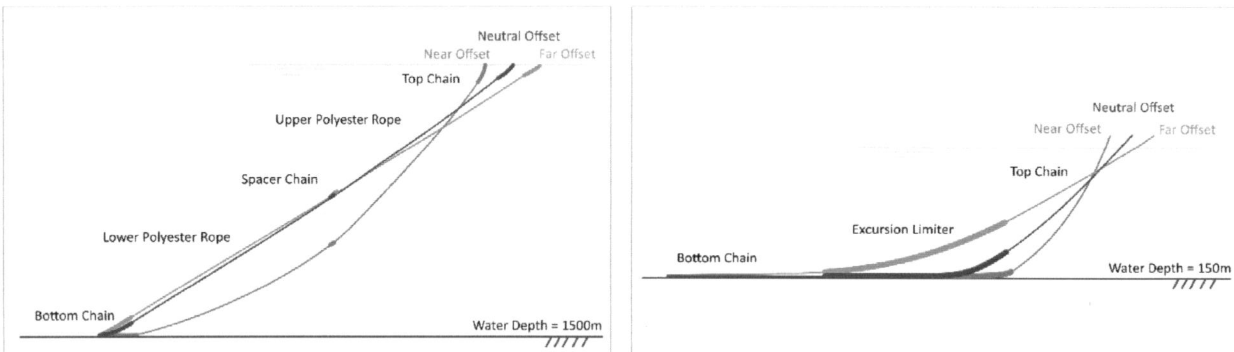

Figure 4: Changes in anchor leg profile for deep (left) and shallow (right) water applications

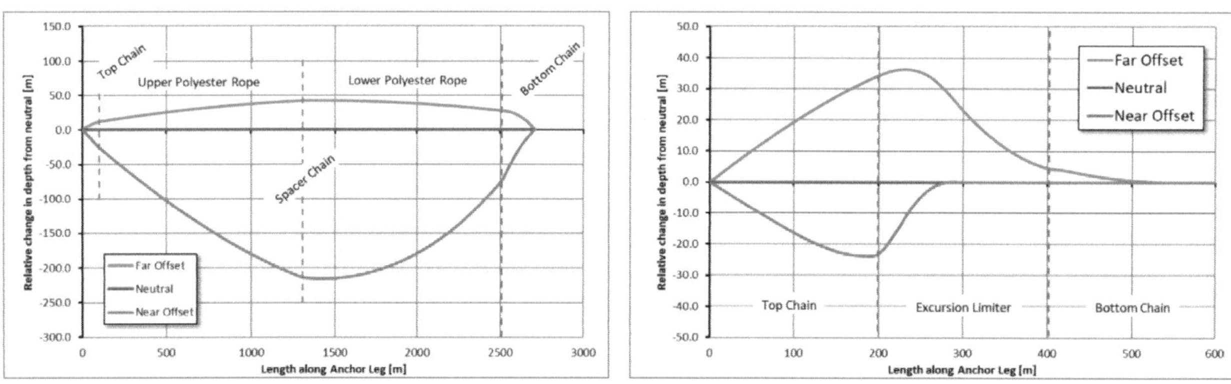

Figure 5: Changes in depth along the anchor leg for deep (left) and shallow (right) water applications

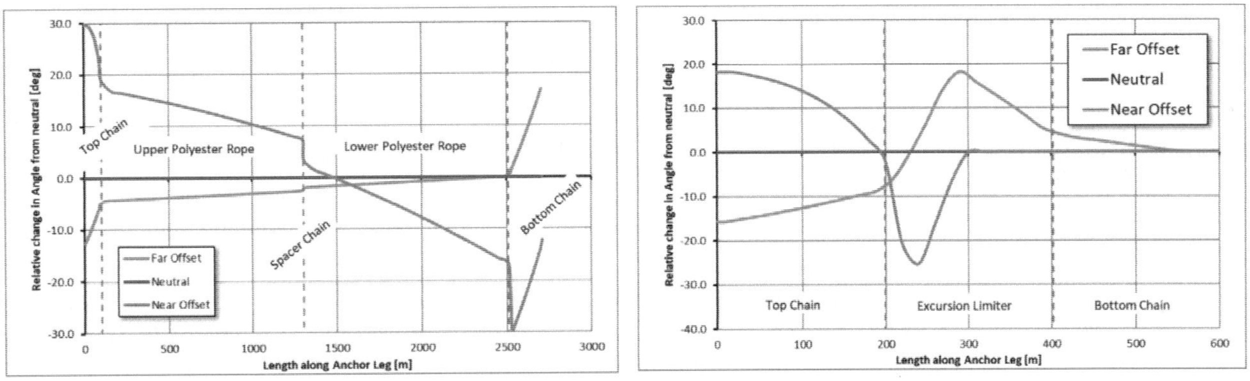

Figure 6: Changes in angle along the anchor leg for deep (left) and shallow (right) water applications

In the examples on the left, a deepwater anchor leg is shown (1500m water depth), which consists of a top and bottom chain, and then two sections of polyester rope, joined by a short spacer chain in between. This anchor leg profile is described as semi-taut, where it transitions from (barely) catenary in neutral and near conditions, to fully taut in far conditions. In this transition, the governing physics describing the profile change from weight (gravity) driven, to axial stiffness driven. The near and far offsets are chosen to be 10% of water depth (i.e. from -150 to +150m).

In the examples on the right, a shallow water anchor leg is shown (150m water depth), which consists of a normal top and bottom chain sections, and then a heavy excursion limiter in the touchdown area, consisting of over-sized chain. This profile is fully catenary driven. The near and far offsets are chosen to be 17% of water depth (i.e. from -25 to +25m).

What can be observed from the above figures is that significant changes in profile occur between near and far offsets. The largest ranges for the angle changes are typically at the top of the anchor leg. In the above examples, both the deep and the shallow water top angle changes are in the order of 30-40 degrees between near and far offsets.

The largest range of depth changes occurs somewhere in the water column. In the deepwater example, the spacer chain in-between the two polyester rope segments provides the largest range of depth. At that location, the depth range is close to 250m between extreme near and far offsets. For the shallow water example, the transition between top chain and excursion limiter provides a depth range of around 55m, making it a suitable location.

ANCHOR LEG MONITORING HARDWARE SETUP

In order to make monitoring systems work reliably throughout the service life of the facility, hardware components shall be robust, reliable and serviceable. With those restrictions, SOFEC considers top angle measurements only feasible on dry anchor leg terminations, such as external turrets, or on spread moored vessels with chain support assemblies above the waterline. Inclinometers can be mounted on chain support assemblies, and can be hard-wired, since they are accessible from on-board without too much hassle (at the very worst, a Rope Access Team may be needed for servicing if the hardware is outboard).

Where above water terminations are not an option (such as internal turrets, or facilities with fairleads in or below the splash zone), SOFEC will typically default to depth sensors, sufficiently deep below the water surface to ensure accessibility by ROV. Cabled solutions are no longer feasible, due to robustness requirements, and the long, exposed distance from the structure. Therefore, the depth sensors will typically come as packages with batteries for power, and an acoustic modem for data transmission. On the floating facility itself, one or more acoustic receivers are deployed below the hull. They can be retrieved through hawsepipes to a dry elevation for servicing. These subsea packages are mostly all built into a single housing, and are periodically retrieved by ROV for fresh batteries, and any recalibration if required. Data collection and especially the acoustic data transfer are the biggest energy consumers, and they need to be managed to ensure a long battery life. Ideally, battery life should exceed the typical ROV inspection frequency of once per 2.5 years, such that battery replacement does not require a separate ROV vessel mobilization.

Resolution plays an important role in achieving reliable data, and sensors shall be selected to provide a good balance between range and resolution. Given the typical 30 to 40 degrees range of top angles, the minimum resolution of an inclinometer should be in the order of 0.1 degrees or better, leading to a range of 300 to 400 points to define the anchor leg profile. Clearly, the finer the resolution, the better the profile is described.

Depth sensor resolution is typically a function of their water depth rating, where shallow water applications will have a finer absolute resolution than deepwater applications. A good range of data points is in the order of 500 points, so for the deepwater example (250m depth range), the depth sensor should have an absolute resolution of 0.5m or better, while in the shallow water case (55m depth range) the resolution needs to be in the order of 0.1m or better.

At the selected locations, the following offset curves are derived, based on either the as-installed configuration of each line, or on updated knowledge from in-service inspections. The difference between the shallow water and the deepwater application is obvious when looking at the non-linearity of both the depth and angle curve for the deepwater example. The physical transition from catenary governed response to axial stiffness governed response is visible as the curves flatten out at far offsets. The curves on the shallow water example remain much more linear across the entire offset range.

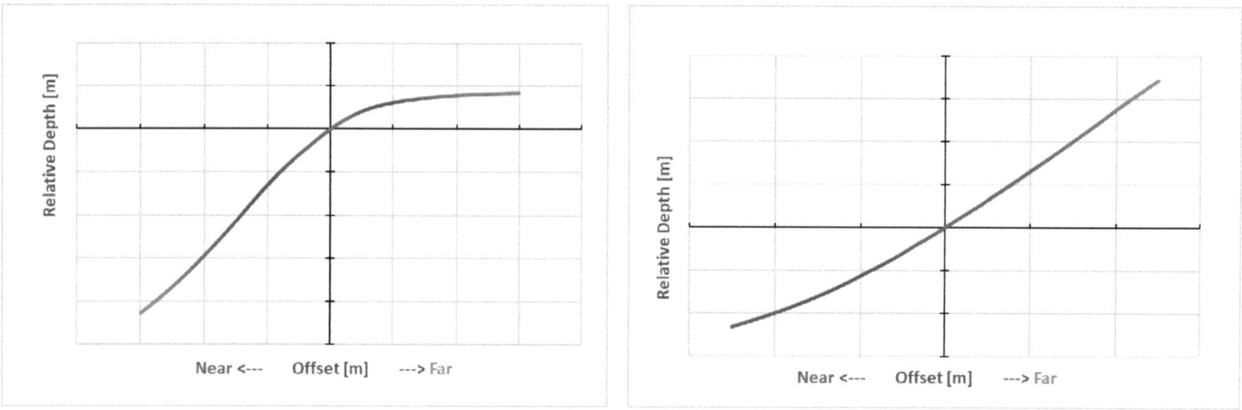

Figure 7: Offset-Depth curves at selected lcoations for deepwater (left) and shallow water (right)

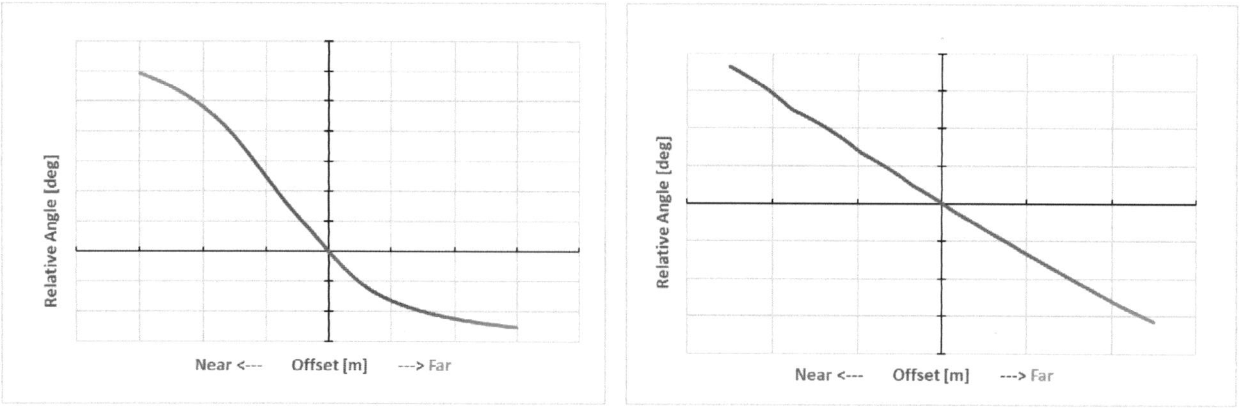

Figure 8: Offset-Angle curves at selected locations for deepwater (left) and shallow water (right)

Now that the angle or the depth at these locations is known given a certain offset, the Anchor Leg Monitoring System can be used to verify that this is actually the case. Using the DGPS, and the offset-depth and offset-angle curves derived for the particular location, the theoretical depth and angle can be calculated, and as long as the measured angle or depth matches the theoretical depth, the line is performing as it is intended to.

Key to automating the detection of anomalous behavior is to define what is an acceptable difference between the predicted profile, and the actually measured value. Threshold values on both the plus and minus side of the predicted value need to take into account sources of uncertainty. Typical sources are: DGPS accuracy, depth or angle sensor accuracy, vessel loading conditions, vessel dynamics (pitching & rolling), line dynamics (the offset curves are derived from static calculations) and wave elevations (a crest or a through over the depth sensor may result in a few meter difference). By taking these sources of inaccuracy into account, the inaccuracies can be expressed as confidence bounds around the offset curves, shown below for the deepwater off-set depth curve as an example.

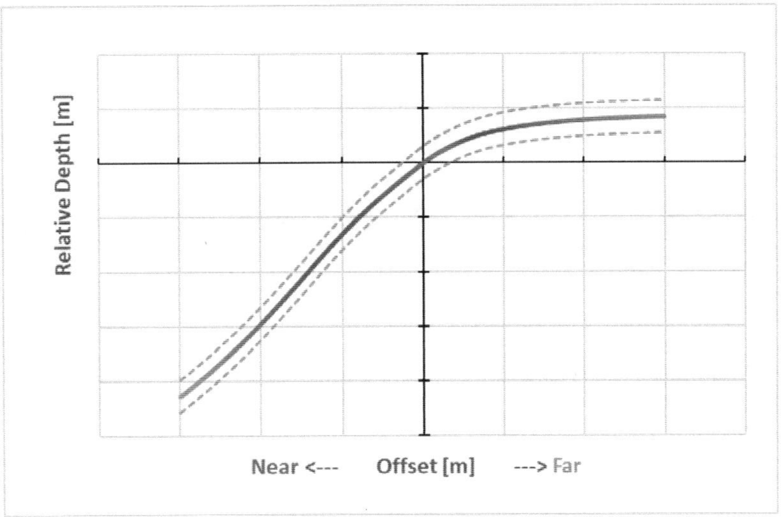

Figure 9: Offset-Depth curve (deepwater) with exaggerated confidence bounds

There are ways of dealing with these confidence bounds. Obviously, the better the accuracy of the Position Monitoring System, the tighter these bounds can be. Also, by monitoring the vessel loading conditions, the whole confidence bound can move up or down, according to the actual draft of the vessel. Vessel dynamics (due to severe weather for instance) can be accounted for by incorporating the feedback from an MRU. Widen the confidence bounds when the vessel experiences large pitching, rolling and or heaving motions, and tighten them when the vessel is calm.

While the above determination and optimization of the confidence bounds can be done as an engineering assessment and applied deterministically, this is an area where Artificial Intelligence can prove itself to be a useful tool. A Machine Learning Algorithm (such as a trained Artificial Neural Network) can deal well with the multitude of input data that needs to be accounted for to minimize the confidence bounds.

Lastly, time-dependent dynamic influences from vessel motions, environmental conditions and line dynamics can effectively be dealt with by averaging the measurements over time, rather than looking at instantaneous values. Typically, when dealing with acoustic communication systems (such as the depth sensors deep down the anchor leg), continuous data from the ALMS is not available anyway, and periodic bursts of statistical data derived from timeseries are transferred to the topsides instead.

DETECTING LINE FAILURE

Let's focus on the depth measurements of a deepwater anchor leg mooring system (spacer chain in between two polyester ropes), even though the principles described below can be applied to angle measurements, and to shallow water applications as well.

Using the vessel position (monitored by the DGPS) and circumstantial data (loading conditions, MRU, environment, etc.), the mean depth of each virtual sensor (red dotted lines) and confidence bounds (the amplitude -or height- of the boxes) are estimated, shown for each line. The figure below shows the schematic for a machine learning application. Within this particular application, this assessment is not made continuously, but based on time averaged windows, matching the time histories of the Anchor Leg Monitoring System. From those actually measured depth time histories, the simple statistics are calculated (mean and standard deviations) and are visually represented by the stick bar shown in the figure below.

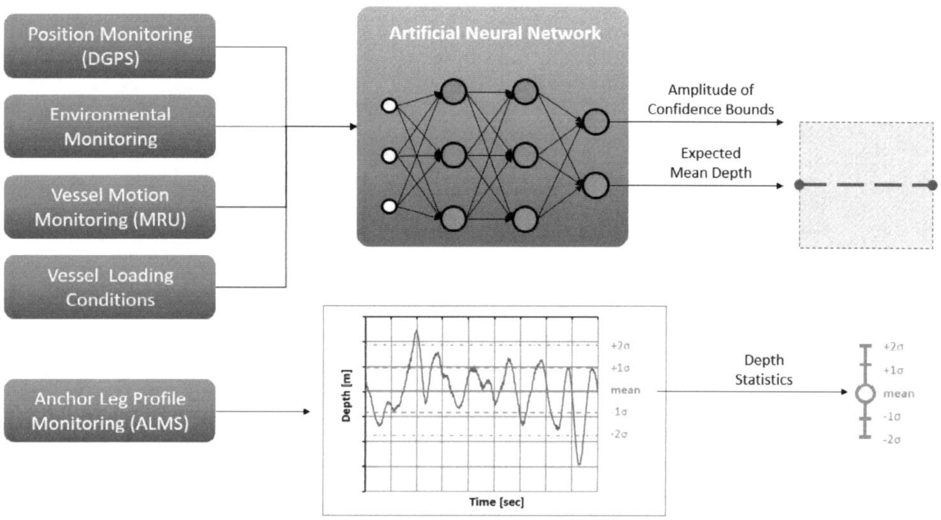

Figure 10: Data Treatment and Visualization

The figures below provide a snapshot of how the statistical data is visualized and interpreted for a deepwater anchor leg system with 10 equidistant lines (10x1 mooring system, with all lines spaced 36 degrees apart). The actual depths have been zeroed immediately after the anchor legs were installed and tensioned, so that all the depths are relative to the as-installed neutral condition. In this example, the turret has an offset towards line #7, so line #7 gets slack, and the opposite line (#2) gets tight. All other lines will be somewhere in between.

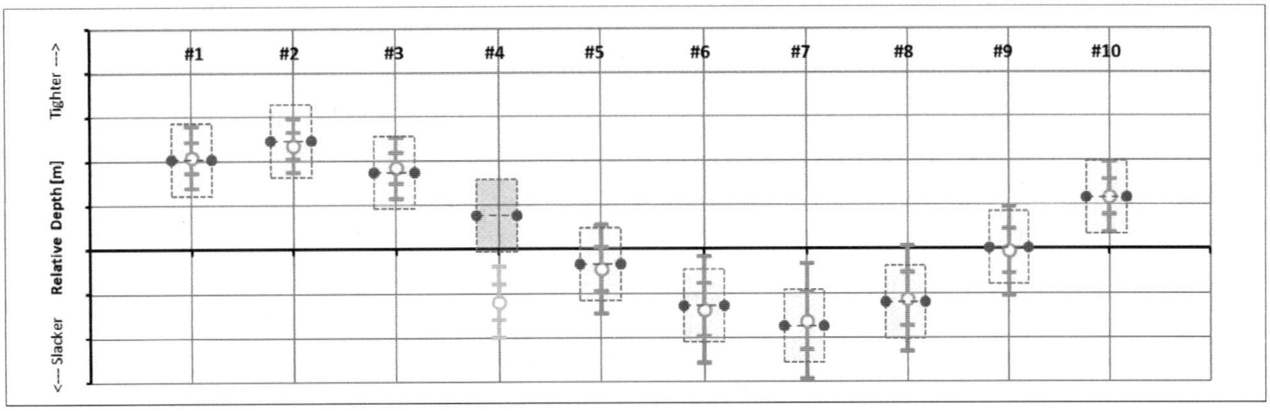

Figure 11: 10x1 mooring system, with anchor leg #4 showing anomalous behavior

In the above example, line #4 is clearly showing anomalous behavior. If the indicated depth of line #4 would be outside the known depth range (so below a relative depth of -220m, derived from Figure 5), this would be a clear sign of line failure, as either the line would be hanging straight down, or the component to which the sensor is attached would have dropped to the seabed. But in this case the line may not have completely failed. An issue like this would likely not be picked up by Artificial Intelligence-based monitoring systems only looking at DGPS data, since the shift in position may not be significant enough to be classed as a failure.

The issue with line #4 in the above example could be an individual polyester rope having elongated (perhaps due to overload), or an anchor having shifted. This Mooring Monitoring System with direct feedback from an ALMS can thus actually show deterioration in individual line performance, even if it is not a complete line failure.

DETECTING UNEVEN LOAD SHARING WITHIN CLUSTERED ANCHOR LEGS

Many permanent mooring systems these days consist of clustered rather than equidistant anchor legs, whether by default (on spread moored systems) or for better performance, redundancy requirements and riser configuration (on single point moored systems). Clustered anchor legs tend to behave very much alike (if properly installed), and share the load reasonably equally between all lines within a cluster. Equal load sharing is an important assumption within the mooring design and often an installation criterion.

While the failure detection algorithm as described above may capture this, it is possible that the confidence bounds are larger than what is required to pick up uneven load sharing, such as shown in Figure 12 below, for a mooring system consisting of 12 anchor legs, in 3 clusters (a 4x3 configuration). The tight clustering of the anchor legs results in all anchor legs within a cluster having very similar mean values as well as standard deviations. If one of the anchor legs shows significantly different behavior than the others within the same cluster, this could be reason to suspect line failure or uneven load sharing between lines within a cluster. Within this example, the predicted depths and confidence bounds are provided by cluster rather than individual line (for reasons of simplicity). In the figure below, line #2 falls within the confidence bounds of the DGPS based depth estimate, but it clearly shows that it is off compared to the other lines within its cluster.

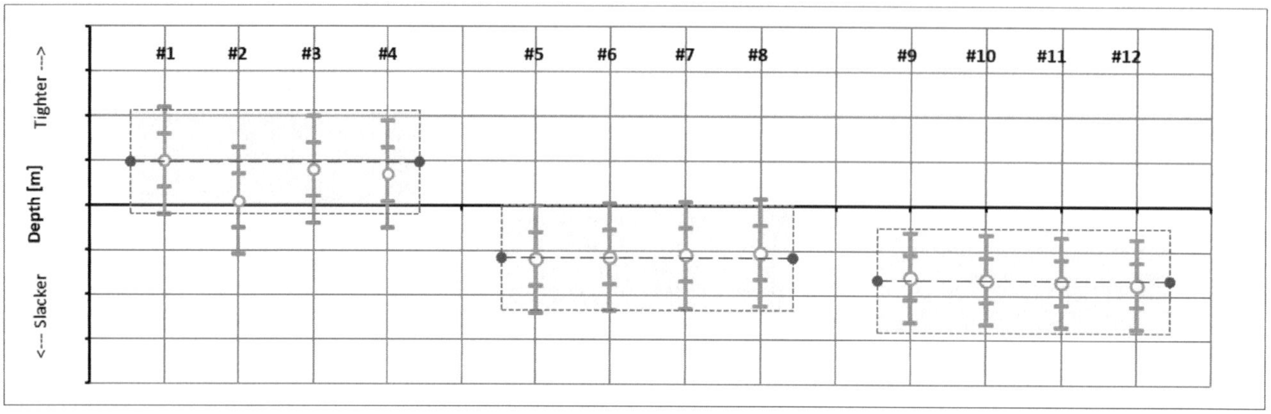

Figure 12: Depth data visualization for Clustered Mooring Systems.

To capture this, SOFEC has developed an algorithm to detect this anomalous behavior and raise warnings and alarms accordingly, and initial results were published in Duggal et al. [15]. This algorithm is referred to as the Cluster Comparison Algorithm. Thresholds that define different levels of warnings or alarms have been tied to the individual mean depth compared to the averaged mean depths of the other lines within the cluster. The allowable range of individual mean depths compared to that average can be defined as fixed values, or as a function of the averaged standard deviations of that cluster. This logic is further visualized in the figure below.

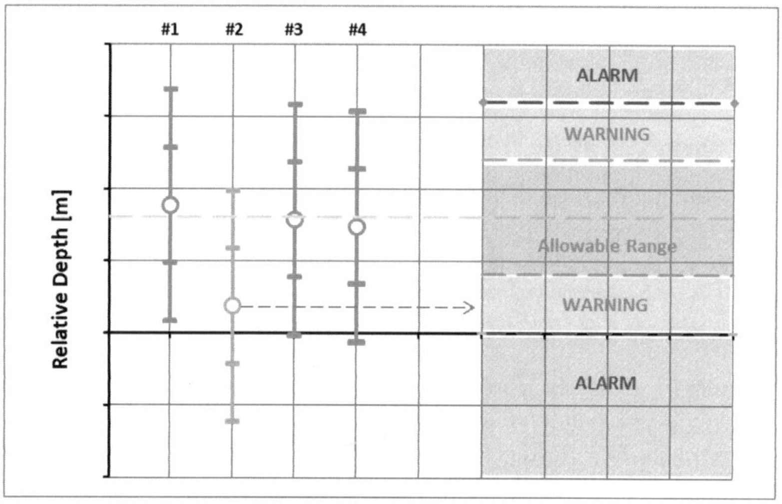

Figure 13: Visualized Example of Cluster Comparison Algorithm

In this example, the warning level was set at the average of the 1-sigma values of the cluster, and the alarm level was set at the average of the 2-sigma values. The appropriate levels are to be defined on a project specific basis.

CONCLUSIONS

Within this paper, SOFEC has presented a practical application of a Mooring Monitoring System, comprising mainly of a DGPS and an Anchor Leg Monitoring System (ALMS). The DGPS is used for monitoring the position of the floating facility, while the ALMS monitors the profile of each of the anchor legs. Based on a thorough understanding of the anchor leg profile, and how this profile changes for changes in floater position, a suitable location can be selected for placing actual sensors on the anchor legs. These are typically either inclinometers at the very top of the anchor legs (if accessible for maintenance), or (battery powered) depth sensors deeper down the water column using acoustic communication for data transfer. SOFEC has applied this successfully on a shallow water FPSO in the North Sea, and is currently implementing it on an ultra-deepwater FLNG offshore East Africa, as well as a deepwater FPSO in West-Austalia.

Individually the DGPS is used for continuous monitoring of the position keeping within specified limitations, while the long-term statistics are used to compare the station keeping performance against the designed conditions. The ALMS by itself provides direct feedback on the anchor leg profile, and on anchor leg condition. Large changes (such as anchor leg failure) are detected instantly, while more subtle changes (such as uneven load sharing between grouped anchor legs in clustered mooring systems) can be identified through a Cluster Comparison Algorithm.

When combining the DGPS and the ALMS, the DGPS can be used for calculating all anchor leg particulars of interest (top tension, angle, position, etc.). By applying a "virtual sensor", the calculated anchor leg profile can be verified by the actual sensor from the ALMS. This methodology can be used deterministically (based on the static offset-profile curve), or with additional input from other monitoring systems (such as an MRU, environmental monitoring and vessel loading conditions), this can be optimized using digital methods such as Machine Learning through Artificial Neural Networks.

Threshold levels can be set for different levels of disagreements between the two sensors, and can be set to warn for suspected anchor leg failure, or for anomalous behavior (i.e. suspected deterioration in individual anchor leg performance due to for instance polyester rope elongation, or anchor shifting etc.).

REFERENCES

[1] American Bureau of Shipping (ABS) (2019): "Guide for Position Monitoring Systems," July 2019

[2] DNVGL, Offshore Standards (2018): "*DNVGL-OS-E301: Position Mooring,*" July 2018

[3] Oil & Gas UK (2014): "*Mooring Integrity Guidelines*", Issue 3, November 2014

[4] Minnebo, J., Aalberts, P. & Duggal, A. (2014): "*Mooring System Monitoring using DGPS,*" OMAE2014-24401

[5] Prislin, I. & Maroju, S (2017).: "*Mooring Integrity and Machine Learning,*" OTC-27866-MS

[6] Sidarta, D.E., O'Sullivan, J., Lim, H-J. (2018): "*Damage Detection of Offshore Platform mooring line using artificial neural network,*" OMAE2018-77084

[7] Jaiswal, V., & Ruskin, A. (2019): "*Mooring Line Failure Detection Using Machine Learning,*" OTC-29511-MS

[8] Gumley, et. al. (2019): *"Application of Machine Learning Techniques as a means of Mooring Integrity Monitoring"*, OMAE2019-96411

[9] Sidarta, et. al. (2019): "*Detection of Mooring Line Failure of a Spread-Moored FPSO: Part 1 — Development of an Artificial Neural Network Based Model,*" OMAE2019-96288

[10] Kyoung, J. et. al. (2019): "*Detection of Mooring Line Failure of a Spread-Moored FPSO: Part 2 — Global Performance Analysis using MLTSIM,* " OMAE2019-96339

[11] Righetti, R. et al. (2020): *"Integrity Assessment of Mooring Systems using Position Measurements,"* OTC-30482-MS

[12] Hageman, R., Aalberts, P., Leeuwenburgh, R., & Grasso, N. (2019): "*Integrity Management of Mooring Systems,*" OTC-29560-MS

[13] Sundararaman, S., Theti, R. & Hill, M. (2018): "*Data Driven Virtual Sensors for Riser Prognostic Integrity Management,*" OTC-28732-MS

[14] Bukowksi (January 2008): "*Christiaan Huygens and the Problem of the Hanging Chain,*" The College Mathematics Journal, Vol. 39, No.1.

[15] Duggal, A., Izadparast, A., Minnebo, J. (2017): "Integrity, Monitoring, Inspection and Maintenance of FPSO Turret Mooring Systems," OTC-27938-MS

[16] Ping, L. & Wang, S. (2016): "*On the Monitoring of Mooring System Performance,*" 21st SNAME Offshore Symposium

SNAME
THE INTERNATIONAL COMMUNITY
FOR MARITIME AND OCEAN PROFESSIONALS

OFFSHORE ANCHORING SYSTEMS WITH TORPEDO PILES

THIAGO ESTEVES MARTINS
DELP ENGENHARIA MECÂNICA S.A.

Proceedings of the 25th Offshore Symposium, February 27th 2020, Houston, Texas
Texas Section of the Society of Naval Architects and Marine Engineers

ABSTRACT

Anchoring systems in deep-water Oil & Gas operations can be very challenging, considering the high loads usually applied and the soft soil to hold the anchor.

The development of offshore anchoring systems with Torpedo Pile is a technological application that can save time and cost during the mooring installation. With a very efficient holding capacity, and using the free-fall energy for its own installation, this technology is widely used in Brazil, by Petrobras. It is estimated that more than 2,000 torpedo piles were successfully installed in Brazil's shore. This Kind of Anchor is field proven, applied to FPSOs, Rigs, Steel Risers, Flexible flowlines and Umbilical cables anchor process, with a large range of holding capacity loads.

This paper presents the advantages to use this technology, considering the overall mooring installation process, and suggest different applications for torpedo piles.

Considering the nowadays increasing requirement for renewable energy, and the large potential for offshore wind power generation, this work brings the proposal to use the torpedo pile technology in the mooring design for the floating wind turbines. With the anchoring efficiency, in addition to the reduced installation time, torpedo pile can improve competitiveness in such a challenged market.

Keywords: Torpedo Pile, Mooring systems, Anchoring systems

INTRODUCTION

Anchoring systems based on Torpedo Pile Technology has been widely used in Brazil by Petrobras, country's major oil and gas operator, since the year of 2005. Flowlines, Steel Risers, Umbilical Cables, and large Floating Production Units are being successfully anchored with this system.

Torpedo pile's design comes from the soil characteristics and load requirements. Through computational programs it is possible to predict pile embedment depth, structural load calculations and inverted catenary configuration in the buried section. A FEA modeling is performed to verify all Stress analysis along pile's body, especially at critical spots (pad eye, plate / wing connection, for example).

The torpedo pile is designed in such way that its kinetic energy acquired by the free-fall gives enough force to penetrate the soil. It is extremely important to have soil data as assertive as possible.

Compared to other anchoring systems, torpedo pile shows a range of advantages, with your simple and effective and field proven design, manufacturability, handling, transportation and above all installation. From the subsea design point of view, by using torpedo pile, which is a taut leg mooring system, you can reduce the radius of anchoring lines compared to a catenary system, helping the subsea layout. Also, by reducing the radius, the anchoring line itself would be shorter.

METHODOLOGY

DESIGN CONCEPT

SOIL PARAMETERS

Soils in deep water offshore locations are normally soft clay. Soft Clays usually have low permeability. in this matter we should consider the undrained shear (Su) as a main parameter to be used in pile design. The shearing stress applied against the soil by the pile itself is the resistance force to be calculated.

One geotechnical model to be used in embedment calculation is the model proposed by True (TRUE, D. G. **Undrained Vertical Penetration into Ocean Bottom Soils**. Ph.D. Thesis, University of California. Berkeley, California, E.U.A, 1976).

In soft clay, especially from deep water locations, the Su has a linear behavior along soil depth. The Equation is:

$Su\ (z) = Su0 + Suk.\ z$ (0)

Where:

$Su0$ is the resistance on the surface and Suk is the resistance growth rate as depth grows, in Kilo Pascal, and;

z is the depth, in meters.

Pile embedment depth can be assessed using True's method, accounting for both strain rate effect and inertia drag. The model starts with Newton's second law as follows:

$$m\frac{d^2z}{dt^2} = W_s - F_{resistance} - F_b - F_{drag} \qquad \text{(Eq.1)}$$

where

m	=	submerged mass of the pile, in kg (kg, lb)
t	=	time, in sec
z	=	embedment depth at the pile tip, in m (m, ft)
W_s	=	submerged weight of the pile in water, in N (kgf, lbf)
$F_{resistance}$	=	upward force resisting movement of the pile through the soil including bearing and frictional resistances, in N (kgf, lbf)
F_b	=	buoyancy weight of the displaced soil, calculated as the soil displaced volume which is limited to the pile volume multiplied by the soil submerged unit weight γ', in N (kgf, lbf)
F_{drag}	=	fluid mechanics drag term, in N (kgf, lbf)

F_{drag} should be considered for very soft viscous clay encountered at soil surface and is calculated using Morison's equation:

$$F_{drag} = \frac{1}{2} v^2 \rho_s A_{tip} C_d$$

where

v	=	pile velocity, in m/s (m/s, ft/s)
ρ_s	=	density of soil, in kg/m^2 (kg/m^2, lb/ft^2)
A_{tip}	=	projected area of the pile, in m^2 (m^2, ft^2)
C_d	=	drag coefficient

The differential equation for the velocity at any given depth or time is solved numerically using a finite difference approach. The predicted embedment corresponds to the depth of pile tip where the velocity of the pile goes to zero.

Another important parameter is soil's sensitivity (St) which represents loss of resistance due to the destruction of its soil. This usually happens when the pile suddenly penetrates the soil, breaking its original structure. Soil sensitivity value is defined by the ratio of original resistance and reduced resistance after disturbing process, as follows:

$St = Su / Su_{(reduced)}$

Empirically this Ratio is often something between 2,0 and 2,7.

TORPEDO PILES

A torpedo Pile is an Offshore Anchor type used for deep water mooring systems. Hydrodynamic designed to be installed by free-fall, with a simple Anchor Handling Vessel Tug (AHTS), this pile works fully embedded in seabed, with the mooring line shaped as a reverse catenary, as shown in figure 1.

Figure 1: Torpedo Pile, courtesy of DELP Engenharia Mecânica S.A.

The reversed catenary also embedded, in addition to the fact the pile ends up in a stiffer area of the soil (as known, shear undrained properties increases directly with soil depth), gives torpedo pile a much higher Holding Capacity, especially in horizontal loads.

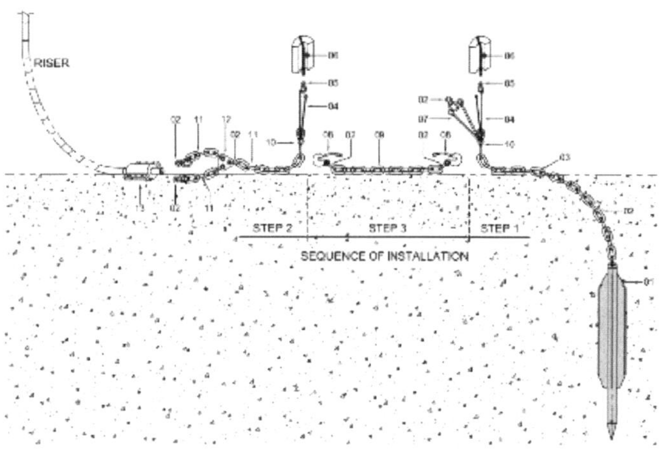

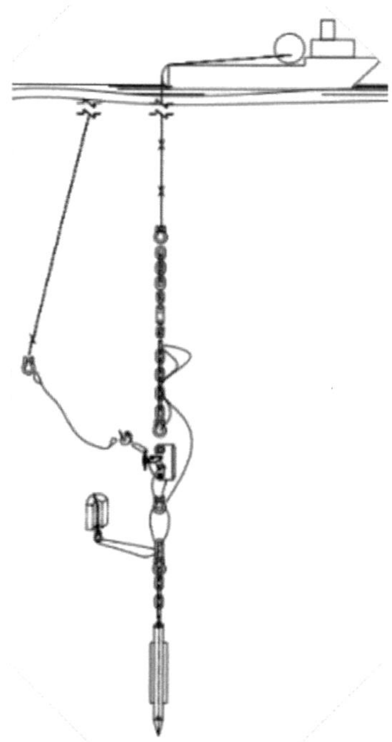

Figure 2: Torpedo Pile Installation Arrangement before and after free-fall.

Design requires inputs from soil geotechnical data and load application. Soil data as Shear Undrained resistance (Su), Sensitivity (St), type of Soil and its layers, in addition to load data of the anchored point are required to specify the proper Torpedo Pile and the installation procedure.

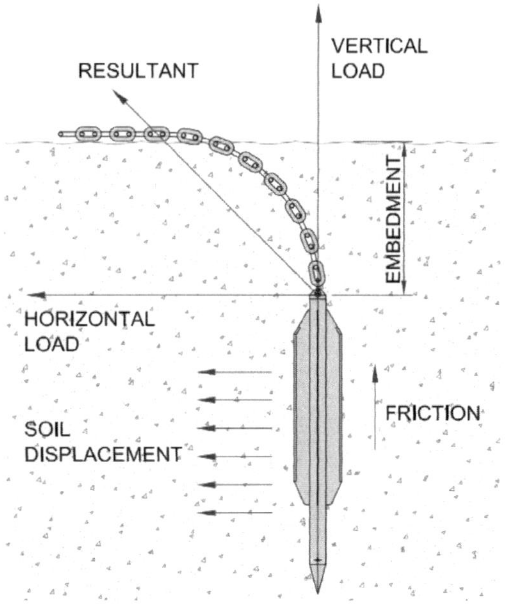

Figure 3: Torpedo Pile Load Analysis

Ultimate Holding Capacity (UHC) is directly proportional to the pile shape and weight. As per the above figure, once the reversed catenary is shaped, pile holding capacity comes from the resultant of vertical and horizontal load. Vertical loads are held for the weight and friction, while the horizontal load is most held by the side area of pile, increased by the wings.

API RP 2A criteria for pile capacity are typically used for loads applied in the axial or horizontal directions. However, the API approach is unable to explicitly account for a mooring line loading angle. The pile capacity can be predicted according to a Finite Element Analysis, as showed in figure 4..

Recently a computational fluid dynamics (CFD) approach has been used to model the complete installation event including: the release and free-fall through the water column, the transition from water into the seabed and the subsequent motion of the pile through the soil.

Pile embedment is perfectly foreseen, according to the method of true (True, D. G., Undrained Vertical Penetration into Ocean Bottom Soils).

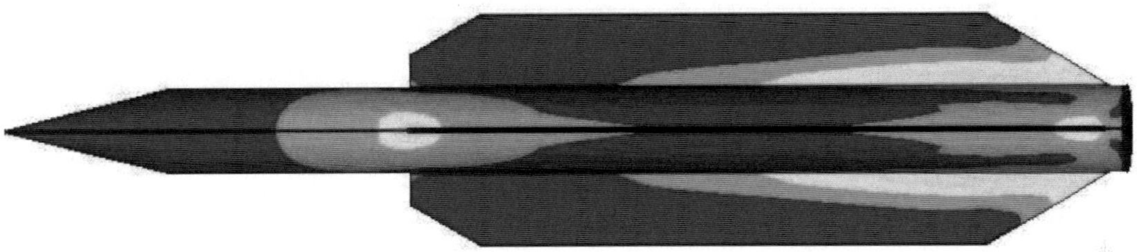

Figure 4: Torpedo Pile FEA exemple.

Nowadays, considering the vast number of data available after so many piles installations, design is well validated.

Important to say that Structural analysis is performed considering the smaller lateral area of the pile, when the tensioning occurs at 45°, between fins, as showed in the figure below:

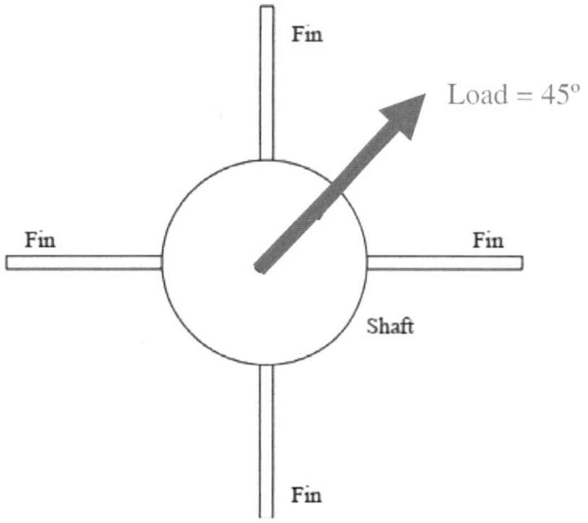

Figure 5: critical load consideration.

The holding capacity analysis is nonlinear. Pile displacement cannot exceed 7% of External pile diameter.

INSTALLATION PROCEDURE

Torpedo pile installation is based on a free-fall method, where the pile is dropped from a calculated distance from seabed to give enough impact energy required for certain penetration. As mentioned before, True's method allows to predict the embedment pile will have.

Installation can be adjusted according to the available vessel, and the most common ways are as follows:

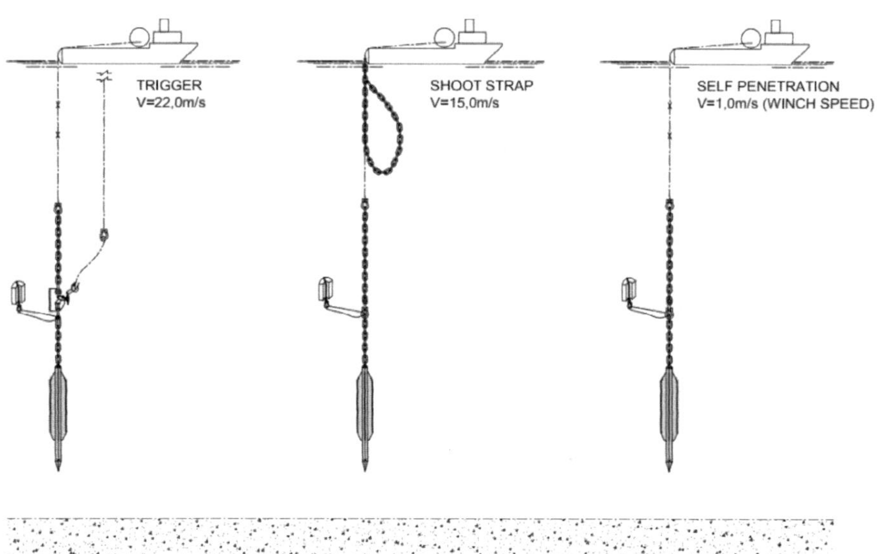

Figure 6: Torpedo Pile installation methods.

a. Mechanical Trigger: higher impact speed, most common for bigger piles, installation line is interrupted close to the pile's pad, with enough length of chain to make the bottom connection.

b. Shoot Strap: medium impact speed, a lace is made on deck and trapped by the shark jaw. Once shark jaw is open, pile falls pulling the entire mooring line. In this case accuracy on the installation spot is higher.

c. Self-penetration: sometimes the installation vessel's crane can't have a sudden variation on the weight, so the pile can be installed according to the which speed. Embedment is lower, but something to be compensate with mass.

Main concern about torpedo pile installation is the final position of it. This is very important to be understood as a very field proven process, so all parameters are well controlled and predicted.

It used to be common to install an instrumented bottle in the pile, in a side housing, to be retrieved after installation. This bottle would give final position and inclination. As discussed before, UHC is a function of the resultant force in the pile's pad eye, so final inclination is a very important parameter. That's why there's a very important characteristic in pile's design: center of gravity should be close to the tip, to assure verticality during the free-fall and embedment.

Estimated pile behavior can be predicted based on a driven test in-situ. Usually this test is performed with the self-penetration method of installation, so the pile may not penetrate to the exact depth as predicted. However, by measuring all parameters from this trial, as embedment, tension to retrieve and speed of fall it is possible to extrapolate results in order to validate soil parameters. Math modeling can be updated with such data in order to validate the installation parameters.

Besides tilting, critical velocity is also an important parameter to assure that final embedment will be achieved. A common curve is presented below, obtained from the instrumentation bottle fixed to the pile. This is a regular flowline pile installed with the shooting strap method.

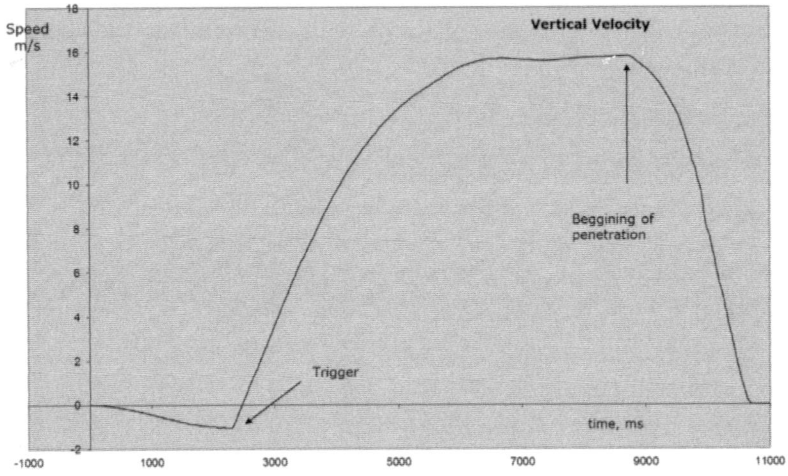

Figure 7: Torpedo Pile post installation analysis

After several installation procedures along Brazilian offshore coast, there was no relevant tilting after embedment, so design was validated. Also, there's a well-established radius of installation's accuracy, as shown in the figure below.

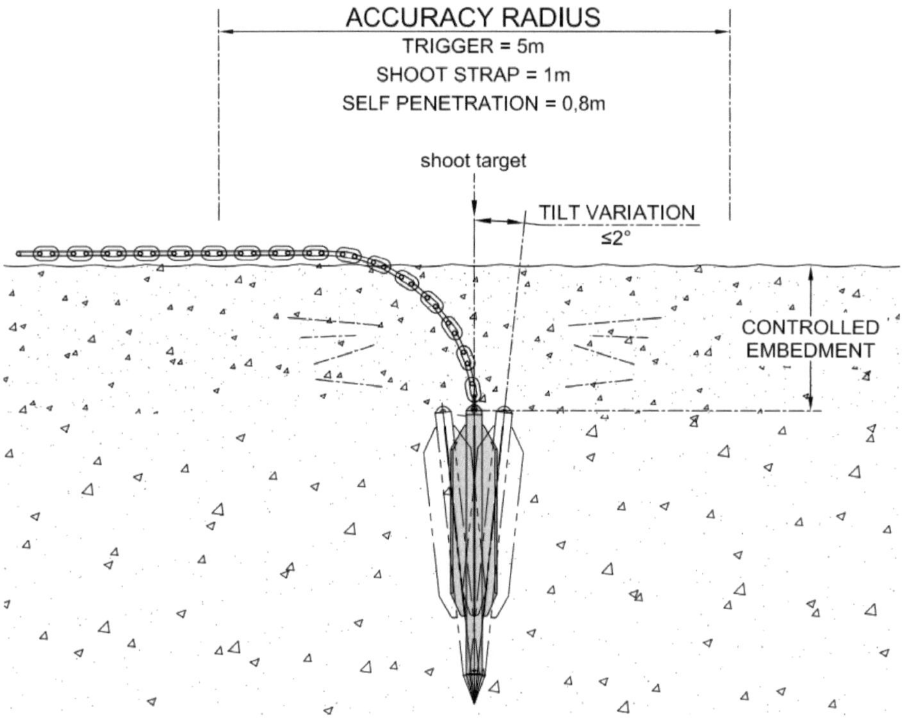

Figure 8: Torpedo Pile post installation analysis

As stated before, load capacity is directly proportional to the angle of resultant force. This is also calculated, so all possible angles as evaluated in order to mitigate any problem. Below there's example of a curve of load capacity *versus* inclination, for a small flowline torpedo pile. Considering the tilting range expected after installation, as the previous figure, there's no big variation on the expected UHC. This small variation is, of course, considered in pile's design, so we can choose the proper pile.

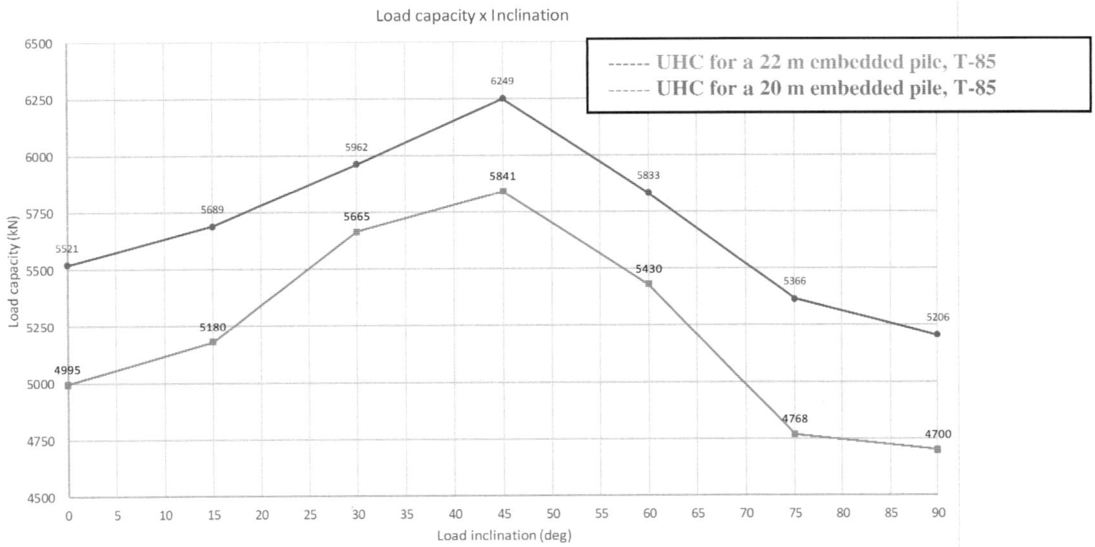

Figure 9: Torpedo Pile curve for load versus inclination

Finally, a very important advantage torpedo pile has is a higher resistance for vertical loads. This allows the pile to stay closer to the floating unit, in a taut-leg arrangement, which means:

a. Optimization of sea bottom layout: anchoring radius are smaller in this type of arrangement, so there's lower risk of interference in subsea layout.
b. Cost savings with the mooring line: This type of system requires shorter mooring lines.

APPLICATION

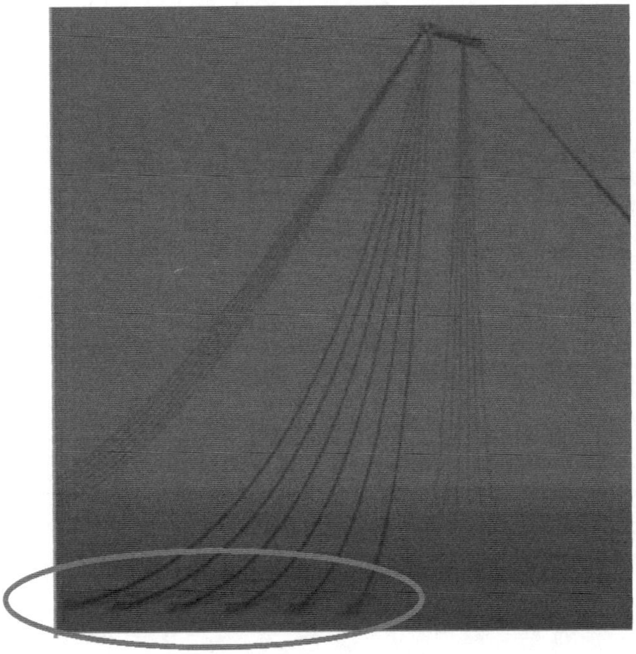

Figure 10: FPSO floating units, by Petrobras Technological report, 2013. Figure shows an FPSO anchored with a spread mooring system, 6 piles in 4 clusters. In the figure pile is already embedded.

Offshore Windfarm (New development)

Offshore wind power or offshore wind energy is the use of wind farms constructed in bodies of water, usually in the ocean on the continental shelf, to harvest wind energy to generate electricity. Higher wind speeds are available offshore compared to on land, so offshore wind power's electricity generation is higher per amount of capacity installed (source: Madsen & Krogsgaard. Offshore Wind Power 2010 Archived June 30, 2011, at the Wayback Machine BTM Consult, 22 November 2010. Retrieved: 22 November 2010).

Concentrated on Europe until nowadays, offshore windfarms are becoming more common around the world. The constant requirement for clean, renewable energy, along with energy demand increasement, will develop this market.

In this scenario torpedo pile becomes a very attractive option to anchor the wind tower, based on the installation effectiveness and design savings.

Required holding capacity of a wind tower is not high as an Oil floating production unit, and also the operational conditions shall not be as severe, so a tail-made pile, with reduced weight, can be specified to optimize this anchoring system.

Below is a sketch of a wind tower anchored with piles:

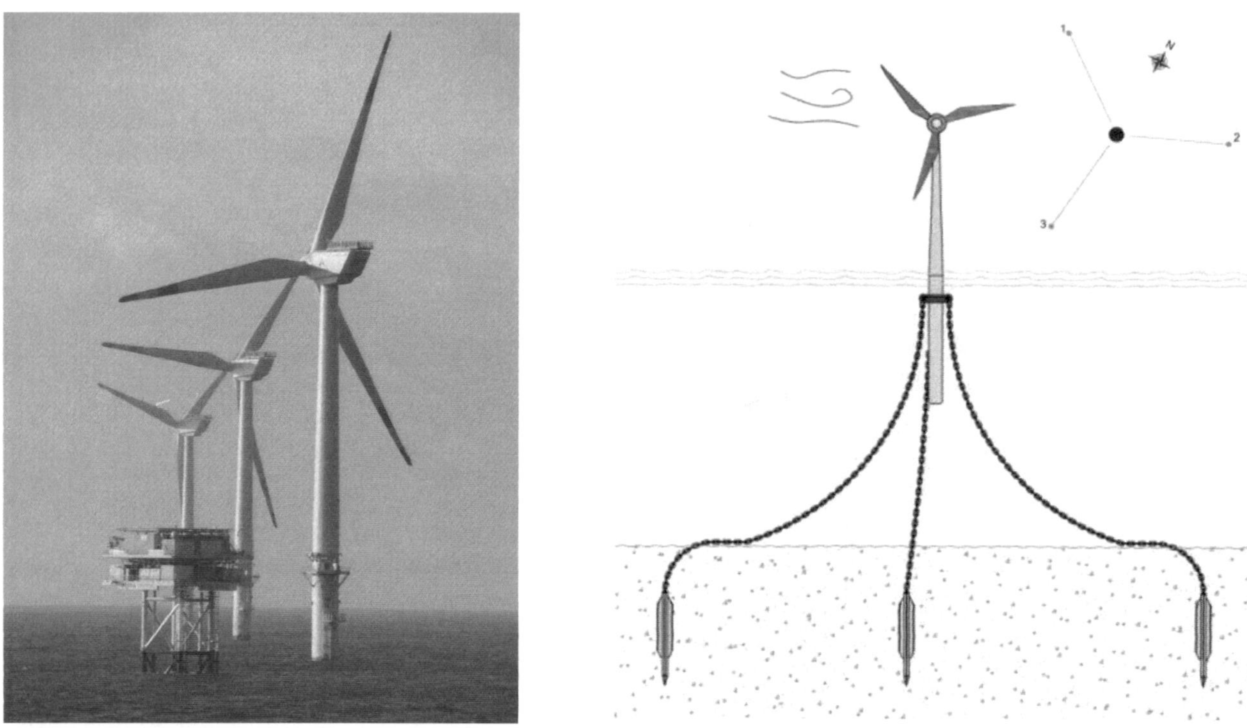

Figure 9: Wind farms anchoring systems

RESULTS

From the economic point of view torpedo pile is a very cost-effective alternative, especially if installation method is considered. Next there's a comparison table between torpedo pile and a suction pile, which is the closest technology. as we can see torpedo pile has advantages in design, fabrication, and mainly installation.

Table 1: Comparison between Suction Piles and Torpedo Piles

	TORPEDO PILE	SUCTION PILE
DRY WEIGHT	90%	100%
HANDLING / TRANSPORTATION	REGULAR TRUCK, NO SIZE RESTRICTIONS	SPECIAL TRANSPORTATION, SOMETIMES NEED TO BE FABRICATED IN HARBOR
INSTALLATION LEAD TIMES	AHTS (1 PILE / DAY)	LCV WITH HCC (1 PILE / 1,5 DAYS)
INSTALLATION COSTS	60% SIMPLER VESSEL, FASTER INSTALLATION	100%
ANCHOR PRICE	60% SIMPLER PILE, EASIER TO MANUFACTURE	100%
OVERALL ANCHORING SYSTEM PRICE	SAVING UP TO 40% IN ANCHORING PROCESS	

This comparison is a general analysis considering a regular anchoring system with 20 mooring lines or more, in any water depth suitable for both anchors, for instance in a range of 1,200 up to 2,400 meters.

CONCLUSION

Oil and Gas market forecast is very positive for the next 20 years, where at least 60 FPSOs should be hired, with a special attention to the Brazilian market.

Offshore wind power market plays an important role in achieving the renewable target in most of the countries around the world. It has been suggested that innovation at scale should deliver significant cost reduction in offshore wind.

Torpedo pile, known as an attractive alternative for offshore anchoring systems, widely used in a large range of soil's types, in several type of applications, is a good option to improve project's competitiveness in any kind of offshore mooring design.

REFERENCES

DELP ENGENHARIA MECÂNICA S.A., available in http://www.delp.com.br/solucoes/oleo-e-gas/ accessed in Dec 16th, 2019.

JUNIOR, C. J. M. **Torpedo Anchor for Deep Water**. Deep Offshore Technology Conference. Rio de Janeiro, RJ, Brasil, 2001.

JUNIOR, C. J. M. **Low Cost Anchor System for Flexible Risers in Deep Waters**. Proceedings of the Annual Offshore Technology Conference. Houston, Texas. Paper 14151, 2002.

Madsen & Krogsgaard. **Offshore Wind Power 2010** Archived June 30, 2011, at the Wayback Machine BTM Consult, 22 November 2010. Retrieved: 22 November 2010

TRUE, D. G. **Undrained Vertical Penetration into Ocean Bottom Soils**. Ph.D. Thesis, University of California. Berkeley, California, E.U.A, 1976.

ABS Guidance Notes On, **DESIGN AND INSTALLATION OF DYNAMICALLY INSTALLED PILES** DECEMBER 2017 (Updated March 2018)

IMPROVING INSPECTION PRACTICES OF OFFSHORE MOORING SYSTEMS FOR OPTIMAL INTEGRITY MANAGEMENT

MENNO VAN DER HORST
CHEVRON

KAI-TUNG MA
CHEVRON

JOERIK MINNEBO
SOFEC

SUE WANG
AMERICAN BUREAU OF SHIPPING

MATHIEU BARGUES
BUREAU VERITAS

ALEX ARGYROS
DNV GL

LEOPOLDO BELLO
VRYHOF

Proceedings of the 25th Offshore Symposium, February 27th 2020, Houston, Texas
Texas Section of the Society of Naval Architects and Marine Engineers

ABSTRACT

In-service inspection plays a critical role in the integrity management of mooring systems of floating production and drilling units, as mooring failures have been a big issue in the offshore industry in recent years. Inconveniently or unfortunately, these offshore mooring systems cannot be easily inspected. Best inspection practices are needed, as they can potentially prevent mooring incidents or pre-emptive replacements caused by degradation mechanisms, such as fatigue and corrosion. Other than the limited guidance provided in API RP-2I through its revision in 2008, there is not much information available on where, what, and how to inspect chain, wire ropes, polyester ropes, and connectors in an offshore mooring system. This paper elaborates on some best practices and recommendations regarding in-service mooring inspection that were learned from field experience and collected data. A documentation of best inspection practices may prove itself useful for operators, contractors, classification societies, and regulatory agencies. Not only may such documentation improve the integrity and reliability of mooring systems, it can also make inspection practices more efficient and effective.

Keywords: Mooring System, Station Keeping, Integrity Management, Inspection, Codes and Standards

1 Background and Introduction

1.1 Reliability Challenges of MODU and Permanent Moorings

Mooring incidents and failures have occurred at an unacceptably high rate in the recent past [4][6][19][21][22]. For permanent moorings, the probability of system (i.e. multi-line) failure was estimated to be in the order of 10^{-3} per year [4][22]. Meanwhile, the reliability record for MODU moorings was even worse. The probability of failure for MODU moorings was estimated to be in the order of 10^{-2} per year [5][7][22]. There were several papers and reports published in recent years that advocate for improving mooring reliability by better understanding some of the failure mechanisms. They all have the goal of improving the reliability to lower the risk level to an acceptable level, as shown in Figure 1.

One of the possible ways to mitigate the low reliability is through improved inspection practices. In-service inspection can be a critical line of defense, as good inspection practices prevent mooring incidents caused by poor condition of mooring components.

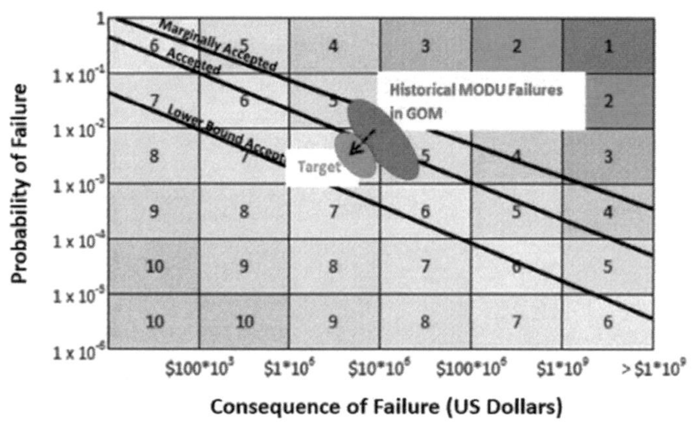

Figure 1: Target reliability needs to be improved, taking MODU mooring as example [5]

1.2 Developing Better Standards for Mooring Inspection

The third edition of API RP-2SK [1] was published in 2005 and is arguably the most commonly used mooring design code in the industry. It was the first mooring design code that provided guidance on several important issues, such as chain corrosion allowance, mooring hardware requirements, clearance criteria, and anchor design and installation. RP-2SK has a companion standard, RP-2I, which is a widely used standard for in-service mooring inspection [2][3]. RP-2SK was used as basis for some other codes, such as ISO 19901-7 published by International Standards Organization (ISO). In comparison to the duo document of 2SK and 2I, ISO 19901-7 Revision 2 has a relatively limited coverage on mooring inspection [14]. There are also several other mooring inspection standards or guidance notes developed by class societies [10][12][13], such as American Bureau of Shipping (ABS), Bureau Veritas (BV), and Det Norske Veritas Germanischer Lloyd (DNV GL).

To a certain degree, the multiple standards caused some confusion for owners, operators and contractors, as they could have different requirements. To resolve the issue of conflicting mooring standards, committee members from ISO and API resource groups initiated an effort under the International Association of Oil & Gas Producers (IOGP) to harmonize ISO 19901-7 and API RP-2SK into one new document. An IOGP Mooring Work Group of more than 30 experts was created, and the group was divided into small panels to each focus on the revision of a topic or chapter. Inspection is one of the topics and is being worked on by the authors of this paper. The preliminary results and learnings are summarized in this paper, which is a compilation of the state-of-the-art practices based on experience of the authors and their colleagues in the industry. The objective of this work is to improve the mooring reliability by compiling and documenting the best inspection practices.

2 Review of Industry Standards

In-service inspection of mooring systems plays a critical role for the integrity and safety of floating facilities. Over the years, the mooring industry has developed codes and standards on inspections of mooring systems by various organizations, see Table 1. They have contributed to the overall safe performance of the mooring systems. However, the industry experiences have indicated that the inspection codes may be further improved and harmonized.

Table 1: Station-keeping standards that cover inspection

Standards	Revision	Title
API RP-2I	2008	In-service Inspection of Mooring Hardware for Floating Structures
DNVGL-RU-OU-0102	2016	DNV GL rules for classification: Floating Production, Storage and Loading Units
DNVGL-CP-0484	2013 (2019)	DNV GL approval of service supplier scheme No 413 Appendix B: 12 Renewal survey examination of mooring chain intended for mobile offshore units (under review)
ABS	2017	ABS Rules for Building and Classing Floating Production Installations
BV NR 493	2015	Classification of Mooring Systems for Permanent and Mobile Offshore Units
ISO 19901-7	2013	Stationkeeping systems for Floating Offshore Structures and Mobile Offshore Units
IACS REC 38	2010	Guidelines for the Survey of Offshore Mooring Chain Cable in Use

Among these standards, API RP-2I is recognized as a founding mooring inspection standard that has been adopted, incorporated, or referenced by many other standards and industry guidance, such as ISO 19901-7. Most Class Rules include API RP-2I's recommendations and have some additional details in terms of inspection schedule and scope, especially for permanent moorings. API RP-2I is even incorporated into regulation by reference within the Code of Federal Regulations (CFR). It covers the inspection of mooring lines, components and connecting jewelry. It was originally written for the inspection of MODU moorings, and largely relied on "in-air" inspections. However, the third edition of API RP-21, published in 2008, incorporated inspection of permanent mooring systems, such as guidance on underwater inspection techniques, but the guidance is rather limited.

2.1 Inspection Schedule Defined in Codes

Class societies, in general, require that the following inspection schedule is followed for permanent moorings:

- Annual survey
- 5-year special survey
- Event inspection after incident or damages occur

Most class societies adopt and accept API RP-2I recommendations for inspection. With the development of integrity management of mooring systems, the offshore industry has also started to adopt inspection frequency based on risk assessments. The UK O&G mooring integrity guidelines [23] are intended to provide a means to better understand and manage mooring integrity of permanent units, introducing the lifecycle approach for undertaking risk review as well as providing general methods for mooring monitoring and inspection. Most class societies accept risk based inspection plans based on mooring integrity management (MIM) on a case by case basis.

For mobile moorings using chain and wire rope, API RP-2I recommends that the inspection schedule should be based on the age, condition, and operational history of the components, see Table 2. For fiber rope inspection and condition assessment, API RP-2I recommends that a plan should be developed by the operator of the mooring system and the manufacturer of the rope in conjunction with the certifying authority to provide consistency in the overall safety assessment for a given installation.

Table 2: Inspection schedule for MODU moorings as recommended by API RP-2I [2]

Component	Number of Years in Service	Recommended Intervals Between Major Inspections *
Chain	0 to 3	36 months
	4 to 10	24 months
	over 10	8 months
Wire rope	0 to 2	18 months
	3 to 5	12 months
	over 5	9 months
	* With a grace period not to exceed 4 months.	

2.2 Inspection Scope Defined in Codes

Scope of Annual Survey

In general, the annual survey is driven by classification societies. It includes documentation review and a visual inspection for the above water mooring components, including chain or wire rope, winches or windlasses, deck sheaves, stoppers, and fairleads or bending shoes. The survey is mainly visual with photographs, video, and inspector's comments that are documented. The survey can also include measurements of component diameter or depth based on the inspection plan. For permanent moorings, all Recognized Class Societies (RCS) require an as-installed survey to be conducted immediately following installation of the mooring system. On top of that, DNV GL requires an expanded 1st annual survey for the first five-year survey cycle. The expanded survey serves as a verification of the as-installed survey, after the mooring system has had time to settle in through all four seasons. It includes the ROV survey from the anchor/pile to the hull connection for all lines and recording any anomalies.

Note that the "as-installed" survey is often denoted as "as-built" survey in codes and standards and in other literature, which creates some confusion. In between the manufacturing and final hookup, there are a lot of operations including transportation and installation that can potentially damage the mooring system. Mixing the terminology can have an impact on whether liability of any damage lies with the manufacturer or installation contractor. To take away the confusion, the following definitions should be used in codes and standards to increase clarity:

- "As-built" refers to the final condition of anchor leg components when leaving the manufacturing facility,
- "As-installed" refers to the final condition of the anchor leg assembly once in operational service (i.e. hooked up to the facility).

Scope of Intermediate Survey

BV requires an intermediate survey every 2.5 years for permanent moorings [13]. The survey is to cover the following: (1) visual inspection of all the lines in their critical areas, (2) visual inspection of selected lines over their full length, (3) dimensional checks of at least one link in the last 5 meters of chains measured from above water level, (4) measurement of chain angle at top, and (5) visual examination of chain links at certain locations.

Scope of Special Survey

For permanent moorings, API RP-2I provides similar recommendations to the as-installed survey with as many measurements as possible to check against the original survey. The survey can be conducted by divers in shallow water or by ROVs in deep water. In addition to addressing the overall mooring performance, the survey should include a detailed inspection of the various mooring components. The survey should be planned in detail to ensure that quantitative information is obtained if required and that the inspectors are provided a list of items to be inspected. Equipment to remove marine growth at the desired inspection locations should be provided. If the mooring system is equipped with load cells, inclinometers, or similar monitoring devices, their data should be recorded along with floater position to enable accurate assessment of the mooring system.

In addition to the scope covered in the annual survey, Class Societies also require the following for the special periodical survey:

- General visual inspection of all the lines over their full length by ROVs or divers.
- Close visual inspection (CVI) with ROV (with cleaning capabilities) of one line from each mooring line cluster. Where lines are not clustered but evenly distributed; at least three mooring lines shall be selected. The selected lines shall be evenly spread, and both normally windward and leeward side shall be included. CVI of selected links should be conducted towards the pile until the chain is buried. Trenches, if any, in proximity to pile/anchors shall be recorded.
- Chain measurements shall be done in the splash zone or, in case of submerged systems, downwards from the hull connector/stopper to check for excessive corrosion. These measurements shall include at least one line from each cluster, or at least two representative lines in case of an equally distributed system. For measurements in the touchdown/thrash zone, sufficient readings shall be taken to ensure that the interlink wear and corrosion rate are mapped. The chains are also to be inspected for loose studs and link elongations.
- The chain tensions are to be checked and where found not in compliance with the specifications, they are to be readjusted accordingly. Excessive loss of mooring line tensions is to be investigated.
- More detailed inspection including MPI shall be required if the ROV or diver inspection reveals defects that are considered as critical, e.g. cracks, severe pitting or wear.

Scope of MODU Inspection

For MODU mooring, API RP-2I provides detailed inspection steps including visual inspection and measurements. For chain measurements, it recommends measuring the chain diameters in two perpendicular directions once every 100 ft and on both sides of each chain link. For wire rope measurements, the inspector should measure the distance of three lay lengths and wire rope diameters in three directions at the beginning, middle, and the end of the portion of the rope being inspected. Internal inspection of the wire rope is recommended for at least one mooring line for internal corrosion by cutting a length of the rope at the end.

2.3 Current Practices for Acceptance/Discard Criteria

The philosophy of API RP-2I is to remove a mooring component with excessive deterioration from service. Based on this philosophy, a criterion of limiting the strength reduction to 10% minimum breaking strength (MBS) was established in the first edition of the document [3]. The criterion has been used by the MODU industry with overall satisfactory results. API RP-2I also includes guidance and recommendations for rejecting wire rope for various broken wire conditions. BV has a similar discard criterion for mooring chains as API RP-2I. They require that the average of the two measured diameters (90 degree apart) is to be greater than 95% of the nominal diameter, and the diameter in any direction is to be greater than the minimum of either 90% of the nominal diameter or 95% of the nominal diameter minus the corrosion allowance [13]. Note that a 5% reduction in diameter is equivalent to approximately 10% of reduction in cross-sectional area and hence in strength.

However, its lack of engineering basis makes the discard criteria of 10% strength loss controversial. More work should be done to gain a better understanding in the failure modes of mooring components and to develop acceptance/discard criteria with a proper engineering basis. For example, advanced approaches for estimating the residual strength and remaining fatigue life of chain have been developed by recent JIPs [15][16][17][18], such as SCORCH, FEARS and FoCCs. In these projects, systematic FEA analyses were performed on heavily corroded chain links. It was shown that the strength and fatigue life reduction of the chain can be estimated with high accuracy based on diameter reduction and local corrosion features.

3 PERMANENT MOORINGS – IMPROVED INSPECTION PRACTICES

3.1 Where to Inspect (Scope)

The newly published API RP-2MIM (2019) [24] makes a clear statement on the main objective for mooring system inspection, which is "to identify anomalies in the mooring system components at the present time to support fit-for-purpose assessment of the mooring system [19]." However, as anchor leg systems can get very large, with many long sections of polyester or steel wire rope, and chain sections with hundreds or even thousands of chain links, it is important to focus the inspection on critical components and locations along the anchor leg. During the design stage of a mooring system, it is important to identify these critical components and areas, where more detailed inspection is required. This ensures that these components are methodologically, and structurally inspected during every future inspection, including the very important as-installed inspection, which serves as the baseline for every future inspection.

Where most Class rules prescribe the interval between mooring inspections (annual, intermediate or special surveys), one should also consider the inspection scope; i.e. how many lines and and how much of each line should be inspected during these different surveys. If the mooring system is only inspected once every 5 years, it is recommended to cover every anchor leg with at least a General Visual Inspection (GVI) from top to bottom. Sections, areas or components that are deemed a higher risk to the integrity of the anchor leg should be subject to detailed cleaning and a Close Visual Inspection (CVI), and if applicable to some quantitative measurements. See Figure 2 for an overview of some historical mooring line failures, which was created in a previous study [4]. The locations of the breaks along the mooring profiles are marked in the figure, which are focused mainly near the top chain, bottom chain, connectors, and wire ropes.

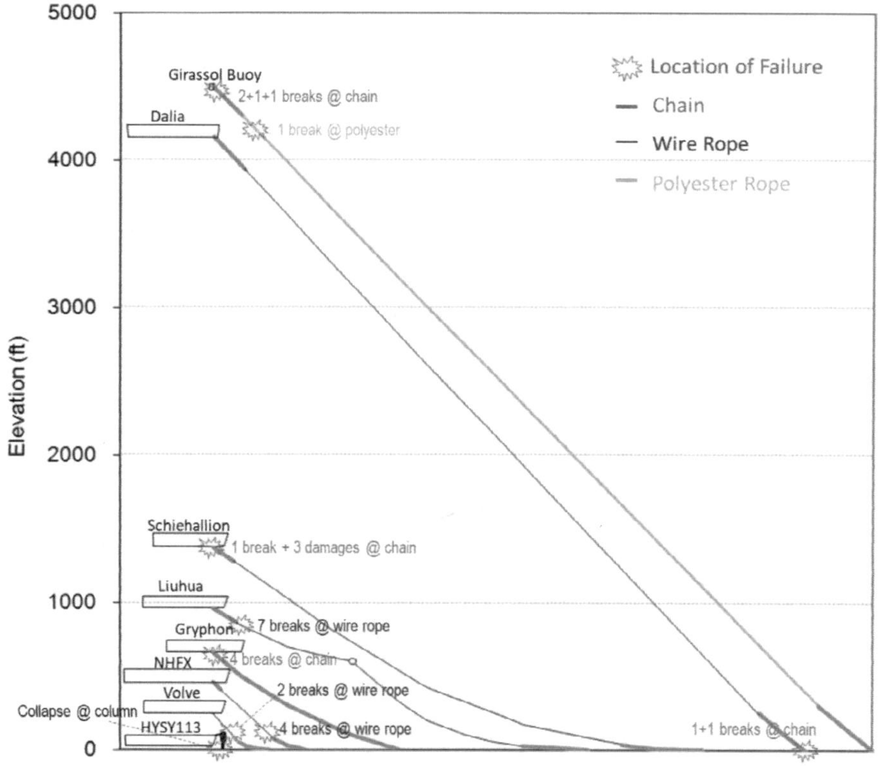

Figure 2: Locations of breaks along mooring line profiles [4]

If the mooring system is inspected more frequently, it may be considered to only perform a CVI on the critical areas of all lines (such as top chains in highly corrosive environments), but reduce the number of anchor legs that receive a complete GVI during every inspection. Especially for mooring systems which include polyester ropes, which have proven to be very robust once in-service, reducing excessive ROV inspections may reduce the risk of inadvertently damaging the ropes. It is not uncommon for large mooring systems to only include half or one-third of the anchor legs in a single inspection, and rotating through the lines during subsequent inspections. Naturally, such scope reduction must consider a good balance between anchor legs in different clusters, and anchor legs at higher risk.

For mooring chains, the most severe degradation is typically observed near the splash zone mainly due to general corrosion, especially the first few links above the waterline. Therefore, it is recommended to perform CVI and quantitative measurements on chain in the splash zone area.

At the transitions between different anchor leg sections (chain to chain, or to polyester or steel wire rope), a large scala of connectors can be found, such as D-type shackles, H-Links, Tri-plates, etc. These types of components should all receive CVI, and for critical components also some quantitative dimension measurements to verify whether any degradation is still within designed tolerances. Specific, special purpose connectors (such as Ballgrab or Balltec type connectors) may have specific inspection criteria from the suppliers of these components.

Other areas of higher risk may be the touchdown area, weight or drape chains, buoyancy modules, clump weights, etc. Depending on the mooring system design, these critical areas are to be identified for inclusion in CVI scope of work.

3.2 What to Look for (Degradation Mechanisms)

To perform an effective inspection, one should have an idea of what to look for. Each mooring component in a permanent mooring system may suffer from several different degradation mechanisms and some degradation mechanisms are more critical than others, depending on material type and environment. The following subsections discuss the most critical degradation mechanisms per mooring component.

<u>Chain</u>

Corrosion is one of the most significant degradation mechanisms for mooring chain. Corrosion reduces the cross-sectional area of the chain links and hence, reduces the strength and fatigue life. In general, there are two types of corrosion: general corrosion (smooth, uniform reduction of chain link diameter) or localized pitting corrosion. Pitting corrosion can lead to large pits that significantly reduce the cross-sectional area locally, but they are easily identified through visual inspection. Pitting corrosion can be categorized as small pitting (<5mm diameter, <2mm deep), medium pitting (5-20mm diameter, 2-10mm deep), and large or mega pitting (>20mm diameter, >10mm deep). General corrosion may be more challenging to detect without taking measurements, but mooring chains are generally designed with a corrosion allowance of at least 0.1 mm/year to 0.8 mm/year, depending on whether the mooring system is in tropical waters. This design assumption can be validated during each inspection by performing diameter measurements.

Figure 3: Examples of general corrosion, pitting corrosion, and other types

Other types of corrosion can be seen in some cases as well, such as Microbially Influenced Corrosion (MIC), Preferential Weld Corrosion, and Galvanic Corrosion. MIC is often related to Sulphate Reducing Bacteria (SRB) and is commonly found in areas where the chain links are in contact with the seabed, or in areas of heavy, active marine growth [15][16]. Preferential Weld Corrosion can be observed as a distinct groove across the butt weld of the chain link. Galvanic Corrosion can occur due to presence of dissimilar metals in close contact or vicinity with the chain sections.

Wear can also be a significant degradation mechanism at the interlink contact areas, especially in environments with a lot of wave action. Interlink wear cannot easily be observed visually, unless extreme, but it can be quantified by measuring the double diameter in the interlink grip area. In some cases, wear can also occur due to contact with the vessel hull, the seabed, or other objects. See Figure 4 for examples of chain links that experienced wear.

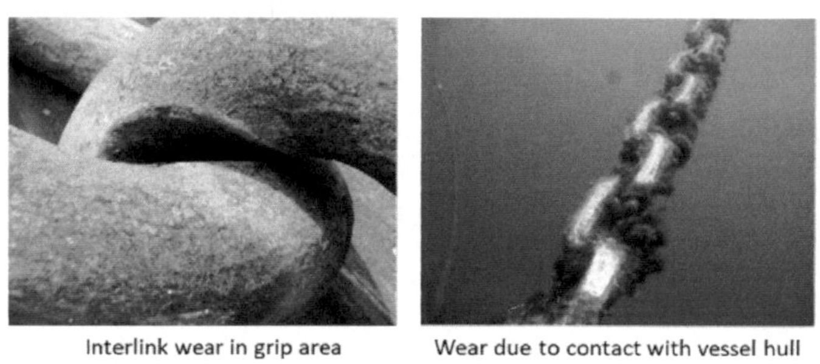

Figure 4: Examples of chain wear

Fatigue is another degradation mechanism that can be critical for the integrity of mooring chains. However, it is almost impossible to detect signs of fatigue damage or make an estimate of how much fatigue life has been consumed by visual in-service inspection. The only physical sign of fatigue damage is the initiation and propagation of cracks. These fatigue cracks can only be detected by nondestructive evaluation (NDE) methods, such as magnetic, dye penetrant, ot ultrasonic testing. For studlink chain, though, one thing that can be detected during visual in-service inspection and that negatively affects the fatigue performance is a loose stud. Hence, it is important to always look for loose studs during inspection of studlink chain.

Wire Rope

Wire ropes may suffer from broken wires, especially at the terminations, which can cause high local stresses that potentially may lead to failure of the entire wire rope. Broken wires can be caused by incorrect fitting of the termination, fatigue, overloading, or mishandling during deployment. In fact, it was shown in [21] that most wire rope failures were caused by damage during the installation phase. When local wire breakage begins, it will usually only get worse, so it always needs to be addressed.

When unsheathed, the wire rope is susceptible to corrosion and wear, which can decrease the breaking strength by reducing the cross-section of the rope and accelerate fatigue by inducing stress concentrations and stress corrosion cracking. External corrosion or wear is a common issue for unsheathed wire rope and can be detected visually. Internal damage, however, is almost impossible to detect. Internal corrosion can occur due to water penetration and internal wear can be caused by friction between individual strands and wires in the rope, particularly when it is subject to bending loads.

For sheathed wire, the initial degradation mode starts with sheathing damage. In case of sheathing damage, the expected life expectancy of the mooring line drops by a factor of two or three due to corrosion exposure. In case of sheathing damage, the integrity of the outer layers of the wire has to be verified before repairing the sheathing.

Wire rope can also be severely deformed due to their slenderness. Kinking, bending, scrubbing, crushing, and flattening are all common wire rope deformations and may result in an uneven stress distribution in the rope. Such severe deformations can usually be detected by visual inspection. API RP-2I provides multiple visual references to these types of degradation modes.

Fiber Rope

Fiber rope damage is mostly caused by contact with sharp objects, work wire, dropped objects, or a hard seabed and can result in cuts or abrasion. The risk of incurring such damage is highest during installation activities or subsea intervention activities, such as the inspection activities themselves. Therefore, it is important to perform a visual inspection of the fiber rope after installation is complete or after subsea intervention activities have been carried out. Most fiber ropes are fitted with an outer jacket. Since the jacket is not a load carrying element in a fiber rope, minor damage to the jacket can be accepted. Internal damage below the jacket can sometimes be visually observed as necking, which is a local reduction in diameter.

To mitigate some of the risk, typically the majority of the rope is subjected to a GVI (with the ROV maintaining adequate distance), but to a CVI at the extremities of the rope, looking in detail at the condition of the spliced eye, polyurethane cover, and the correct seating in the thimble (see Figure 5). Reducing the number of anchor legs to be inspected every inspection also reduces this risk.

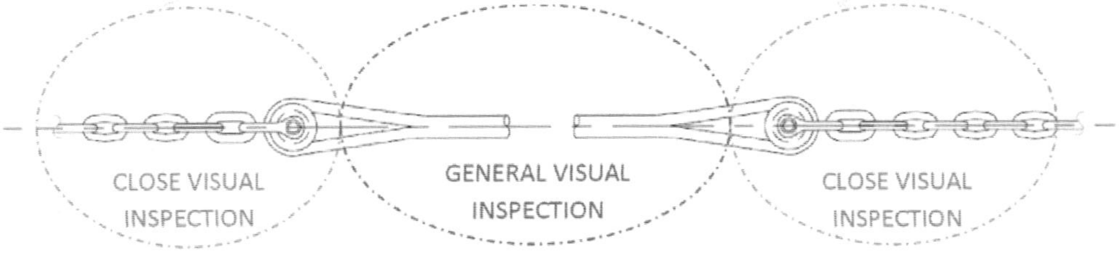

Figure 5: CVI and GVI locations on polyester ropes

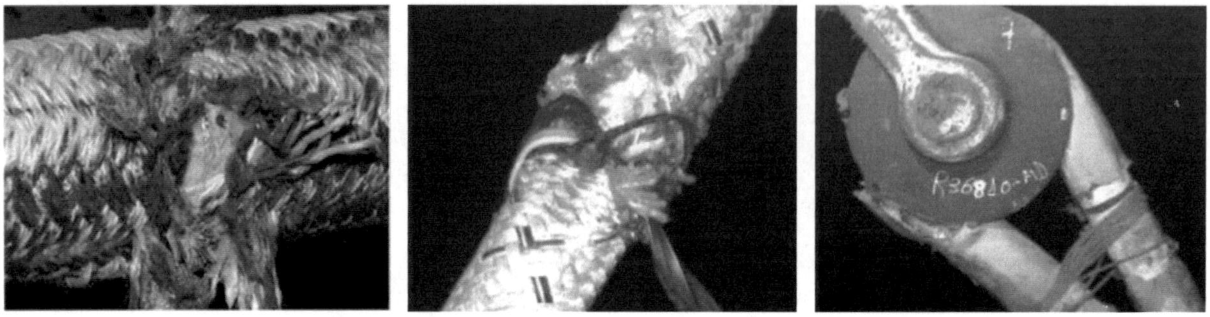

Figure 6: Different types of rope damage – (left) damage on jacket, (middle) foreign object stuck in jacket, and (right) polyurethane cover cut by thimble and lashing material

Connectors

Most permanent mooring systems use connectors such as thimbles, D-shackles, H-links, or tri-plates to connect various chain, wire rope, and anchor components. Some connectors are fitted with anodes to provide corrosion protection, in which case the remaining volume of the anode should be assesssed during each inspection or cathodic protection (CP) readings can be taken. Whichever kind of connector is used, every connector should be subject to a detailed cleaning and a CVI, with a specific focus on ensuring that the retaining mechanism is functioning. Especially for D-shackles, there have been several cases where pins have come apart due to failure of the retaining mechanism. Therefore, it is recommended to visually inspect all connectors for anomalies and to ensure that all retaining hardware is intact. In addition, connectors should be inspected for corrosion damage between the threads of the pin and the nut, and wear measurements should be taken, if possible, to assess remaining strength of the component. It is also recommended to check whether the connector is still oriented in its design orientation, such that it does not experience any side loading for example.

3.3 How to Inspect (Methods)

Three main resources are available for performing the actual offshore inspection of anchor leg systems. These are Rope Access Teams (RAT), divers and Remotely Operated Vehicles (ROVs). Each of these resources operate in a specific area of the anchor leg.

For mooring systems which are terminated above water (and part of the anchor leg and the connecting hardware is dry), inspection is typically performed by RAT inspectors. These highly skilled inspectors can go outboard of regular pathways, and work at height or even suspended with abseil equipment. RAT inspectors can cover the full scale of GVI, CVI, and quantitative measurements, and their typical scope includes outboard chain support assemblies, chain stoppers, top chains, pocketed fairleads, and bending shoes.

Air divers are typically used for inspections within the first 30m under the surface, as this is out of the typical working range from most ROVs. Similar to RAT inspectors, air divers can cover the full scope of GVI, CVI, and quantitative measurements, typically applied to underwater connecting structures (chain support assemblies, fairleads, etc.) and chain or wire segments within this range of water depths. Divers can also more easily get into congested areas, such as internal turrets, where all the anchor legs, risers, and umbilicals come together within a very small footprint.

Below 30m from the surface, mooring inspections fall rigidly in the world of ROVs. While saturation divers could be used to inspect the full anchor leg system on shallow water moorings, the high cost and risk of saturation diving typically favors a split scope between air divers and ROV inspection. ROVs are very efficient performing GVI work, especially over long lengths of polyester or steel wire ropes, or long bottom chain sections. CVI can definitely be performed as well, but may be more time consuming due to cleaning requirements, and the array of specialty tools that are required (such as rotating brushes, flailing tools, high pressure water jets, or cavitation blasting). Also, ROVs typically remain unattached from the mooring line (perhaps on chain sections they can grip onto a chain link), and any relative motion between the anchor leg and the ROV may slow down detailed visual inspections and other detailed work. Equipped with the right tools (often custom made, or provided by specialist companies), ROVs can perform a large range of quantitative measurements as well, including chain diameter measurements, CP readings, 3D scanning of components, etc.

Example: Chain Inspection Methods

Chain sections throughout the anchor leg should be inspected by GVI during each major inspection campaign, independent of their location along the anchor leg.

Specific locations of a chain section should be subjected to a CVI, where a set of 3 links is thoroughly cleaned, completely removing any marine growth and exposing the bare metal. It should be noted that the choice of these locations should take into close consideration the potential for accelerated corrosion, since cleaning the links to bare metal may temporarily increase the corrosion rate [16]. Especially in tropical water areas with high general corrosion rates, one may want to consider changing these CVI locations for each inspection, to prevent the same links from being exposed to this higher corrosion rate over and over again.

Finally, quantitative measurements are to be taken only at specific locations where the wear and/or corrosion rate is expected to be the highest. The easiest and most widely used method to measure the chain diameter is with a caliper. This method provides instant results and is reasonably accurate when used properly. However, when taking caliper measurements underwater by ROV, diver, or rope access technician, it can be challenging to take accurate measurements, an using a Vernier caliper with large jaws. For example, if the caliper is not held exactly perpendicular to the chain link, an error will be introduced that suggests that the diameter is larger than it really is. Therefore, it is advised to instruct the inspector to be as precise as possible when taking caliper measurements, to take multiple measurements at the same locations to gain higher confidence in the measured value, and to do a sense check on whether the measured value falls within the expectations based on prior measurements or based on the nominal diameter of the chain. ROV operated calipers need to also be properly calibrated.

A chain link's geometry may seem uniform, but this is never the case in reality, especially for used chains. The manufacturing tolerances, flash weld area, nonuniform corrosion, and interlink wear all contribute to a chain link's nonuniform geometry. This nonuniformity makes it challenging to measure the diameter of a link. Therefore, it is recommended to take multiple measurements per chain link that are taken in particular areas on the link. The minimum recommended measurement locations per chain link are shown in Figure 7. As can be seen, it is recommended to take measurements of the double diameter in the interlink grip area and of the single diameter in the straight bar and/or crown area if possible. Note that the measurement in the straight bar area should be sufficiently away from the center to avoid the flash weld area and that two measurements are recommended per location in two perpendicular orientations to account for any ovality. As a bare minimum, at least two measurements should be taken at each location and orientation to increase the confidence in the measured data.

To gain an accurate picture of degradation over time, it is recommended that repeat measurements are taken on the same chain links and same locations on the links, where practically possible. Reliable marking methods should be considered to enable such repeat measurements.

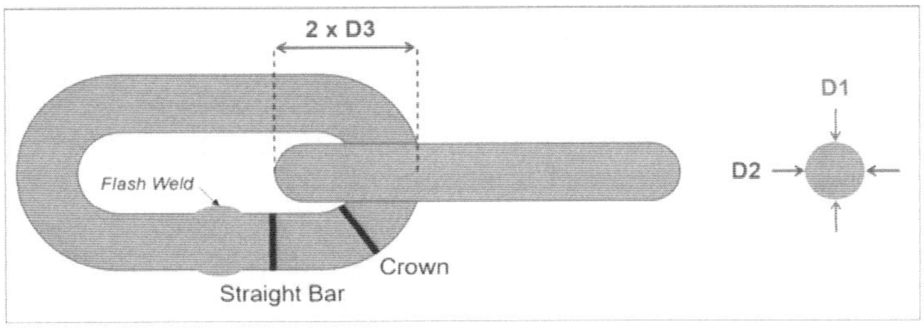

Figure 7: Minimum recommended chain link diameter measurement locations

3.4 Inspecting the Anchor Leg Profile

The above sections have focused mainly on the inspection of individual components and segments, but equally important is the inspection of the anchor leg profiles in order to determine that they are still providing the restoring characteristics that they were intended to provide, and that anchor legs within a cluster share the load equally (preventing a single taut line in a cluster of slack lines). While some anchor leg systems are more prone to changes in their profile (especially systems with polyester ropes), other factors may contribute to changes as well, such as shifting of drag anchors, or an extreme offset event picking up previously unrecovered slack (due to poor installation for instance).

On shallow water mooring systems, the top angle of the anchor legs is typically a good indicator of the profile and can be measured by simple inclinometers (preferred to use a long baseline, taking the angle over multiple links to prevent distortion from chain link surface effects). This can be done either using RAT inspectors for above water anchor leg terminations or using divers or an ROV in underwater terminations.

Additionally, good catenary profile measurements can be taken by ROV over the depth of the water column. For instance, an ROV can take an accurate position fix next to the anchor leg every 10 or 20m in depth. Measured depth, and position can be used to create a profile, and can be compared to as-built or design profiles. For deepwater mooring systems, especially the ones using polyester or steel wire ropes, taking depth and position fixes using an ROV can easily be combined with the CVI requirements for the connectors, since the ROV needs to move in close to those components anyway. Taking these depth and position measurements at the top and bottom of each polyester rope provides data to assess any potential elongation.

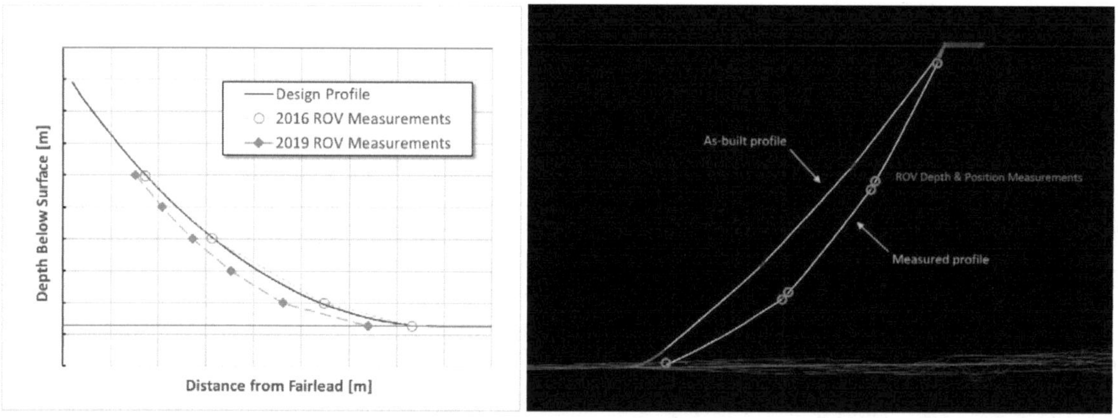

Figure 8: Examples of catenary profile mapping by depth increments (left) or at connector locations (right)

With any profile mapping measurement, whether it is angle or depth, it is extremely important to capture the floater position, heading, and loading conditions as well and as accurately as possible. As the vessel moves around in wind, waves and current, the line profile changes. These changes may impact consecutive measurements on a single anchor leg.

It is recommended that with every individual angle or depth measurement, the following parameters are documented at the exact same time as the measurement:

- Floater position - UTM coordinates, using a GPS with differential level accuracy (1.0m or better)
- Spread moored facilities: any point of reference, typically center of the vessel.
- Single point moored facilities (turret moored FPSOs): turret centerline
- Floater heading (especially for spread moored facilities as the heading may also impact the location of the fairleads, and thus the line profile)
- Vessel draft, and trim levels

With the above situational floater data included with the profile measurements, a large portion of the measurement uncertainties can be removed during the assessment of the profile.

4 TEMPORARY MOORINGS – IMPROVED INSPECTION PRACTICES

4.1 Where to Inspect (Scope)

The inspection practices of MODU and permanent moorings can be quite different. MODUs are typically moored temporarily on location. The mooring system is retrieved periodically and is therefore accessible for direct inspection in dry conditions. On the contrast, permanent mooring systems are normally not retrieved during their service life, so all inspections must be performed in-situ and rely primarily on diver or ROV inspection. As a result of the difference, deterioration of permanent mooring components can be more concentrated in a few locations, such as splash zone, fairlead/stopper and seafloor touchdown point. Deterioration of MODU mooring components is normally more evenly distributed over the entire mooring legs. However, for temporary moorings in deepwater, wire ropes seem to be one of the most problematic components, while chain and connectors contribute to a smaller share of the problems [5][22]. On the other hand, most of the failures in harsh environments such as the North Sea are on chains and connectors due to fatigue or handling damage. Critical spots in the mooring lines of a MODU are shown in Figure 9 [5].

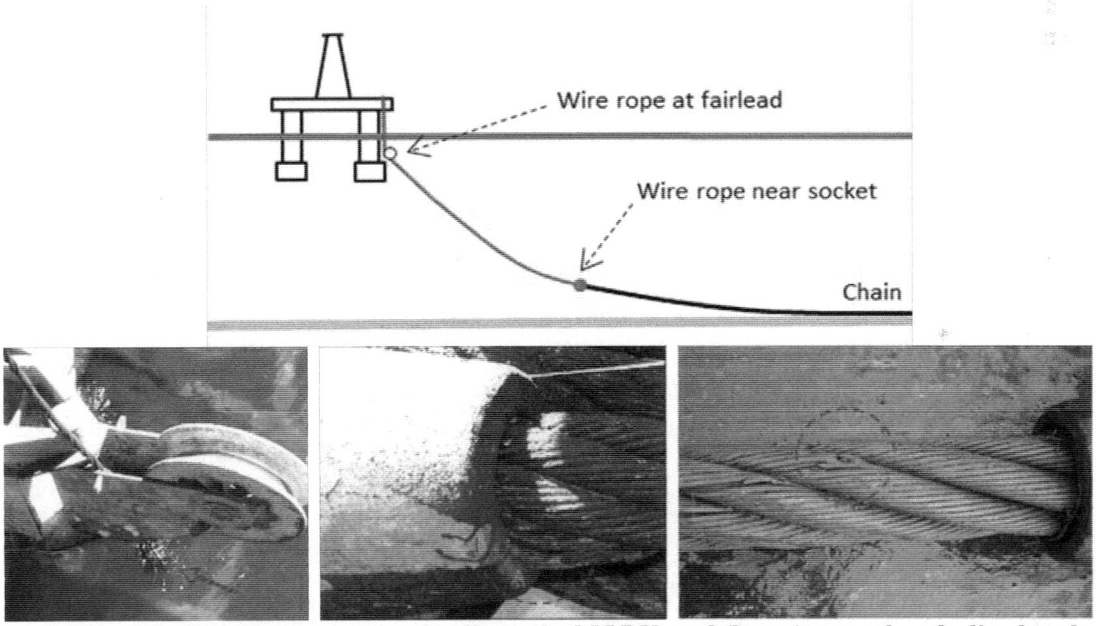

Figure 9: (Upper) Critical spots in mooring lines of a MODU, and (lower) examples of a line break and broken wires

4.2 What to Look for (Degradation Mechanisms)

The inspection and maintenance requirements are critical to maximizing the life-cycle and level of assurance of the MODU mooring components. Each component has specific maintenance requirements that help ensure that performance expectations and anticipated service life are achieved. The inspection protocols for each type of mooring component allow for condition monitoring, grading of equipment, and early identification of increased probability of failure (i.e. discard criteria). The assessment criteria should follow Class Rules guidelines as well as manufacturers' recommendations for best handling, preservation, maintenance, and operation of each product.

Each mooring component may be subjected to one of the three inspection levels. A good example of recommended inspection items for each component type is shown in Table 3. Owners and operators are encouraged to develop a similar table for optimal mooring integrity management.

Table 3: Example of recommended inspection items for each component type

	Level 1 Minimum Scope	Level 2 Enhanced Scope	Level 3 Maximum Scope
Chain	• Missing/loose studs • Cut damaged • Deformation	• Items in Level 1 • Dimensional Survey	• Items in Level 1 & 2 • MPI on 5-20% of the links with Dimensional Survey • D1/D2 defined dimensions • Loose-stud inspection
Wire	• Broken wires • Cracks • Deformation • Corrosion	• Items in Level 1 • "Broken wire per strand" inspection and documentation • Lay length check to confirm no crush or corrosion.	• Items in Level 1 & 2 • MPI of Sockets • Internal Core Check of Cut Section • Dimensional Survey • Break Test if possible
Kenters	• Cracks • Loose pins • Lead Plug	• Items in Level 1 • Groove Check (open the Kenter) • Dimensional Survey	• Items in Level 1 & 2 • MPI (critical areas defined by manufacturer)
D-Shackle	• Deformation • Cracks • Missing parts • Bolts, nuts	• Items in Level 1	• Items in Level 1 & 2 • MPI of critical areas (defined by manufacturer)
Swivels	• Defects • Cracks • Functional check (if possible)	• Items in Level 1 • Functionality (360 degrees check) • Lubrication	• Items in Level 1 & 2 • MPI of critical areas (defined by manufacturer) • Inspecting the inner surfaces and replacing gaskets and bearings if necessary
Anchors	• Deformation • Nuts, bolts, cotter pins • Anchor Shackle • Cutter Points • Plugged Soil	• Items in Level 1 • Grip Check • Dimensional Survey	• Items in Level 1 & 2 • MPI of main welds, high stress areas, anchor shackle and pins • Fluke/Shank Assessment • Anchor Disassembly • Dimensional Survey

4.3 When and How to Inspect (Schedule and Methods)

The inspection schedule for MODU chain or wire rope follows roughly the framework defined in Class Rules, but is also based on the age, condition, and operational history of the mooring components and type of operation as previously shown in Table 2. The inspection intervals may be modified based on the actual condition and previous inspection history of the mooring component.

In temporary moorings, the system is designed such that it can be easily moved from location to location. That implies that the MODU moorings can't be overly robust due to weight limitations. They are used in different locations and by different parties including drilling contractors, operators, and rental companies. Therefore, the inspection frequency and scope may be considered as a function of the Component Condition (including history service, maintenance and repairs) and the anticipated Working Environment (e.g. environmental loading, corrosion, etc.). Owners or operator may set the inspection frequency and scope dependent on these two criteria according to Figure 10. Such a well-planned strategy may yield a more optimal inspection result. Note that Component Condition may be determined by auditing of the suppliers, their planned maintenance system, and consistency with their equipment records. Working Environment may be determined on a per location basis as a result of the mooring analysis and design.

Working Environment

Component Condition		Mild	Severe
	Good	Level 1 Minimum	Level 2 Enhanced
	Poor	Level 2 Enhanced	Level 3 Maximum

Figure 10: Suggested inspection requirements as function of Component Condition and Working Environment

Level 1 – Minimum Scope

- When: Minimum scope is performed whenever there is a rig move. Each component shall go through a list of items in Level 1 inspection.
- How: Mostly a visual inspection is carried out. Marine growth and/or debris should be removed to give a better visual inspection of the components. Inspection is performed offshore with a competent trained inspector.

Level 2 - Enhanced Scope

- When: Accordance with interval-matrix for the given component, or when there is a failed component detected in Level 1 inspection.
- How: Inspection can be performed offshore (or onshore) with a competent trained inspector with experience dealing with the component. When offshore, the mooring line is typically inspected with the assistance of a workboat. The chain in the chain locker can be paid out fully and then examined by an inspector standing close to the windlass, while the chain is slowly taken back into the chain locker. A "go-no-go gauge" can be used to check the chain diameter; and checks should be performed at every certain distance, say for example every 35 meters. At the same time, the workboat may pick up the anchor and moves slowly toward the vessel. The advantage of this method is that it requires no dock facilities. There shall be a trackable form to fill out that documents the stated and checks on critical areas per each component type. It includes gross dimensional checks, disconnecting components where applicable, function testing, etc. If there are loose studs in the chain identified either by rattling sound or when hammering the chain, the links should be marked and noted in the report.

Level 3 - Maximum Scope

- When: Accordance with interval-matrix for the given component, or when there is a failed component detected in Level 2 inspection.
- How: Can only be performed onshore with a highly trained inspector with substantial training in dealing with the component. The drilling vessel is taken into a dock, and the chain is laid out on a dry surface for inspection. In this manner, the entire chain can be thoroughly cleaned and carefully inspected, and the connecting links and anchor shackles can be examined by magnetic particle inspection (MPI). There shall be a trackable form to fill out that documents the stated and checks on critical areas per each component type. The inspection shall be performed at a qualified location with an approved scope defined in accordance with the requirements for the components. It should include full MPIs, full lubrication, and internal inspections.

5 A LOOK INTO THE FUTURE

5.1 More Measurements to Establish Corrosion Trend

The industry is undergoing a rapid and large-scale digitalization. This may create opportunities to conduct more in-service measurements and to analyze the data to obtain valuable information, such as corrosion rates. This information can then be used to validate design assumptions, gain better understanding of failure modes, and for more rational decision making.

For example, Figure 11 shows measured data on chain diameter loss due to corrosion of a mooring line in a catenary shallow water mooring system in West-Africa. The mooring line was retrieved on the back deck of an anchor handling vessel and diameter measurements were taken in a few different sections: in-air, splash zone, water column, and ground chain. Note that the black dots in the graph represent real measurement data and the grey dots are data points created by trend analysis in combination with engineering judgment. The data was translated into corrosion rates by the using Equation 1.

$$Corrosion\ rate = \frac{(D_{nominal}+\Delta)-D_{measured}}{Service\ life} \quad \textbf{(1)}$$

Where $D_{nominal}$ is the nominal chain diameter, Δ is the average manufacturing tolerance, and $D_{measured}$ is the average of two diameter measurements 90 degrees apart and on the straight side bar location of the chain (see Figure 7).

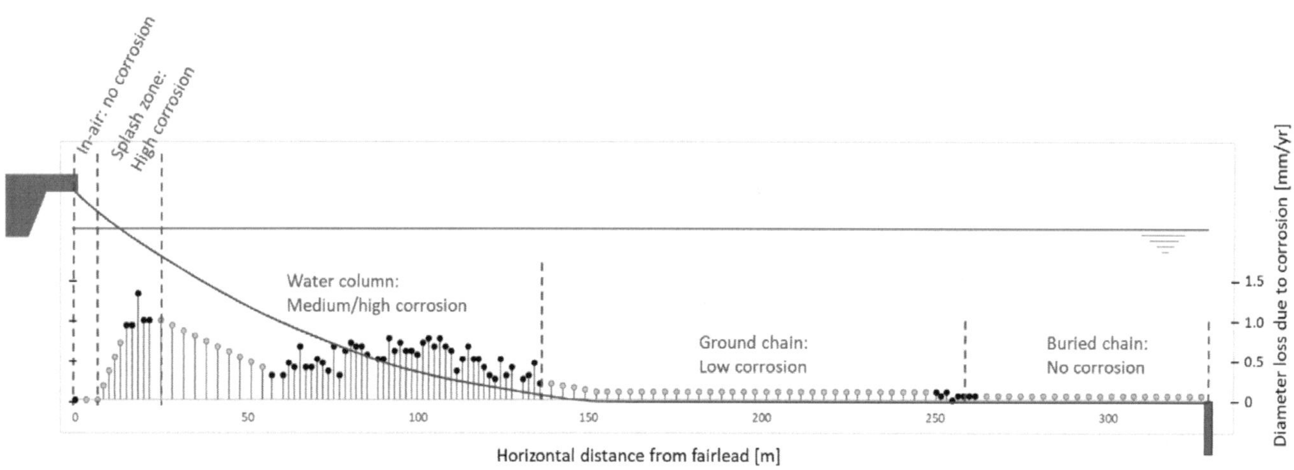

Figure 11: Trend of diameter loss due to corrosion along a catenary mooring chain – based on real measurement data (black dots)

One can see a clear difference in corrosion rates between the different sections along this mooring line. No corrosion is observed on the section in-air, high corrosion in the splash zone, medium/high corrosion in the water column, low corrosion on the ground chain, and no corrosion again on the buried chain near the anchor pile. Note that there were different kinds of corrosion observed as well. The chain in the splash zone showed general uniform corrosion, while the chain in the water column showed both general and pitting corrosion.

5.2 Certification or Training Program for Inspectors

Offshore mooring is a relatively young discipline compared to other established ones, such as hull structures and marine equipment system. Inspectors that work in this field are constantly on a learning curve to catch up with the new lessons learned from mooring systems deployed in deeper water or harsher environment. The competency of inspectors is of high importance as offshore mooring systems can be complex and novel.

One solution to is to create a certification or training program for the personnel working in this discipline. Class societies may be the most suitable organizations to create such a certification program, which could be similar to those for other disciplines like hull structure, marine system, etc. Alternatively, an academy may be created either by Class societies or third parties to provide training courses for inspectors. The advancement in competency will increase the effectiveness of inspections and thus improve the mooring reliability. It may also reduce the overall cost of mooring integrity management.

5.3 Promising New Technologies for Mooring Inspection

A more advanced chain measurement method, which can even be used for measuring any dimension on any object, is underwater 3D scanning using photogrammetry. This method uses still images taken with a high-resolution ROV mounted camera to reconstruct a 3D model of a chain link, connector, or any other object. The 3D model can then be used to query any dimensions such as the diameter of a chain link at various locations. Even better is to obtain the cross-sectional area directly, since that is directly linked to the strength of the chain link. Such a 3D model can also be imported into Finite Element Analysis (FEA) software to obtain a better prediction of the mooring chain's remaining strength and fatigue life, with its actual degradation taken into account [15][16][17][18]. Note that it is of importance to clean all marine growth off the chain before doing 3D photogrammetry, or the marine growth will show up in the 3D model as part of the component's geometry. Also, the accuracy of 3D reconstructions can vary significantly among various service providers and currently there exists no standardized methodology to quantify and compare their accuracy levels.

More frequently these days, especially in certain parts of the world, and from certain operators, there is a strong push to eliminate divers for planned inspection purposes from a safety perspective. The advancement of technology is now allowing for unmanned inspections near the surface area, and many solutions have been or are being developed for hull inspections, such as hull crawlers, or mini-ROVs. Similar solutions for mooring inspection, however, are still in a development stage, and cannot yet deliver inspection results to the same level of detail and accuracy as divers can. This is mainly related to the level of cleaning and getting access into tight spots (such as around underwater fairleads, or near/inside internal turrets).

5.4 Better Ergonomics and Marking System for MODU Mooring Inspection

Typically, inspection of mooring equipment for MODUs happens in open air, on the floor and with little tools that allow for easy handling. It is recommended that appropriate facilities and tools are provided and set up to facilitate inspection in a more ergonomic manner. Figure 12 shows a good example of an ergonomic inspection bench for inspecting Kenter links, which are critical components.

Figure 12: Examples of an ergonomic work bench for inspecting Kenter links

MODU moorings require extensive logistical efforts with significant amounts of equipment being mobilized and demobilized in parallel. Marking and tagging systems should include visual identification as well as machine-readable marking through Radio Frequency Identification (RFID) or Quick Response (QR) codes. This will enable the use of handheld devices, see Figure 13, to assist as equipment gets mobilized, deployed, retrieved, and demobilized. This system can notify an operator trying to pick equipment from storage with overdue maintenance/certification jobs.

Figure 13: Surveyor using a handheld device to read equipment information

6 CONCLUSION

State-of-the-art mooring inspection practices were compiled and summarized by a team of authors who represent owner, operator, contractor and classes societies. The industry has experienced a high number of mooring incidents and pre-emptive replacements in the past years and it is the authors' vision that improved inspection practices is one of the keys to achieving increased reliability of mooring systems. Hence, this work provides guidance to mooring practitioners on what, where and how to inspect chain, wire ropes, polyester ropes, and connectors for the various degradation mechanisms that are encountered.

The state-of-the-art practices were presented regarding in-service mooring inspection, which were learned from field experience and collected data. It is recognized that the inspection strategies for permanent mooring and temporary mooring are very different. Temporary moorings for MODUs are frequently retrieved and redeployed, which gives the opportunity to regularly inspect each component in-air. A methodology was proposed in which the inspection frequency and scope may be considered as a function of the Component Condition level (including history service, maintenance and repairs) and the anticipated Working Environment (e.g. environmental loading, corrosion, etc.). It is expected that such a risk-based methodology will yield optimal inspection results. Permanent mooring systems, on the contrary, are normally not retrieved during their service life. Therefore, all inspections must be performed in-situ and are done primarily by Remotely Operated Vehicle (ROV), diver, and Rope Access Team (RAT). As a result of the mooring system staying in operation for prolonged periods, deterioration of permanent mooring components is often more concentrated in a few locations, such as the splash zone, at the fairlead/stopper, and near the touchdown point. Guidance is given on where, what and how to inspect the different components of a permanent mooring system, such that the efforts more focused on the most critical failure modes and locations while minimizing valuable offshore inspection time. This work reflected on future developments as well, like novel inspection methods using digital technologies, certification programs for surveyors, and more ergonomic inspection practices.

Finally, it was identified that there are gaps between the codes, standards, and practices of individual organizations. To address the gap, the findings of this work will be presented and submitted to the IOGP Mooring Work Group for review and potential adoption in the revised and harmonized mooring standard. It is understood that harmonization of multiple codes and standards has never been an easy task. It is particularly challenging for mooring design standards, as offshore mooring is a relatively young discipline that is still constantly evolving. Therefore, this work can hopefully serve as a foundation for the harmonization effort and it is expected to be incorporated into Revision 3 of ISO 19901-7 and the next revision of API RP-2I in the next years as well.

7 NOMENCLATURE

ABS	American Bureau of Shipping
API	American Petroleum Institute
BV	Bureau Veritas
CFR	Code of Federal Regulations
CP	Cathodic Protection
CVI	Close Visual Inspection
DNV GL	Det Norske Veritas Germanischer Lloyd
FEA	Finite Element Analysis
GVI	General Visual Inspection
IOGP	International Association of Oil & Gas Producers
ISO	International Organization for Standardization
JIP	Join Industry Project
MIM	Mooring Integrity Management
MODU	Mobile Offshore Drilling Unit
RAT	Rope Access Team
RCS	Recognized Class Societies
ROV	Remotely Operated Vehicle
SCORCH	Seawater Corrosion of Steel Wire Rope and Chain (JIP)
FEARS	Finite Element Analysis Residual Strength (JIP)
FoCCs	Fatigue of Corroded Chains (JIP)

8 REFERENCES

[1] API RP-2SK, "Design and Analysis of Stationkeeping Systems for Floating Structures", 3rd edition, October 2005.

[2] API RP-2I, "Recommended Practice for In-Service Inspection of Mooring Hardware for Floating Structures", 3rd edition, April 2008.

[3]. API RP-2I, "Recommended Practice for In-Service Inspection of Mooring Hardware for Floating Drilling Units", 1st edition, 1996.

[4] K.T. Ma, A. Duggal, P. Smedley, D. L'Hostis, H. Shu, "A Historical Review on Integrity Issues of Permanent Mooring Systems", Proc. Offshore Technology Conference, OTC 24025, 2013.

[5] K.T. Ma, R. Garrity, K. Longridge, H. Shu, A. Yao, T. Kwan, "Improving Reliability of MODU Mooring Systems through Better Design Standards and Practices", Proc. Offshore Technology Conference, OTC 27697, 2017.

[6] J. Stiff, "How Reliable are Reliability Calculations – Illustrated with Stationkeeping Examples", Proceedings of the Offshore Structural Reliability Conference, September 16-18, 2014, Houston.

[7] A. Ku, J. Stiff, B. Huang, F. Puskar, "MODU Mooring Reliability Analysis for Hurricane Conditions in the Gulf of Mexico", Proc. Offshore Technology Conference, OTC 20147, 2009.

[8] C.T. Kwan, "Mooring Design Standards – The Past, Present, And Future", Proceedings of the 20th SNAME Offshore Symposium, Houston, Texas, February 2015.

[9] ABS, "Guide for Building and Classing Floating Production Installations", 2009.

[10] ABS, "Rules for Building and Classing Floating Production Installations", January 2018.

[11] DNV, "Draft Revision of POSMOOR Rules", Det Norske Veritas, April 1998.

[12] DNV GL, "Position Mooring", DNVGL-OS-E301, Edition July 2015.

[13] BV, "Classification of Mooring Systems for Permanent and Mobile Offshore Units", Rule Note NR 493 DT R03 E, Bureau Veritas, December 2015.

[14] ISO, "Stationkeeping Systems for Floating Offshore Structures and Mobile Offshore Units", ISO 19901-7, International Standard, 2nd edition, 2013.

[15] E. Fontaine, A.E. Potts, K.T. Ma, A. Arredondo, R.E. Melchers, "SCORCH JIP: Examination and Testing of Severely-Corroded Mooring Chains from West Africa", Proc. Offshore Technology Conference, OTC 23012, May 2012.

[16] E. Fontaine, J. Rosen, A. Potts, K. Ma, R. Melchers, "SCORCH JIP - Feedback on MIC and Pitting Corrosion from Field Recovered Mooring Chain Links", Proc. Offshore Technology Conference, OTC 25234, May 2014.

[17] K. Ma, O. Gabrielsen, Z. Li, D. Baker, A. Yao, P. Vargas, M. Luo, A. Izadparast, A. Arredondo, L. Zhu, N. Sverdlova, I. Hogsaet, "Fatigue Tests on Corroded Mooring Chains Retrieved from Various Fields", 38th International Conference on Ocean, Offshore and Arctic Engineering, OMAE 2019-95618, June 9-14, 2019.

[18] S. Wang, X. Zhang, C.T. Kwan, K. Ma, Z. Li, D. Baker, et al., "Assessing Fatigue Life of Corroded Mooring Chain", OTC Conference, May 2019.

[19] C. Carra, T. Lee, K. Ma, A. Phadke, D. Laskowski, G. Kusinski, "Towards API RP 2MIM - DeepStar Guidelines for Risk Based Mooring Integrity Management", Deepwater Offshore Technology, October 2015.

[20] K. Ma, et al., "Life Extension of Mooring System for Benchamas Explorer FSO", Proceedings of the 19th Offshore Symposium, Society of Naval Architects and Marine Engineers (SNAME), Houston, Texas, February 2014.

[21] E. Fontaine, A. Kilner, C. Carra, D. Washington, K. Ma, A. Phadke, D. Laskowski, G. Kusinski, "Industry Survey of Past Mooring Failures, Pre-emptive Replacements and Reported Degradations for Mooring Systems of Floating Production Units", Proc. Offshore Technology Conference, OTC 25273, May 2014.

[22] K. Ma, Y. Luo, C. Kwan, and Y. Wu, "Mooring System Engineering for Offshore Structures", Gulf Professional Publishing, Elsevier, June 2019.

[23] UK O&G, "Mooring Integrity Guidelines", Issue 2 October 2014.

[24] API RP-2MIM, "Mooring Integrity Management", 1st edition, September 2019.

SNAME
THE INTERNATIONAL COMMUNITY
FOR MARITIME AND OCEAN PROFESSIONALS

A SEQUENTIAL APPROACH TO INTERNAL STRUCTURAL LOADS DERIVATION OF FOWT SUPPORT PLATFORMS

DIMITRIOS TSIROPOULOS
SBM OFFSHORE

Proceedings of the 25th Offshore Symposium, February 27th 2020, Houston, Texas
Texas Section of the Society of Naval Architects and Marine Engineers

ABSTRACT

The purpose of the present study is to demonstrate and verify a sequential approach towards deriving internal structural loads and performing the complete structural analysis of large volume FOWT support platforms, through utilisation of commercially available rigid body hydrodynamic and aero-elastic wind turbine design software. The platform's response, initially predicted by the coupled analysis software, is recreated in the rigid-body hydrodynamics software through the application of the wave elevation and modified tower base loads to an equivalent, segmented, model. With the coupled analysis load output "re-tracked" onto the segmented model, internal structural loads can be calculated.,

Keywords: SNAME, Offshore Symposium, coupled analysis; FOWT; structural design; internal loads; large volume platform; sequential; multibody

DEFINITIONS AND ABBREVIATIONS

ANSYS	General-purpose Finite Element analysis program
AQWA	Potential flow simulation model by Ansys
Bladed	Aero-hydro-servo-elastic simulation model by DNVGL
C-A	Aero-hydro-servo-elastic Coupled-analysis simulation
CoG	Centre of Gravity
CPU	Central Processing Unit
DLC	Design Load Case
DOF	Degree of Freedom
EDW	Equivalent Design Wave
EoM	Equation of Motion
FAST	Aero-hydro-servo-elastic simulation model by NREL
FE	Finite Element
FFT	Fast Fourier Transform
FOWT	Floating Offshore Wind Turbine
H/I	Hydrodynamically Interacting
H_S	Significant Wave Height
LC	Load Case
LRFD	Load and Resistance Factor Design

MBAHM	Multi-body Advanced Hydrodynamic Model
RAO	Response Amplitude Operator
RNA	Rotor-Nacelle Assembly
SBAHM	Single-body Advanced Hydrodynamic Model
TB	Tower Base
T_P	Peak Wave Period
QTF	Quadratic Transfer Function
WTG	Wind Turbine Generator
θ	Vessel's pitch angle
φ	Vessel's roll angle
ψ	Vessel's yaw angle

INTRODUCTION

The derivation of structural design loads for FOWT support platforms under combined wind and wave loading requires the integration of wind turbine aerodynamics, floating platform hydrodynamics, structural dynamics, and mooring line dynamics. Each of these subjects are a complex problem on their own and the currently available C-A software has its origins in the fixed wind turbine technology, which focuses on deriving structural loads in the tower and not the substructure. Moreover, the commercial and academic software for analyzing FOWTs has predominantly advanced aerodynamics but only basic hydrodynamics capabilities.

The main shortcoming, however, of the available C-A software – which the present study shall address – is the inadequacy of performing advanced hydrodynamic analysis for large volume FOWT support platforms. With the majority of the fixed offshore structures comprising of slender bodies, the hydrodynamic loading of beam structures is well catered for through Morison's theory; yet there is very little capability of performing the detailed hydrodynamic analysis of large volume floaters. The complete structural analysis for the hull of a FOWT is thus characterised by a lack of dedicated commercially available software to generate hull loading as well as a lack of standards and recommended practices for performing the analysis and validating the numerical model.

In the absence of such software, experimental tests have been carried out to study the dynamic responses of various FOWT designs (Coulling 2013 & Duan 2016). Nonetheless, model testing is a rather expensive analysis method, which – most often – comes with the caveat of Reynolds number dissimilarity between the model and full-scale structures. As a result, custom-made numerical modelling tools have been developed by researchers to tackle the complex full-scale FOWT analysis. Two discrete types of such tools can be identified. One that emphasizes the hydrodynamic loading and motion response of a FOWT's support structure, while simplifying the wind turbine aerodynamic loading (Nematbakhsh et al 2014 and 2015, and Tran & Kim 2015); and another that attempts to analyse the complex fluid flow of combined wind and waves through a complete system, two-phase CFD simulation that accounts for the wind turbine aerodynamics, floating platform hydrodynamics as well as the influence of one to another (Quallen 2014). The former approach has the disadvantage of the simplified aerodynamic model; while the latter, although rather accurate when performed immaculately, is greatly expensive in terms of CPU power and time demands along with being rather complicated in its setup.

The main purpose of the present study is to demonstrate and verify a sequential approach towards deriving internal structural loads and performing the complete structural analysis of a large volume FOWT support platform, through utilisation of commercially available rigid body hydrodynamic and C-A design software. This approach allows one to perform a fully coupled aero-hydro-mooring structural design analysis of a FOWT support platform and its main benefits are:

- The use of established and commercially available advanced hydrodynamic and C-A software packages;
- The potential for coupling between such software packages;
- The drastically reduced time and CPU power demands compared to: time domain pressure mapping FE structural analysis; and CFD analysis.

C-A software, such as FAST or BLADED, may be used to simulate the aero-hydro-servo-elastic response of the floating platform and wind turbine, taking into account both wind loading and wave frequency hydrodynamic loads. The approach to the C-A is to make use of the obtained results (TB loads) in combination with a segmented advanced hydrodynamic model for two primary reasons:

- To appropriately distribute the hydrodynamic loads between the hull's sectional planes which are critical for the structural design below the TB interface level; and
- To model hydrodynamic loading that also includes load components required, but not modelled in the C-A due to limitations in the available software.

Hydrodynamic software, such as AQWA, shall be used to combine the C-A TB loads with hydrodynamic loading. Termed the Multi-body Advanced Hydrodynamic Model (MBAHM), this model shall be segmented at the critical section planes and used to generate global hull loads required for the structural design.

The verification of the MBAHM against an equivalent Single-body Advanced Hydrodynamic Model (SBAHM) will be examined through a case study, outlined in detail. The de-risking of this modelling approach is vital for the reliability of the outcome of such an analysis, as this type of geometrical segmentation constitutes an unconventional type of hydrodynamic analysis that may challenge the functionalities of most commercially available hydrodynamic software.

METHODOLOGY

FULLY COUPLED MODELLING STRATEGY

In determining the global loading for the hull design of a FOWT support platform, a fully coupled modelling strategy is proposed. A sequential approach is regarded necessary to generate a simulation where wind turbine aerodynamics, floating platform hydrodynamics and mooring line dynamics are coupled in the time domain, whereby the internal structural loads can be derived. A high-level description of the modelling and analysis steps required to attain this is presented below and graphically illustrated in the flowchart of Figure 2. The details of each step are outlined in the following subsections.

Step 1. In the potential flow hydrodynamic software (e.g. AQWA), derive the hydrostatic and frequency domain hydrodynamic parameters (added mass, radiation damping, Froude-Krylov and diffraction force RAOs) from the rigid, SBAHM of the FOWT's hull.

Step 2. In the C-A software (e.g. FAST), derive – for each LC – the time series of the wave elevation, global motions and loads at the TB interface level, using the abovementioned hydrodynamic parameters to represent the FOWT's hull.

Step 3. In the potential flow hydrodynamic software, recreate the derived global motions of the FOWT – in the time domain and for each LC – using a rigid, MBAHM. This is achieved through the application of the C-A TB loads and wave elevation time histories. Before the TB loads can be applied to the MBAHM – as an external loading time history –, they require to undergo modification so that any reproduced load components are removed, thus avoiding double counting. If necessary, include potentially important load components for the hull design, which could not be modelled in the C-A software, e.g. current, 2nd order difference and sum frequency wave loads, mooring line dynamics. For each section plane of the MBAHM's hull, extract the global loads' time series per LC.

Step 4. Finally, determine the maximum sectional loads[1] and adopt a EDW approach (DNVGL 2015). In a general-purpose FE structural analysis program (e.g. ANSYS Offshore), generate sectional load RAOs and subsequently, derive the equivalent regular wave to replicate the first order wave load contribution to the total sectional load.

[1] The specifics of the methodology for determining maximum sectional plane loads from time histories may vary between the relevant codes and fall outside the scope of this study.

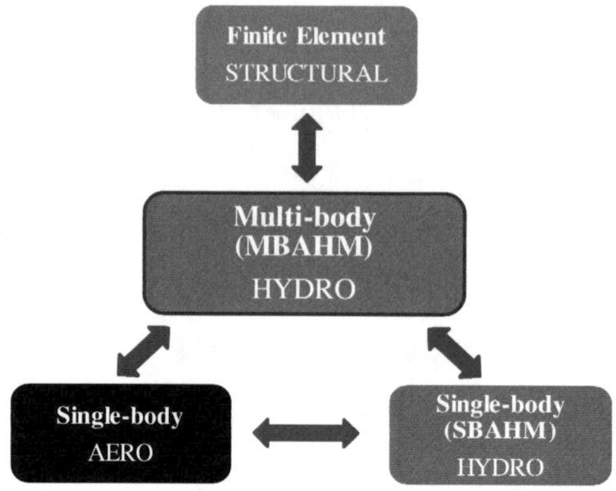

Figure 1: Required verifications between the MBAHM and the rest models employed in the sequential approach

The following terms and abbreviations are used throughout the present study to refer to the models employed in each step:

Step 1: *Single-body*; *SB*; *SBAHM.*
Step 2: *C-A*; *Aerodynamic*; *aero-hydro-servo-elastic.*
Step 3: *Segmented*; *multi-body*; *MB*; *MBAHM.*
Step 4: *Structural*; *Finite Element; FE.*

This multifaceted interface, between primarily the MBAHM and the rest three models (C-A, SBAHM and FE) requires a four-part verification process, as graphically illustrated in Figure 1. Of the four depicted verifications, the one between the two hydrodynamic models – MBAHM and SBAHM – is considered of the outmost importance as well as the most complex. This is due to the segmentation approach employed being rather unconventional for a hydrodynamic analysis; with adjacent bodies rigidly connected and no water allowed to infiltrate between them. As such, a case study of this verification process will be thoroughly outlined in the following section.

The surface mesh used for the hydrodynamic modelling should be consistent across all analysis models. This will ensure passing of wave pressure loads can be undertaken efficiently between software packages.

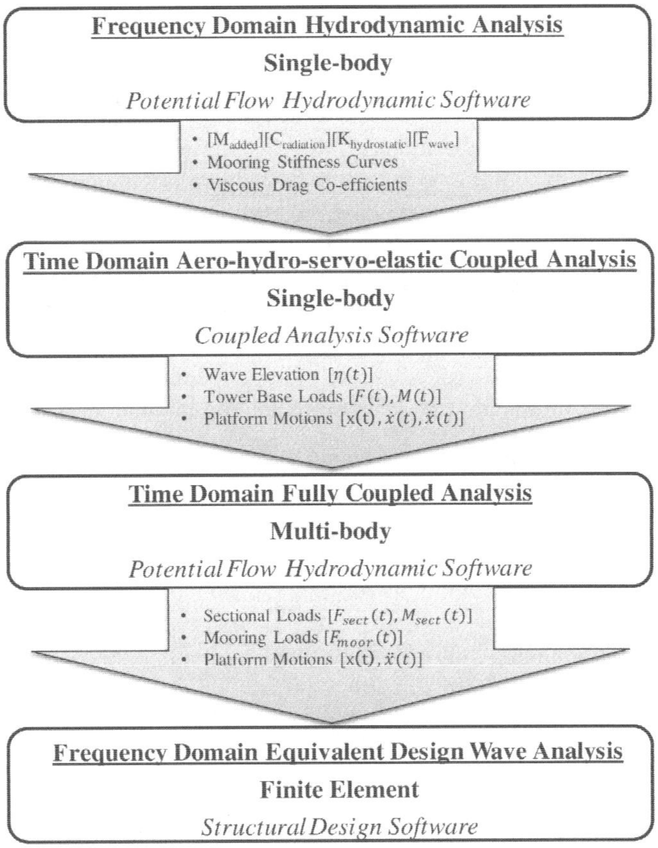

Figure 2: Proposed sequential approach to structural EDW analysis based on a fully coupled modelling strategy

Single-body Advanced Hydrodynamic Model

The computation of the frequency domain hydrodynamic parameters, required by the C-A, should be carried out using commercially available hydrodynamic software. Hence, through the radiation-diffraction analysis of a SBAHM, the following inputs for the C-A shall be generated:

- 1st order hydrodynamic parameter RAOs, i.e. added mass, radiation damping, Froude-Krylov and diffraction forces;
- 2nd order dimensional wave drift and sum frequency QTF.

Moreover, the linear hydrostatic stiffness matrix should be calculated, together with the mooring system in the form of individual mooring line stiffness curves.

If required, a methodology for modelling viscous damping e.g. equivalent Morison diameters and drag coefficients should also be provided.

For model verification purposes, it is recommended that the following data are compared between the SBAHM and C-A model:

- Platform motion RAOs;
- Mooring system stiffness at equilibrium;
- Decay tests with and without the mooring system.

Aero-hydro-servo-elastic Coupled Analysis Model

Commercially available, C-A software should be used to simulate the coupled aero-hydro-servo-elastic response of the platform in time domain, accounting for all wind loads and wave frequency hydrodynamic loads. Utilising the hydro-static and -dynamic parameters derived from the SBAHM, the main output time series from the C-A should be:

- Forces and moments at the TB interface level;

- Wave elevation;
- Global motion responses.

This approach is also applicable for when the C-A software is directly coupled to the advanced hydrodynamic tool.

Multi-body Advanced Hydrodynamic Model

Section planes of the structural design's interest are defined and a hydrodynamic model, split into the relevant bodies, is created. The purpose of the MBAHM is to transform the C-A output into design loading for the hull, a process termed load "re-tracking". This is attained by coupling the aerodynamic, hydrodynamic and dynamic mooring loading in the time domain. A hydrodynamic model that is segmented, contrary to the single-body one, is deemed necessary so that the hydrodynamic loading is appropriately distributed between bodies on each side of the defined section planes (Figure 4).

The hydrodynamic and mooring line dynamics can be directly simulated in the potential flow hydrodynamic software, but the aerodynamic forces and moments need to be applied as an external, time series loading. The validity of applying the C-A aerodynamic loads onto the MBAHM relies upon recreating the same platform response – in the time domain –. This is because changes in the platform's speed alter the wind speed that the turbine sees, which in turn, result to change in the turbine's aerodynamic load, i.e. the "aerodynamic damping" load[2]. From a typical wind turbine thrust curve, it becomes apparent that the aerodynamic damping is a rather non-linear phenomenon, since the aerodynamic force changes irregularly with changes in the relative wind speed (Figure 3). This is especially pronounced around the rated wind speed, where the thrust/wind-speed gradient changes sharply from positive to negative. Hence, for the derived time history of the aerodynamic damping to be directly applied onto the MBAHM, it is crucial that the – 1st order, higher speed – platform motions are correctly recreated.

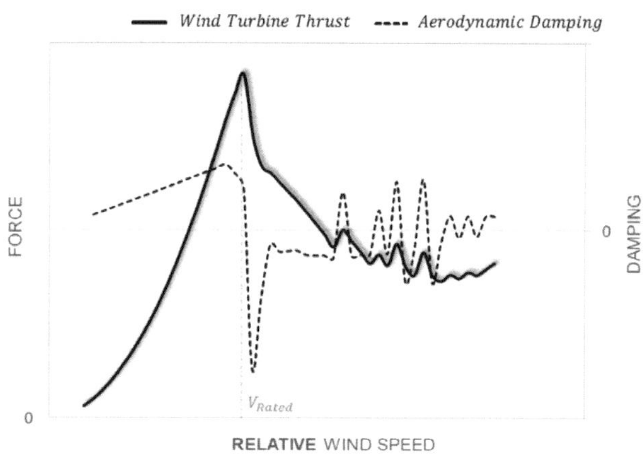

Figure 3: Typical thrust curve of a wind turbine and its associate aerodynamic damping

The wave elevations generated in the C-A are directly applied onto the MBAHM to generate the hydrodynamic loading. Nonetheless, before the derived time histories of the C-A TB loads are applied onto the MBAHM, load components that shall be reproduced are removed to avoid duplication. These replicated components arise from the presence of the WTG, i.e. tower and RNA, mass inertia in both the C-A model and the MBAHM. In particular, the loading that would otherwise be double-counted are the low frequency (LF) and wave frequency (WF) inertial and gravitational (static and dynamic[3]) forces, and moments, due to the WTG's mass inertia components.

Effectively, the LF loads are recalculated in the hydrodynamic software as response to the LF motions of the platform, whereas the WF loads are recalculated from the applied wave elevation. With the inertial loads being proportional to the platform's accelerations – in the relevant DOF – and the gravitational loads to the displacements, the removal of the desired LF and WF loads can be achieved as described in the following process:

[2] As this effect is linked to the velocity term in the EoM, it is comparable to damping, hence the term aerodynamic damping.

[3] Dynamic gravity arises from the structure's inclination as a function of time, $\alpha(t)$, resulting to the time-dependent $g \cdot sin\alpha(t)$ gravitational load.

- For each displacement and acceleration time series component derived in the C-A, filter out the high frequency (HF) components – the cut-off frequency can be taken as ~0.33Hz, i.e. 3s minimum wave period equivalent, subject to the site's metocean data–;
- Use these filtered time histories of the platform's displacements and accelerations to calculate the WTG's gravitational and inertial loads, respectively, – at LF and WF only –; and finally,
- Subtract these time series loads from the original TB ones.

Thus, the modified load time series will be acquired, which when applied to the MBAHM, in conjunction with the wave elevation time history, shall generate the same motion responses as derived in the C-A.

Regarding the HF TB loads, these may remain unaltered as they are not accounted for anywhere else in the analysis. However, these are derived in the C-A as internal loads and yet are applied as external loads onto the MBAHM. Hence, there shall be a discrepancy between the applied and the reacted loads at the TB interface level due to the presence of the WTG mass inertia in both the MBAHM and C-A models. This issue will be analysed in more detail before conclusions are drawn and alternative ways to remedy it shall be proposed.

For model verification purposes, it is recommended that the motions between the MBAHM and C-A models – for at least one LC from each DLC group – are compared.

Once the verification of the models is completed, potential load components that cannot be modelled in the C-A software but which are still important for the platform design, e.g. current, drift and sum frequency wave loads, mooring line dynamics, can then be included in the MBAHM.

FE Structural Model

The general approach to deriving the global loads to be assessed in the structural analysis can be based upon a conventional EDW concept used for assessing floating structures. The main modifications to the standard offshore structure approach is the inclusion of wind load and mooring loads, which although they are traditionally neglected due to their small impact on global loading and relative phasing with the first order wave loads, their impact on a FOWT hull design cannot be dismissed in most cases.

In this approach, sectional load RAOs are calculated from an FE structural model and an equivalent regular wave and set of point loads for TB and mooring loads are derived that allow reproduction of the design load in each load section or global response. The design loads will need to be split into static and environmental loads to be compatible with a LRFD approach.

For each LC examined, it is proposed that the responses presented in Table 1 are maximised. Hence, the corresponding loading set for each maximised response is provided for the structural analysis of the floating platform.

Table 1: Maximised responses for global load analysis

Type	**Response Maximised**
Section Plane Loads	F_Z, F_R^*, M_X, M_Y, M_Z, M_R^*
Global Motions	$\varphi, \vartheta;\ \ddot{x}, \ddot{y}, \ddot{z}, \ddot{\varphi}, \ddot{\theta}, \ddot{\psi}$
Point Loads	Mooring Line(s) Tension, TB Loads

Subscript "R" denotes the resultant X and Y load components

The advantage of this approach is that it allows the wind and wave loads to be simultaneously evaluated in the time domain, while it avoids the requirement to perform a, complex and time expensive, stress analysis.

In terms of model verification, the sectional loads at static equilibrium as well as in – a series of – regular waves can be compared between the FE structural model and the MBAHM.

RESULTS

MULTI-BODY ADVANCED HYDRODYNAMIC MODEL VERIFICATION

An example case study of verifying a MBAHM against its equivalent SBAHM is presented in this section. The analysis was performed in the hydrodynamic software AQWA. A verification test matrix is derived and its importance to be followed in similar analyses is well demonstrated.

Multi-body Modelling

To acquire the hull's internal loads at cross sections, the single-body hull geometry needed to be segmented into a multiple bodies one, where each body was bounded by the adjacent load section plane(s) (Figure 4). The model's segments were articulated with fixed constraints – at the relevant section planes – and thus the whole platform behaved as a rigid body.

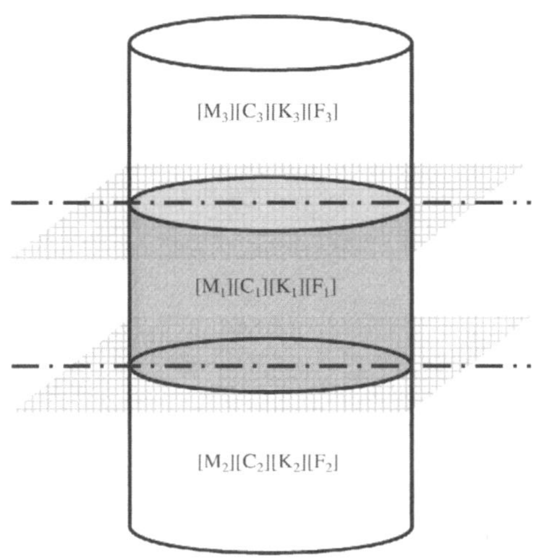

Figure 4: Typical adjacent section planes that bounded individual bodies (shown in blue colour) of the MBAHM model

The multi-body model was defined whereby the portion of the hull between load sections was represented by individual mass and hydrodynamic properties. The derived hydrodynamic properties in AQWA included interaction effects, hence ensuring that the influence between adjacent bodies was captured. In this case, the complete hull was segmented into twenty H/I bodies.

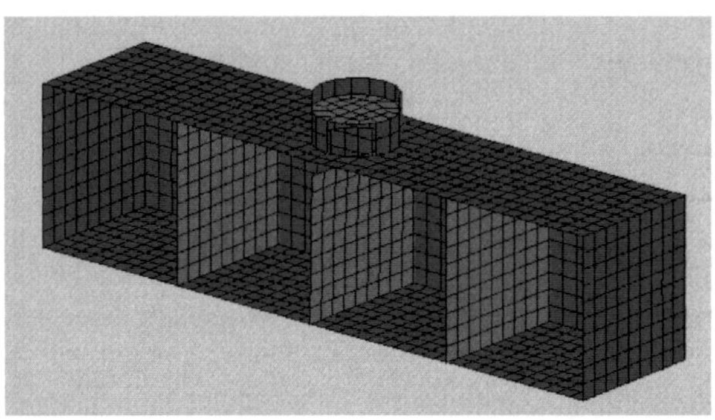

Figure 5: Discretised, non-diffracting lids enclosed submerged diffracting bodies; orange colour for non-diffracting and blue for diffracting panel elements

Non-enclosed Diffracting Bodies, with more than one of their sides open, caused numerical inaccuracies – these were evident in the initial static inclination tests – hence, non-diffracting, dummy lids were manually created to close the open sides of all the submerged bodies in AQWA. This modification resolved the issue but it also required that the buoyancy force and the linear hydrostatic stiffness matrix for each diffracting body were defined manually.

Morison Elements, where required, were appropriately defined for each body and thus, the relevant slender body drag and inertia loads were correctly distributed between the section planes.

Verification Tests

Due to the unconventional configuration of adjacent H/I bodies being rigidly attached to each other and no water allowed to infiltrate between them, it became imperative to verify that the multi-body model behaved the same as the equivalent single-body one.

The sought-after output from the analysis of the multi-body, MBAHM in AQWA was the time-domain, six DOF loads at the required sections (articulation loads) that would occur from a variety of sea states and their corresponding sets of external aerodynamic loads, as derived from the C-A software.

To attain reliable results from AQWA, it was essential to ensure that the motions of the MBAHM would match these of the equivalent SBAHM, for any examined case. Therefore, a verification test matrix (Table 2) was compiled, which comprised of a series of tests deemed appropriate to verify the static and dynamic parameters governing the models' motions for a wide range of cases.

Mechanical Mass & Inertia were the first properties of the MBAHM to be successfully verified, ensuring that the resultant total mechanical mass and inertia properties were identical to the SBAHM's structure. Although these parameters were user-defined, verification of all the variables involved in the equation of motion was considered necessary.

Linear Hydrostatic Coefficients were also user-defined and verified successfully by assembling the resultant 3x3 matrix of the 20 H/I bodies. Where needed, free surface effects – in bodies that contained liquid compartments – were also accounted for by correcting the rotational hydrostatic stiffness of any individual body. The resultant displacement and CoB showed excellent agreement with the SBAHM's values.

Added Mass & Radiation Damping Coefficients were assembled for the MBAHM and successfully verified against the SBAHM ($<0.1\%$ deviation). The assembly of the resultant 6x6 added mass and radiation damping coefficients for each frequency was done in two steps:

First, since each body was interacting hydrodynamically with all the rest, the H/I matrices needed to be added on the non-H/I coefficients for each body. Thus, the 6x6 hydrodynamic matrices for each body were first assembled and then, the resultant added mass and radiation damping matrices of the combined twenty bodies were calculated.

Froude-Krylov & Diffraction Forces were also assembled for the MBAHM and the resultant values were successfully verified against the SBAHM ($<0.1\%$ deviation) in all 6 DOF and for multiple wave directions.

Additional Linear Diagonal Damping is derived for SBAHM – in this case through CFD analysis, but could have also been determined through experimental tests or empirical formulae – to account for damping effects calculated neither by potential theory nor Morison's equation e.g. viscous and eddy drag effects from heave plates. With the derivation of internal structural loads being the main objective of the segmented, MBAHM, it was essential to properly distribute the additional damping loads among the individual bodies. Hence, the following process was employed in order to assign the SBAHM's additional linear damping values in heave, roll and pitch to each of the H/I bodies:

1. The single-body's additional heave damping was proportionally distributed, relative to the vertically projected area of the individual bodies.
2. The additional heave damping component, assigned proportionally to each individual body, resulted to an additional roll and/or pitch damping component, dependent on the dy and dx distance, respectively, between the overall CoG and that of any individual body. Ideally, the resultant, heave-imposed roll and pitch additional damping should match the corresponding values of the single-body.

3. In case of any observed deviation in the resultant additional rotational damping (overestimate of ~10% in this analysis), it should be corrected via means of negative, or positive, roll and pitch diagonal damping values assigned proportionally to the relevant bodies.

The validity of the method employed and the resultant damping values in all 3 DOF of interest was confirmed through decay tests' comparison (Figure 9) between the two models, both with and without the additional linear damping.

Pushdown, Static Tests were performed in AQWA Librium on both the single- and multi-body models through the application of varying extra weight at the overall CoG. The key parameter compared was the heave displacement (Figure 6), which showed excellent agreement between the two models for a wide range of additional weight applied. Moreover, no coupling in any other DOF was observed in both models, as expected.

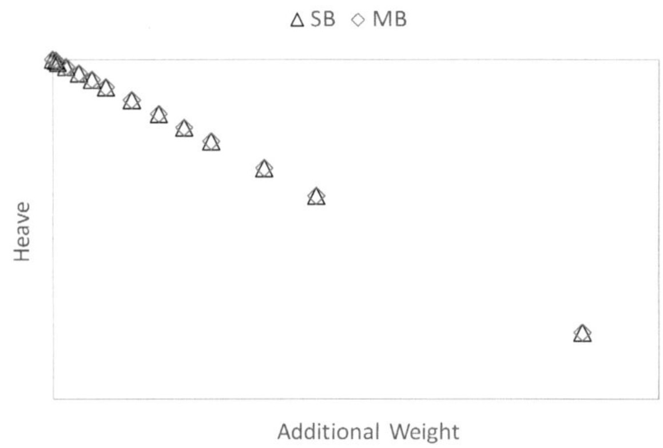

Figure 6: Heave displacement comparison between the single-body (SB) and the multi-body (MB) due to the static application of extra weight at the hull's CoG

Inclination, Static Tests were performed through the application of extra weight at a range of distances from the hull's CoG to induce a range of inclination moments. Unlike the pushdown tests, this additional weight was fixed in magnitude and its location varied depending on the magnitude and direction of the induced moment required. Making use of the model's symmetry, four directions of inclination were examined – 0, 30, 60 and 90 degrees from the model's longitudinal centreline – and the results between the single- and the multi-body showed excellent agreement in the obtained heel and trim. An example comparison of the models' inclination angles for the 30 degrees case is shown in Figure 7.

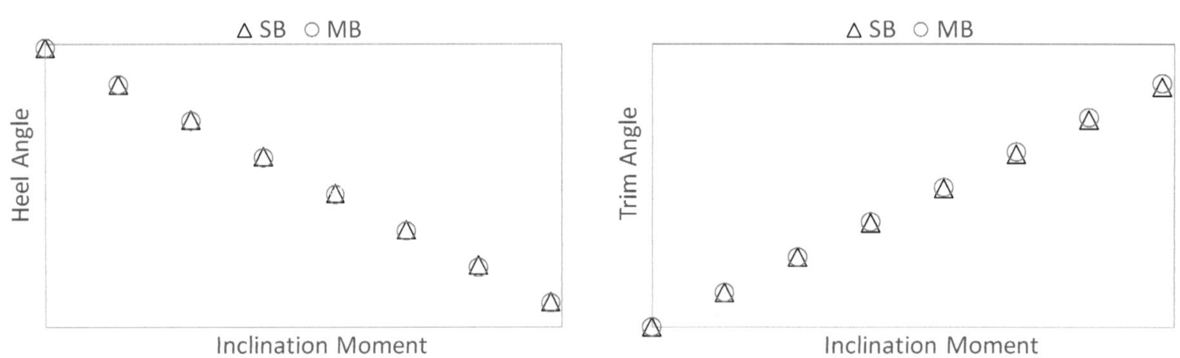

Figure 7: Heel and trim inclination angle comparison between the SB and the MB due to the static application of an ascending moment about the hull's CoG,

Decay, Dynamic Tests were performed separately for surge, sway, heave, roll, pitch and yaw motions by means of offsetting the hull's initial position on the corresponding DOF only. The hull was then left to freely oscillate (for a minimum of four periods) in the time-domain and its motions were recorded. For heave, roll and pitch, decay tests were performed with and without inclusion of the frequency-independent, additional linear diagonal damping, which was previously discussed. These tests were necessary to verify the method employed for the distribution of the hull's, single-value additional damping onto the individual bodies of the MBAHM. As can be seen in the plots of Figure 8 and Figure 9, excellent agreement was obtained between the SBAHM and MBAHM motions for all 6 DOF decay tests; thus, providing confidence regarding the rigid-body natural periods of the segmented model and consequently, the resultant damping, added mass and stiffness coefficients.

White Noise, Dynamic Tests were deemed necessary to compare the RAOs of the MBAHM to those of the SBAHM, because, at the time of writing, the resultant RAOs of all the H/I bodies combined cannot be directly output by the frequency domain analysis performed in AQWA Line. However, with all the rest components of the equation of motion verified (mechanical and added mass, radiation damping, hydrostatic stiffness, Froude-Krylov and diffraction forces), confidence on the expected RAO motions existed.

An identical, unit amplitude white noise wave spectrum was applied in both models for a duration of 3600 seconds (3-hour sea state). Only one wave direction (0 degrees) has been investigated in terms of the RAOs verification and, due to the model's geometry, motions only in surge, heave and pitch were induced. The Power Spectral Density (PSD) and consequently the RAOs were determined by means of FFT of the platform motions time series (signal). It should be noted that depending on the nature of the analysis, it is recommended that RAOs are verified for all directions and DOF of interest.

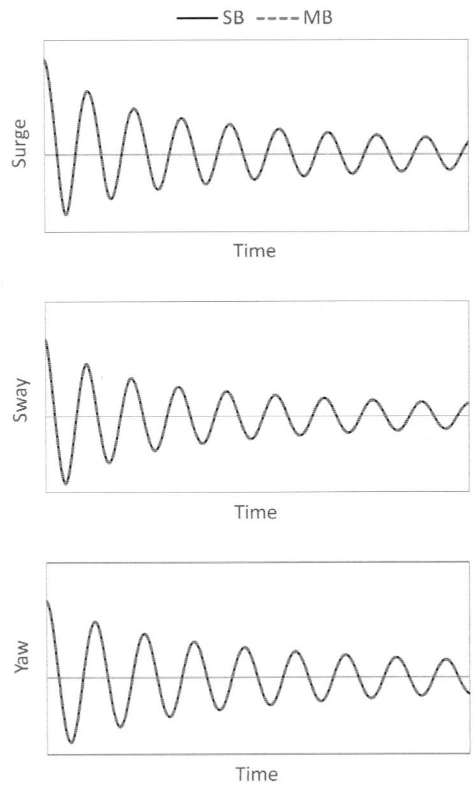

Figure 8: Comparison between the SB and the MB models' motions during surge (top), sway (middle) and yaw (bottom) decay tests

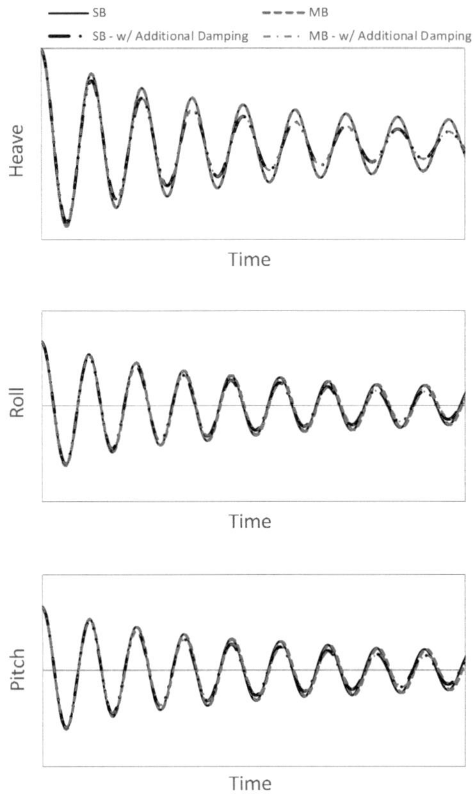

Figure 9: Comparison between the SB and the MB models' motions during heave (top), roll (middle) and pitch (bottom) decay tests, with and without additional linear diagonal damping

Comparison of the obtained RAOs for both models is shown in Figure 10. The agreement between the two models' motions was very good in all DOF examined, with the only noteworthy deviation observed at the peak of the pitch RAO, where the MBAHM's value was found to be ~4% less than the single-body's equivalent. This implied a difference in the pitch damping of the MBAHM, which, nonetheless, was not confirmed by the decay tests. This, in turn, suggested that the 3-hour duration of the white noise tests was possibly insufficient for a pitch natural period significantly higher (~3 to 5 times) than a characteristic ~10s wave.

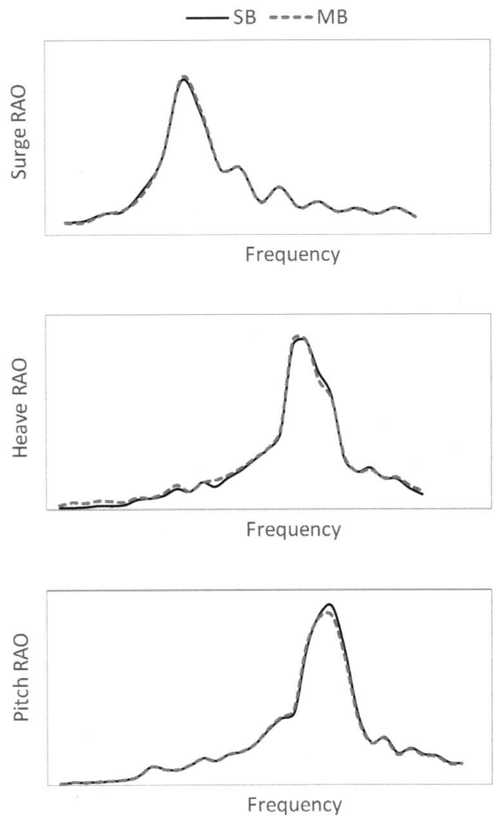

Figure 10: Comparison of the SB and the MB model RAOs derived from white noise wave tests for 0 degrees; surge (top), heave (middle) and pitch (bottom)

Table 2: Single- VS Multi-body model verification test matrix

Verification Test	Type	Software	Examined Model Parameters
Mechanical Mass/Inertia	Spreadsheet Calc	-	6x6 Mass/Inertia Matrix
Linear Hydrostatic Stiffness	Spreadsheet Calc	-	3x3 Linear Hydrostatic Stiffness Matrix
Added Mass	Spreadsheet Calc	-	6x6 Added Mass/Inertia Coefficients Matrix
Radiation Damping	Spreadsheet Calc	-	6x6 Radiation Damping Coefficients Matrix
Pushdown	Static	Librium	Hydrostatic and Mooring Stiffness in heave
Inclination	Static	Librium	Hydrostatic and Mooring Stiffness in roll and pitch
Decay	Dynamic	Drift	Natural Frequencies in all 6 DOF
White Noise	Dynamic	Drift	Displacement RAOs in surge, heave and pitch
Irregular Wave (1yr, 50yr & cut-off)	Dynamic	Drift	Motions in Time-domain in 6 DOF
Irregular Wave & External Force	Dynamic	Drift	Motions in Time-domain in 6 DOF

Irregular Wave, Dynamic Tests were performed for three sea states of interest:

1. Omnidirectional 1yr design wave;
2. Omnidirectional 50yr design wave;
3. Cut-off wind speed wave.

Although the white noise wave tests provided confidence with regards to the MBAHM's RAOs, it was also necessary to ensure good agreement with the SBAHM's motions for a number of critical ULS design waves. Moreover, the small difference observed in the pitch RAOs could be further investigated to determine its influence in the predicted motions of the segmented platform model.

The motions for cases 1 and 3 showed excellent agreement in all 6 DOF between the two models (representative plot of pitch motion for case 3 shown in Figure 11).

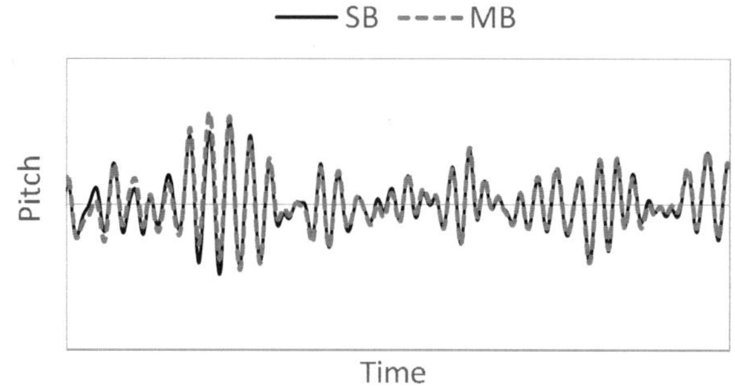

Figure 11: Representative snapshot of the SB and MB model's pitch motion in time domain for the cut-off wind speed wave

Nonetheless, the following issue with the MBAHM's motions was noticed for the 50yr case: instances of excessive pitch motion were observed, which also triggered excessive, and unrealistic, roll motion (Figure 13). To further investigate this critical issue, sensitivity studies with regards to the wave's T_P and H_S parameters were performed.

No link with the wave's T_P, and hence the wave's steepness, was found. Instead, the studies revealed that the problem lied with the value of H_S being significantly high, and thus, very slightly damped natural frequency modes of one, or more, of the individual bodies were triggered. Therefore, an upper bound of H_S was determined (~7 metres), beyond of which the MBAHM's motions would become problematic. A representative record of the bounding wave case's motions in all DOF is given in Figure 12.

As it becomes apparent, these irregular wave tests revealed an unexpected behavior of the segmented model which was not otherwise obvious; despite the confidence attained from the series of the previous tests. It should be noted that the observed problematic behavior of the multi-body model for H_S greater than seven metres would not necessarily hold true for all hull geometries. Hence, it is strongly recommended that all the tests outlined in this verification matrix are conducted independently for any hull geometry. Moreover, it is proposed that white noise wave tests of greater than a unit's wave amplitude are also conducted to attempt and reveal potential irregularities occurring in sea states of considerably high H_S.

Although further work to investigate this matter is required, the proposed remedy to such issues is to artificially increase the total damping value of the relevant individual bodies, and in the relevant DOF, either at the system's natural frequency or frequency-independently. This increase should be done gradually and – through a sensitivity study – an acceptable level of additional damping can be determined, which resolves the issue without compromising the reliability of the obtained motions and hydrodynamic loads.

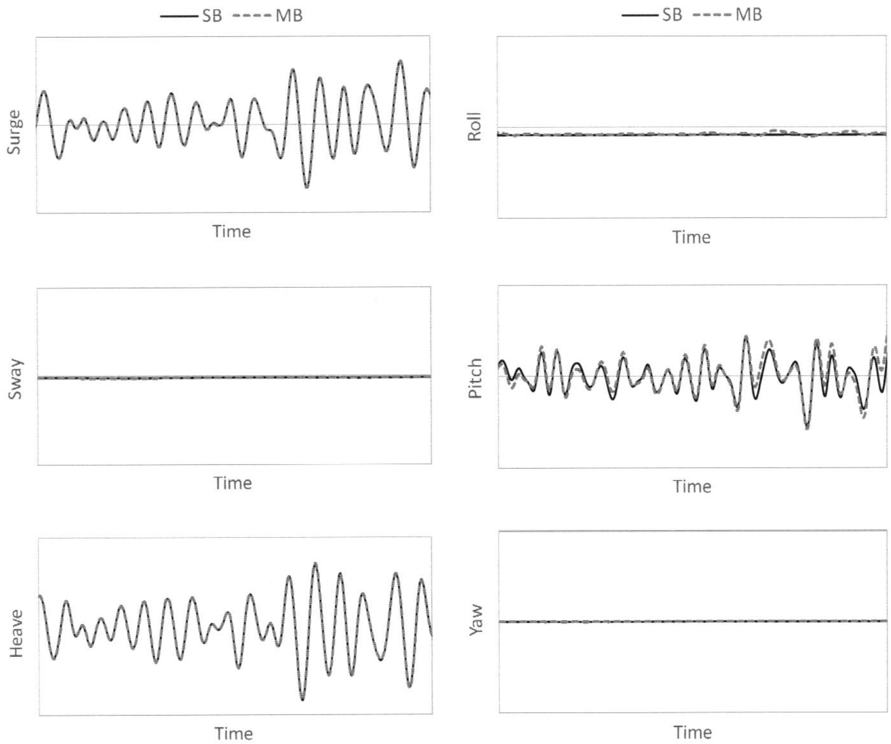

Figure 12: Representative snapshot of the SB and MB models' motions in all DOF, in time domain for the bounding H_S = 7m

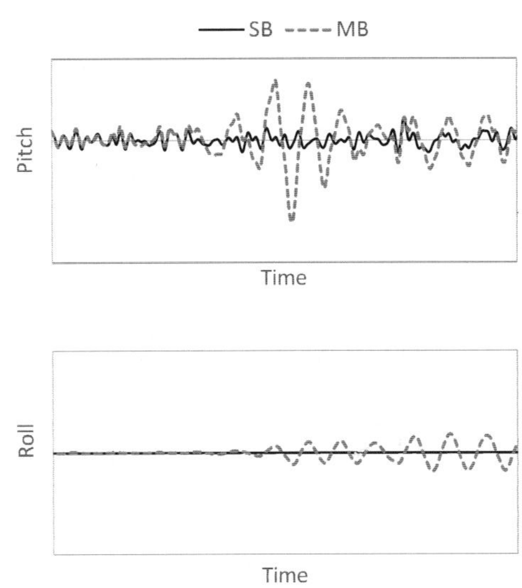

Figure 13: Time interval where roll and pitch motions of the MB model became erroneously excessive

Combined Irregular Wave & External Force, Dynamic Tests were the last type of tests to be completed in the proposed multi-body modelling verification matrix. Effectively, these tests were an exact replication of the pursued analysis to be performed onto the MBAHM to extract the sectional design load time series. The external force would be representative of the aerodynamic, C-A loads and the irregular wave would be representative of the corresponding's load case design wave. It is proposed that at least one design wave is tested from each DLC category.

The MBAHM motions showed excellent agreement with the SBAHM for all cases examined, e.g. normal operation, extreme sea state, emergency shutdown. As raised in the last section, the maximum H_S of the tested waves was restricted to an upper limit of ~7 metres. Comparison plots between the motions of the two models are shown in Figure 14 for an example (emergency shutdown) design scenario.

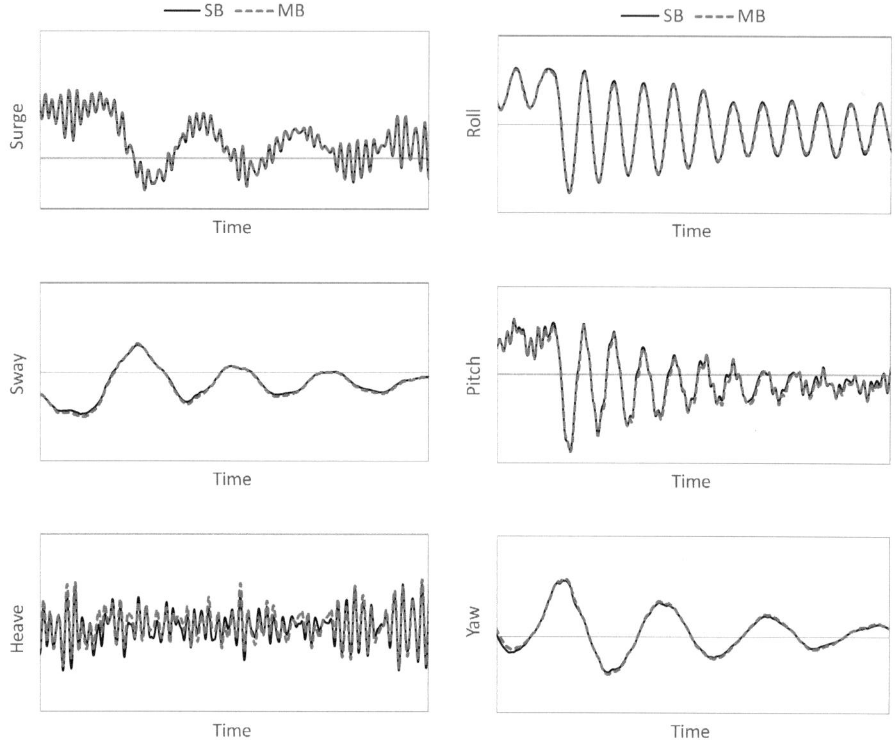

Figure 14: Representative snapshot of the SB and MB models' all DOF motions in time domain for a turbine shutdown case

HIGH FREQUENCY LOADS RE-TRACKING

It is essential that the ability to recreate the platform's response, as predicted by the C-A software, in the MBAHM is verified before – any additional loads are included, and – the sectional loads are extracted. LF and WF load components included in the C-A TB loads and replicated in the (coupled aero-hydro-mooring) analysis of the MBAHM are removed, as previously discussed. HF TB loads are, nonetheless, applied unaltered onto the MBAHM. The intact application of the HF TB loads as external loading gives rise to the following issue that needs to be taken into consideration.

The HF loading, as output from the C-A software, is the loading that acts on the floating substructure at the interface with the WTG, as a result of turbine operation, wind turbulence, etc. Hence, an equal and opposite loading, coming from the substructure, shall also exist on that TB interface. Based on the proposed modelling approach, this internal loading is applied to the integrated MBAHM as external loading. However, with the MBAHM also including the WTG's mass inertia, not all of this HF applied load will get transferred down into the substructure as expected, but instead, part of it will go up into the tower. This can be seen in Figure 15, where the reduction in the reacted load at the TB interface level of the MBAHM (with WTG mass inertia included) is quite significant compared to the applied, C-A load. The discrepancy will be dependent on the ratio between the structure's mass inertia above and below the TB interface – with the WTG's mass normally being an order of magnitude less –.

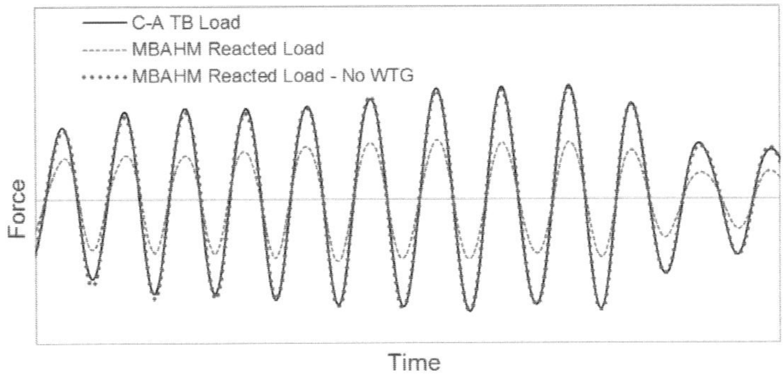

Figure 15: Representative HF TB load comparison between the C-A and two MBAHMs, with and without the WTG's mass inertia

In order to remedy this issue, occurring in the MBAHM, the following is proposed:

- Identify and quantify the difference between the applied and the reacted HF loads;
  - If the difference is deemed negligible then proceed with applying the original unaltered HF TB loads;
- If the quantified discrepancy is important, then proceed by comparing the magnitude of the HF loading to that of the LF and WF;
  - If the HF loading is deemed negligible, for the limit state design considered, then proceed with applying the original unaltered HF TB loads;
- Finally, if the HF discrepancy is considered an important loading component for the structural design, then the following two alternative modelling methods are recommended:

1. Adjust the HF loads applied to the MBAHM in order to artificially increase the reacted loads at the TB interface level, such that they match the original C-A loading. This increase will need to be performed based on the mass inertia ratio between the structure above and below the TB.

2. Create a second, identical MBAHM which has the mass inertia of the WTG removed. In effect, this will lead to two simulations for each LC examined, the loads of which would then need to be combined. The second model should be used to derive the HF loading only. An example case verifying such an application in terms of the HF loading is shown in Figure 15.

Whichever of the above methods is selected, it is imperative that the motions between the C-A and the MBAHM are compared and verified. Moreover, considering that it is unlikely for the platform's motions to be affected by the HF loading (due to high inertia), the final internal HF loads in the time domain should also be verified between the two models.

CONCLUSION

To overcome the lack of relevant dedicated software and recommended practices, a sequential approach for deriving internal structural loads of large volume FOWT support platforms was presented and verified in detail.

Commercial software packages were combined in a sequential manner to produce the coupled aerodynamic, hydrodynamic and dynamic mooring loads acting on the floating platform in the time domain. Four discrete steps and the associated models were required to perform the global structural analysis.

The requirement for deriving loads at sectional planes crucial for the floater's structural design, deemed a multi-body modelling method necessary. This allowed for the hydrodynamic loading to be properly distributed across the individual bodies composing the floater's hull.

The importance of performing a thorough verification of the multi-body hydrodynamic modelling approach on a case-by-case basis was demonstrated through an example study. A verification test matrix was necessary to be followed to substantiate critical modelling features between the two, multi- and single-body, hydrodynamic models.

The key aspects that could compromise the sequential approach's validity were identified across all four steps and models employed; with emphasis given to the transformation of the C-A aerodynamic loads into design loading for the hull. Where required, recommendations were made that need to be adhered to so that the reliability and accuracy of the analysis's outcome is not compromised.

To conclude, the sequential approach is rather versatile as it can be performed through any combination of commercially available C-A, potential flow theory and FE software packages. The appropriate verification studies will ensure that the reliability of the produced results remains high, while the required timescales for performing such a fully coupled analysis compare excellently with the current industry practice. Finally, the integration of the approach's main sequential steps into a single process through multidisciplinary design optimization platforms, will result to vast reductions in the overall solution time and minimum data exchange risks.

ACKNOWLEDGEMENTS

I would like to offer my sincere thanks to Aaron Burden and Ben Smith for being great role models and always offering their support and guidance to me.

REFERENCES

Coulling A.J., Goupee A.J., Robertson A.N., Jonkman J.M., Dagher H.J., *Validation of a FAST semi-submersible floating wind turbine numerical model with DeepCwind test data*, J. Renew. Sustain. Energy, 5:2 (2013): 29

DNVGL, *Rules for Classification of Ships – Part 3: Hull – Chapter 4: Loads*, Oct (2015)

Duan F., Hu Z., Niedzwecki J.M., *Model test investigation of a spar floating wind turbine*, Mar. Struct., 49 (2016): 76-96

Liu,Y., Xiao Q., Incecik A., Peyrard C., and Wan Decheng., *Establishing a fully coupled CFD analysis tool for floating offshore wind turbines*. Renewable Energy, 112:Nov (2017): 280-301

Nematbakhsh A., Bachynski E.E., Gao Z., Moan T., *Comparison of wave load effects on a TLP wind turbine by using computational fluid dynamics and potential flow theory approaches*, Appl. Ocean Res., 53 (2015): 142-154

Nematbakhsh A., Olinger D.J., Tryggvason G., *Nonlinear simulation of a spar buoy floating wind turbine under extreme ocean conditions*, J. Renew. Sustain. Energy, 6:3 (2014): 20

Quallen S., Xing T., Carrica P., Li Y., Xu J., *CFD simulation of a floating offshore wind turbine system using a quasi-static crowfoot mooring-line model*, J. Ocean Wind Energy, 1:3 (2014): 143-152

Tran T.T., Kim D.H., *The coupled dynamic response computation for a semisubmersible platform of floating offshore wind turbine*, J. Wind Eng. Industrial Aerodynamics, 147 (2015): 104-119

SNAME

THE INTERNATIONAL COMMUNITY
FOR MARITIME AND OCEAN PROFESSIONALS

TRANSPORTATION FUELS FROM THE SEAS POWER

AUTHOR: ROY ROBINSON
AFFILIATION: EXCIPIO ENERGY, INC.

AUTHOR 2: GEORG ENGELMANN
AFFILIATION: EXCIPIO ENERGY, INC.

Proceedings of the 25th Offshore Symposium, February 27th 2020, Houston, Texas
Texas Section of the Society of Naval Architects and Marine Engineers

ABSTRACT

Offshore Renewable Energy (ORE) already has great promise as a primary producer of electricity, but the real advantages off ORE are:

- *The amount of power per square kilometer*
- *Access to an unlimited supply of water*
- *Positive impacts on wildlife compared to onshore systems*
- *Potential for ORE platforms to fulfill multiple functions*

Recent advances in electrical separation of water into hydrogen, carbon capture from air, gas to liquids conversion, and floating ORE platforms open up an entirely new area for development. This paper details the technical requirements and economic and social benefits of large floating ORE platforms located far out to sea making either syngas or eco-fuels. Developing these systems decouples the platforms from a requirement to transmit power directly to shore and frees them to be located where the resources are best. The fuels made from these platforms will be carbon neutral if burned, and carbon negative if used for feedstock to the petrochemical industry.

Imagine instead of having to rebuild the entire transportation grid, we could convert to a carbon neutral renewable energy derived hydrocarbon? You can effectively fill up your 1965 Mustang with wind and wave energy, drive to the airport and fly to your destination in a jet powered by eco-fuel and know you did it with zero carbon load!

Keywords: SNAME, Offshore Symposium, Offshore Renewable Energy, Offshore Wind, Power to Fuel, Wave Energy, Tidal Energy, Ocean Thermal Energy Conversion, OTEC, Floating Wind, Hydrogen Economy, Hydrogen Fuel

INTRODUCTION

Imagine instead of having to rebuild the entire transportation grid, we could convert to a carbon neutral renewable energy derived hydrocarbon? You can effectively fill up your 1965 Mustang with wind and wave energy, drive to the airport and then fly to your destination in a jet powered by eco-fuel and know you did it with zero carbon load!

The simple fact is the technology to accomplish this exists, and the basic principles are simple. Split water into hydrogen and oxygen, pull carbon dioxide out of the environment, combine the carbon and the hydrogen and you get a hydrocarbon not based on fossil resources.

Hydrocarbons made in this way have several advantages:

- They are pure in that they contain none of the common trace elements in oil-derived fuels
- They can be used to replace mined hydrocarbons in petrochemical processes
- The process can be tailored to produce hydrocarbons ranging from methane gas to heavy oils.
- They can be used to store and transport renewable energy from high resource, but unpopulated, areas to urban centers that need the energy; often using existing oil and gas transport networks.

The stumbling block has been the cost of these fuels could not compete with cheap natural gas and direct renewable electric power on a one to one basis.

This paper hopes to show that the cost dynamic is changing, that simple bbl-to-bbl cost comparisons that have been the focus of many papers ignores some key advantages of power to fuel systems, and that building offshore renewable energy powered industrial parks can be a near term and long term solution yielding a green energy future.

DEFINING THE MARKET

Some of the significant stumbling blocks for shifting to a fossil fuel free, carbon neutral/negative economy are:

- Replacement of fossil fuel-based propulsion systems, especially for aviation.
- The societal costs of replacing trillions of dollars of infrastructure, equipment, and industries.
- The continuing need for hydrocarbons in the pharmaceutical and petrochemical industries.
- Limitations and negative environmental impacts of battery technologies.

How Much Fuel?

To put some numbers to, and get an idea of, the scale of the problem. Per the U.S. Energy Information Administration (EIA) world consumption of oil is around 100 million barrels per day, and still growing at about 2% per year. The total global energy consumptions in 2017 14,500 MTOE and slated to grow to 18,400 MTOE by 2050. Currently 79% of all energy used is from fossil fuels.

Using US EIA Data (Ref 3) in terms of use of fossil fuels the breakdown is almost evenly divided as follows:

- Transportation 3600 MTOE
- Petrochemical 3800 MTOE
- Electrical Power 3700 MTOE
- Other (heating oil, coke production etc.) 345 MTOE

The total of 11,445 MTOE is the energy equivalent of 100,000 TWh. Even at 100% efficiency that would mean more than 15,000 new 1 GW sized power plants. In fact, the number would be many times that. Figure 1 gives a breakdown of the global energy consumption in 2017.

To replace the current global market for transportation fuel of 3600 MTOE will require 1.3 billion tonnes of hydrogen per year. That represents 11.3 billion tons of pure water per year. By comparison New York City uses 1.4 billion tonnes of fresh water per year.

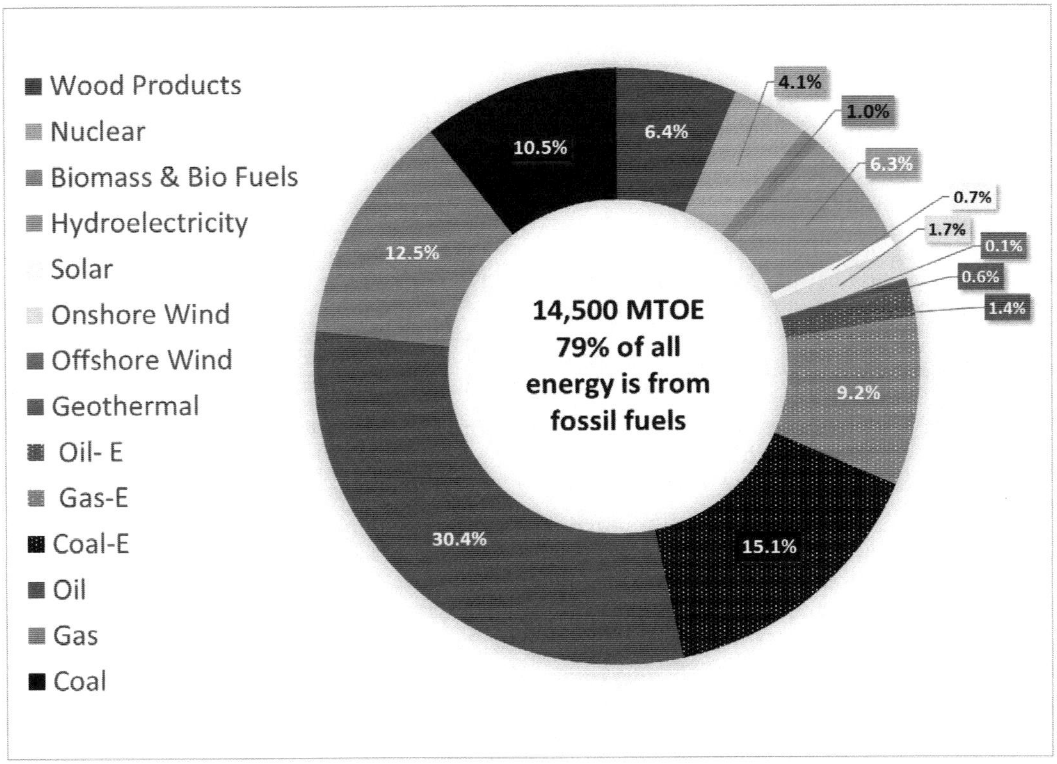

Notes:
1. Oil-E, Gas-E, and Coal-E denote used to generate electrical power
2. MTOE = million tons of oil equivalent = 11.63 MWh
3. Sources, Ref 2 and 3.

Figure 1: 2017 Global Energy Use

The Hidden Costs of Replacing Infrastructure

When accounting for the benefits of switching to renewable energy sources, particularly in transportation, the economic and social costs seem too often overlooked or ignored. To completely do away with hydrocarbon-based fuels means every car on the road will have to be replaced or undergo expensive retrofitting. Aircraft will require complete ground up redesign and because of decreased energy density will have less payload capacity.

The entire value chains that support internal combustion engines and turbines from the smallest weed-wacker to the largest gas turbine will need as a minimum retooling, or more likely replacement. The new plants are likely to be on average more automated than their legacy predecessors, not located in the same regions, and will require new skill sets.

The distribution networks for oil and gas built up over 100 years will lose all utility and need replacement.

In short, the costs will be many trillions of dollars, the effects on job markets will be disruptive, and it is not clear that the proposed replacements will be better in the long term. Unless the governments of the world decide otherwise and give massive subsidies to average people to replace the existing equipment, the majority of the economic burden will fall on the lower end of the income scale as people are forced to pay for new hydrogen or battery powered transport.

Again, putting the scale of the problem in hard numbers, globally there are approximately:

- 30,000 commercial airliner and military planes (Ref 11)
- 500,000 private aircraft (Ref 8)
- 30,000 helicopters (Ref 8)
- 1.4 billion cars (Ref 10)

- 325 million commercial vehicles (Ref 9)

All of the above are designed to run on hydrocarbon fuels, just 3 of which make up 90% of the fuel use per the US EIA:

1. Diesel
2. Gasoline
3. Kerosene/Jet Fuel

The good news is all of the above can be made using power to fuel technology, the problem is that as a practical replacement for fossil fuels the Power-to-Fuel industry is in its infancy.

USING RENEWABLE ENERGY FOR TRANSPORTATION

There are five primary methods being considered for using the energy produced by renewable energy sources for transportation.

1. Batteries – Lithium Ion, Flow Batteries
2. Burning hydrogen directly, stored and transported either as a liquid or high-pressure gas
3. Using fuel cells power by hydrogen or methane
4. Conversion of H2 and N2 to ammonia to be used as a fuel
5. Conversion of CO2 and H2 to fuels, which is this paper's focus

Batteries have limitations in their application, particularly for aircraft. Added to that the batteries currently used require rare earth metals that restricted in their supply, are difficult to recycle, can burn in the event of an accident, have a limited life span, and are slow to recharge when compared to filling a gas tank.

Burning hydrogen directly can be attractive, but it would require a complete re-tooling of the transportation infrastructure and the engines that run it. Hydrogen also has some costly material challenges and cannot, as is sometimes claimed, be used directly in most existing pipeline systems. Hydrogen also has a power density problem. Hydrogen Promoters often state a kilogram of hydrogen has the heat energy of 1 gal of gasoline, however they leave out that even in liquid form a kilogram of hydrogen is takes 3.8 gallons of volume. Put another way you can hold 4 times the times the energy in a tank using gasoline than you can in the same tank filled with liquid hydrogen. And the gasoline tank does not need to be pressurized or chilled. It is the main stumbling block in direct use of hydrogen for aviation.

Fuel cells are more promising than direct combustion, but to date are too expensive and very sensitive to the purity of the activation gases.

Ammonia has also been considered as a replacement fuel for hydrocarbon-based fuels. The drawbacks here are pure ammonia is a very volatile and caustic liquid and would require the same level of re-tooling as either batteries or pure hydrogen as fuels. In controlled environments like petrochemical plants, ammonia is used safely, but tanks of pure ammonia being driven around as todays cars are raises some serious safety concerns. Ammonia used as fuel is not like the ammonia used in cleaning your home. While less likely than gasoline to ignite, pure ammonia is much more dangerous than gasoline when you consider that it is the fumes and contact with the liquid ammonia that are the hazard. If the ammonia tank in a car ruptures the released ammonia will give off toxic fumes that can kill a person in just one or two breaths. If it comes in contact with skin, it literally dissolves the skin and flesh off the bones.

Lastly, we get to power-to-fuel. If it can be made to work efficiently at the very least it would ease the transition to a hydrogen-based economy, and it would more importantly allow the world to go carbon neutral much more quickly. It can also serve as a transition phase if going to a hydrogen-based economy remains the goal. It will allow the retirement and replacement of legacy hydrocarbon infrastructure as it wears out, rather than ripping out equipment still in its prime.

Onshore vs Offshore

So, given that power-to-fuel is worth pursing, why would an offshore solution be preferable to an onshore one? Why not build a power-to-fuel plant attached to every onshore wind and solar farm that converts any excess energy (energy not sold to the grid) to fuel?

Water Access and Discharge

The first and most obvious point is access to water. To replace one years-worth of fuel consumed by the worlds current transportation needs would require more than 8 times the freshwater used by New York City in a year. Less obvious is the ability to dilute the discharge. The water that goes into the electrolyzer must be purified meaning that regardless of the source there will be a waste stream. Offshore for every 4.5 kg of water that goes into the hydrogen separation unit, 5.5 kg of brine are discharged. If for example an onshore plant was set up along the Mississippi river, then it is probable that the wastewater will require treatment before it would be allowed back into the river. Any waste products or chemicals in it would have been concentrated by the separation process.

The brine discharge offshore would also be an environmental problem if discharged in large quantities from a single location, such as if an FPSO type processing vessel is used with power supplied by offshore energy. For this reason, among others the authors have assumed a distributed model, with each energy platform making its own freshwater and hydrogen. Only the CO2 capture and reformation will take place at a central location.

Size and Efficiency

Onshore sites are more restricted in size than their offshore cousins and are becoming harder to permit near population centers. Onshore it is becoming more common to include solar power on the same site, but the energy density is still low compared to offshore. This is in part because the size of offshore wind turbines is on average larger than onshore. At sea, besides offshore wind, the possibility of capturing wave energy, tidal energy, or ocean thermal energy means the power per installation can be many times that of an onshore site.

Once the power is made and converted to hydrogen, transport offshore will be simpler and safer than it is onshore. While the capital cost of offshore pipelines is higher per mile than onshore pipelines, the permitting and routing are usually much simpler, with an overall economic advantage to the offshore solution. The installed lines are less likely to be damaged than onshore pipelines and if damaged the risk to people is much lower offshore.

While view shed (covered below) is an issue, most large urban centers are located near a coast. This means that the renewable energy sites, even if placed over the horizon can be within 25 miles of the final consumer.

Given a power to fuel solution, both onshore and offshore sites can transport the produced clean fuels using existing systems, but offshore systems (ships for liquids and long high-pressure pipelines for gas) are not open to onshore production.

Recent developments pushing offshore renewables to floating solutions will eventually lead to offshore turbines becoming even larger. Wind turbines to 25 MW are being designed, and these are not practical for current fixed wind designs.

View Shed

View Shed, defined as the visual impact on the local community, is a real issue in wind farm siting. Onshore and near shore this is often the issue that kills a potential permit. Installations out of sight of land effectively remove that as an issue. Considering the speed at which the world must progress to meet its climate goals, and the quantity of power that will be required the only real answer to view shed issues is to move the energy farms far enough out to sea.

Wildlife Impacts

The wildlife impacts of onshore renewables are more negative than offshore installations providing the offshore installations are located far enough offshore. Onshore bird kills and bat kills are a concern the industry is working hard to correct and should not be seen as a reason not build a wind farm, but offshore the issue is reduced or eliminated.

Any sailor will tell you that one sign you are getting near shore is you start to see birds. In the open sea unless you are in a migration route there are no birds or bats.

Noise pollution is also an issue, and turbines make noise offshore and onshore, onshore the effects on local wildlife and domestic stock is under study, and offshore the noise impacts on marine mammals are likewise under study. To date the position of groups such as the World Wildlife Federation (Ref 6) are that the positives of clean carbon free energy outweigh and possible negative impacts that wind farms may have.

What is also clear is offshore there are real benefits to marine life caused by renewable energy facilities. Within the confines of the site commercial fishing is restricted or prohibited. The hulls and moorings of floating systems, and the legs and piles of fixed systems offer habitat locations.

Even oil and gas platforms act as artificial reefs in places like the Gulf of Mexico. At its peak there were 4500 platforms in the Gulf of Mexico (now down to 1700), but once offshore renewables go global and floating that number will be dwarfed. To meet the 15,000 GW of installed power using offshore wind turbines would, allowing for a 50% capacity factor, and 15MW per turbines, mean 1.7 million offshore installations worldwide.

That would create habitat on the scale of the Great Barrier Reef, onshore that many turbines would have a measurable negative impact on global bird and bat populations.

Multiple Use

Unlike onshore renewable energy sites, one being used for offshore energy can host multiple technologies, often not related to energy. This is especially true for floating facilities. Activities that can be added on the same facilities to increase the positive economics of an offshore renewable energy site are:

- Aquaculture
- Ocean Mining
- Blue-tourism/floating hotels
- Marine research
- Bio-fuel production
- Marine Rescue and Monitoring (coast guard)
- Weather stations
- Communications relays
- Big Data Storage

Summarizing

Offshore:

1. Using a distributed solution ensures minimal environmental impact
2. Unlimited water supply
3. No view shed issues if placed more than 12 miles offshore
4. Overall impacts on wildlife populations are positive
5. Can often be located nearer to population centers.
6. Can host other industries such as aquaculture or ocean mining.
7. The amount of renewable energy per installation is much higher offshore than onshore

Onshore:

1. Depletes local water supply
2. Potential pollution issues
3. Takes up useable land,
4. Often far removed from population centers
5. View shed and local tourism issues
6. Bird and bat kills are a continuing problem
7. Noise pollution can be an issue depending on location
8. Would require new onshore pipelines, which are more difficult to permit than offshore

9. Distributed solutions are not an option

THE POWER TO FUEL CYCLE AND ECONOMICS

The study of power-to-fuel has taken on a renewed interest since the signing of the Paris Accords. There are multiple studies underway and various national programs to encourage research into the technological components. While the systems are very different the basic functions and objectives remain the same. Enough of an advance in the technology has occurred that the construction of offshore prototype facilities is already being proposed for power-to-hydrogen plants, and power-to-fuel will not be far behind.

Figure 2 (Ref 1) below shows a basic cycle, mass, and energy balance. It is not tremendously efficient in terms of value, producing roughly 8.4 bbl of liquid fuel per day for every MW of power. The interesting thing is even at that low output and ignoring allowance for things like carbon credits or tax incentives; the result is a nearly a break-even facility. As the technology advances, offshore facilities based on a cycle like the one below will be profitable without subsidy within a few years.

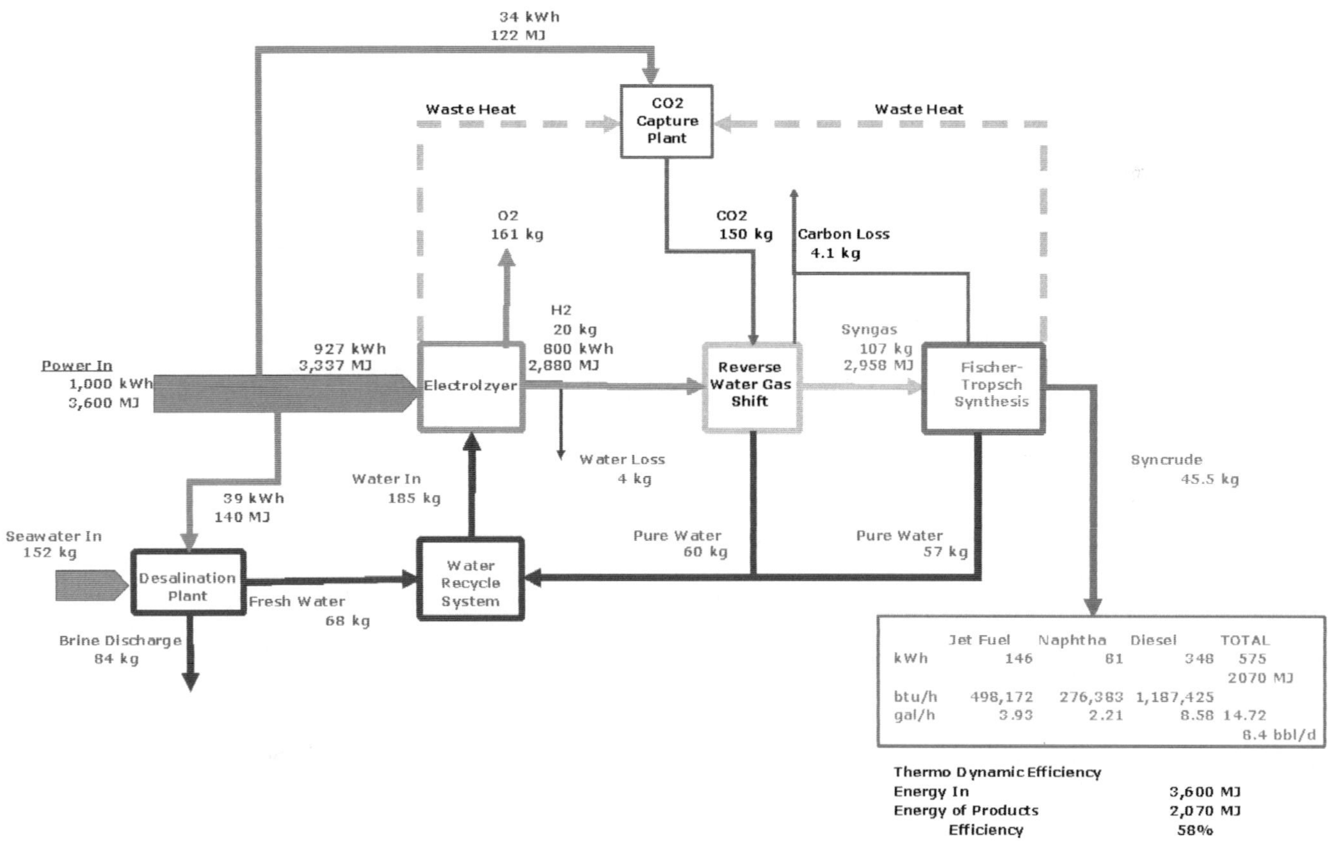

Figure 2: Energy and Mass Balance, Simplified from Reference 1

The economic model is based on using a floating platform with multiple energy sources generating 30 MW of power, with the deck capacity to support the power to fuel process equipment shown in Figure 2 above.

Many papers have gone into great and often misleading detail of how much a gallon of econ diesel will cost to produce. There are many unknowns, and the technology is evolving at a pace that is astonishing, so predicting and exact break-even cost of production is at best a moving target with a very large diameter bullseye.

This paper is not trying to justify the economic value of power to fuel, but neither can the economics be ignored. To that end a simplified model was run, using the following assumptions:

1. Realized diesel price 1.5 times the current wholesale price ($1.95 per gallon, Ref 12) on the assumption there would be a market for premium carbon neutral fuel.
2. No tax breaks or incentives, including carbon credits were applied
3. An assumed overnight cost of $100 Million for a 30 MW, floating, renewable energy, Power-To-Fuel plant was assumed. The is based on what would be possible if the plants are mass produced.
4. A 30-year design life, and 1year fabrication and installation time were assumed.
5. A Capacity Factor for the renewable energy of 65%, and a plant availability of 95%.
6. An OPEX of $2 million per year per platform
7. Decommissioning was assumed to be negligible.

Model Inputs

Item	Case 1	Case 2	Units
Number of Platforms	10	10	
Diesel Production Rate	8.4	8.4	bbl/MW
Technology Factor*	1	1.64	
Platform Nominal Power	30	30	MW
Capacity Factor	0.65	0.65	
Diesel Dec 2019 per EIA	$82.00	$82.00	per bbl
Price Premium	1.5	2	
Daily Production	2520	2520	bbl/d
Availability	0.95	0.95	
Annual OPEX per platform	$2	$2	MM
CAPEX Per Platform	$100	$100	MM

* Current process efficiency is 58%, assumed maximum is 95%

Table 1, Economic Model Inputs

Case 1: Current Technology

	PROJECT YEAR (Values in million US$)					
	0	**1**	**2**	**3**	**……..**	**30**
CAPEX	-$1,000				……..	
OPEX		-$20	-$20	-$20	……..	-$20
Revenue		$70	$70	$70	……..	$70
					……..	
Total	**-$1,000**	**$50**	**$50**	**$50**	**……..**	**$50**

IRR **3%**

Table 2, Current Technology IRR

Case 2: 2050 Technology

	PROJECT YEAR (Values in million US$)					
	0	**1**	**2**	**3**	**……..**	**30**
CAPEX	-$1,000				……..	

OPEX		-$20	-$20	-$20	....	-$20
Revenue		$153	$153	$153	....	$153
					....	
Total	**-$1,000**	**$133**	**$133**	**$133**	....	**$133**
IRR	**13%**					

Table 3, Estimated 2050 IRR

Even at the low rate given in Figure 2 of 8.4 bbl/d/MW the IRR is 3%. A sensitivity using what we termed a Technology Factor, defined as the positive effect on production in terms of bbl/d/MW of advancing technology gave the following result. It would require an improvement from the current thermodynamic efficiency of 58% to 95% (TF=1.64) and raising the price premium to 2 to achieve a marginal IRR of 13%.

In short it can be done, it can make money, but it is not cheaper than mining fossil fuels. That really is not the question. Leather can be produced for much less of the current cost of manufacture if discharge water is not treated, workers toil in unsafe conditions, child labor is used, and hides are sourced unethically. Society however chose to take the path that is a better long-term solution, even though it means cowboy boots will cost more. Producing clean fuels instead of mining hydrocarbons is the same thing. As a society we must chose the sustainable, if more expensive path.

CONCLUSION

While there are still technical and financial hurdles to overcome the obvious long-term benefits of offshore production of carbon neutral and carbon negative fuels using renewable energy has the potential to become the primary source of transportation fuels. No other solution offers the versatility, potential knock on benefits, and minimal societal disruption that offshore renewable based power-to-fuels does. In one technology the answers to energy storage, energy transport, and low-cost low impact production are found.

The economics can be improved by taking advantage of:

- Shared use activities such as aquaculture and ocean mining
- Sale of additional product gases such as oxygen
- Tax incentives
- Carbon Credits
- Improved power generation and
- Improvements in the power-to-fuel component technologies.

It may not be possible to make fuels generated using renewable energy as cheaply as we can from mined fossil fuels, but as a society we must chose the sustainable path. There are other options such as hydrogen, but power-to-fuel is the sustainable option that will be the least disruptive and costly, while still meeting those sustainable goals.

ACKNOWLEDGEMENTS

The Authors wish to thank Elias Greenbaum, President of GTA Inc, Knoxville, Tennessee, and Mahdi Fasihi, Research Assistant at Lappeenranta University of Technology, Finland for their assistance with the technical details of converting water to fuel.

GLOSSARY

MTOE Million Tons of Oil Equivalent – the energy equivalent of a million tons of oil

TWh Terawatt Hours – one terawatt produced for 1 hour, the are 8760 hours in a year.

LCOE Levelized Const of Energy – metric used to compare different energy sources.

FPSO Floating Production, Storage, and Offloading vessel, often a converted tanker.

ORE Offshore Renewable Energy – offshore wind, wave energy, tidal and ocean current energy, ocean thermal energy conversion

OTEC Ocean Thermal Energy Conversion – running a turbine powered by the temperature difference between the surface water and the deep ocean. Requires a minimum 15 deg Celsius difference.

IRR Internal Rate of Return

EIA Energy Information Administration of the US Department of Energy

UNIT CONVERSION

kWh	Btu	J	MTOE
1	3412	3.60E+06	8.60E-05
2.93E-04	1	1055	2.52E-08
2.78E-07	9.48E-04	1	2.39E-11
1.16E+04	3.97E+07	4.19E+10	1

REFERENCES

1. Mahdi Fasihi, Dmitrii Bogdanov, Christian Breyer; Techno-Economic Assessment of Power-to-Liquids (PtL) Fuels Production and Global Trading Based on Hybrid PV-Wind Power Plants; 10th International Renewable Energy Storage Conference, IRES 2016, 15-17 March 2016, Düsseldorf, Germany.
2. BP Statistical Review of World Energy, 2018, 67th Edition.
3. US Energy Information Administration, INTERNATIONAL ENERGY STATISTICS website, 2019.
4. International Energy Outlook 2019 with projections to 2050.
5. US Energy Information Administration, U.S. No 2 Diesel Wholesale/Resale Price by Refiners (Dollars per Gallon), Jan to Sep 2019.
6. WWF Position on Wind Power, June 2004.
7. Alternative Fuels Data Center Fuel Properties Comparison.
8. Dr. Thomas Schiller, Michael Maier, Martin Büchle; Truck Study 2016 The truck industry in transition, 2017 Deloitte.
9. 2016 General Aviation Statistical Databook & 2017 Industry Outlook, www.GAMA.aero.
10. carsguide.com.au How Many Cars are There in the World
11. www.airliners.net
12. US Energy Information Administration, Petroleum Marketing Monthly, December 2019.
13. US Energy Information Administration, Global Transportation Energy Consumption: Examination of Scenarios to 2040 using ITEDD September 2017.
14. Manuel Geotz, Jonathan Lefebvre, Friedemann Mors, Amy McDaniel Koch, Frank Graf, Siegfried Bajohr, Rainer Reimert, Thomas Kolb; Renewable Power-to-Gas: A technological and economic review, 2015, Renewable Energy, http://www.elsevier.com/locate/renene.

THE INTERNATIONAL COMMUNITY
FOR MARITIME AND OCEAN PROFESSIONALS

RELIABILITY SENSITIVITY ANALYSIS OF DROPPED OBJECTS HITTING ON THE PIPELINE AT SEABED

XIAOCHUAN YU, GONG XIANG , HANQI YU* , HAOZHAN MENG, YI LI, LINXIONG LI
UNIVERSITY OF NEW ORLEANS

Proceedings of the 25th Offshore Symposium, February 27th 2020, Houston, Texas
Texas Section of the Society of Naval Architects and Marine Engineers

ABSTRACT

Dropped objects are often the principal cause of accidents in the oil and gas industry and contribute significantly to the total risk level for offshore and onshore facilities. In this paper, the risk-based approach for assessing pipeline protection against accidental external loads is briefly introduced. Secondly, a limit state function based on the impact energy absorbed is proposed to calculate the probability of failure (P_f). Thus, the formulas to perform the reliability sensitivity analysis can be developed based on this function. Further, the Monte Carlo (MC) method is applied to the reliability sensitivity analysis, in which the mass (M), collision area (A), the volume (V), added mass coefficient (C_a) and drag coefficient (C_d) of the objects are assumed to follow a certain distribution, and how each of them may affect the probability of failure will be investigated.

Keywords: Dropped Objects, Probability of Failure, Reliability Sensitivity Analysis, Monte Carlo (MC) Method

1. INTRODUCTION

Accidentally dropped objects from a platform deck or from a supply vessel or during a hoisting operation may cause fatal casualties. DNV presents a risk-based approach for assessing pipeline protection underwater against accidental dropped objects [1]. ABS guide specifies the requirements for an onboard dropped object prevention program in one guide [2]. It provides the design requirements for equipment primary securing, secondary retention methods and securing for specific equipment, etc. For the risk exposure to deeper water structures (either fixed jackets or hull systems) sea equipment, this needs specialized techniques to track the trajectory of dropped objects and to estimate the subsequent likelihood of striking additional structure and equipment as well as predicting the consequences of such subsequent impacts [3].

Both numerical and experimental investigations have shown that the accidentally falling drilling pipe followed different paths. In addition, the neglect of axial rotation may cause some errors in simulating the trajectory in real experiments [4]. A computer program, called DELTA, has been developed to calculate the trajectories in two dimensions (2D) of dropped drill casings [5,6]. These studies confirmed that drop height above waterline and the initial drop angle were key parameters influencing the horizontal velocity. Reliability-based impact analysis successfully established the relation between impact velocity and the probability of its exceedance. Katteland and Øygarden evaluated the risk related to dropped objects in offshore oil and the gas industry [7]. They established four distributions to describe the spreading of dropped objects from an offshore installation at deep water. Majed and Cooper presented nonlinear dynamic simulations of dropped objects for a detailed and accurate assessment of dropped object trajectories by incorporating detailed 3D hydrodynamic models of complex geometries [8]. The experimental data from two more recent experimental tests can be found in [9, 10]. After comparing them with the results from a simplified method from [1], it can be concluded that "the present recommended methodology for use of calculation by DNV is generally conservative and in some case's not conservative at all [10]. The simplified method gives the initial estimate since it's based on a general category rather than a specific object hydrodynamics. The authors have extended the 2D equations of motion presented in [4] into 3D, and further developed the in-house tool Dropped Objects Simulator (DROBS), which can be used to simulation the Motion Dynamics of Cylindrical Objects in three dimensions (3D) more accurately [11 - 15].

In the offshore oil and gas development, it is mandatory to develop a dropped-objects plan and make it available to Bureau of Safety and Environmental Enforcement (BSEE) upon request [16]. However, it's difficult to estimate the size of a dropped object and its hitting probability in reality. Therefore, a comprehensive risk-based approach for assessing pipeline protection against accidental external loads has been presented in [1], and the procedures have been summarized in the flowchart shown in Figure 1. One of the examples about how to apply this approach into the risk assessment of submarine pipeline can be found in [17], in which the reliability sensitivity analysis has been performed.

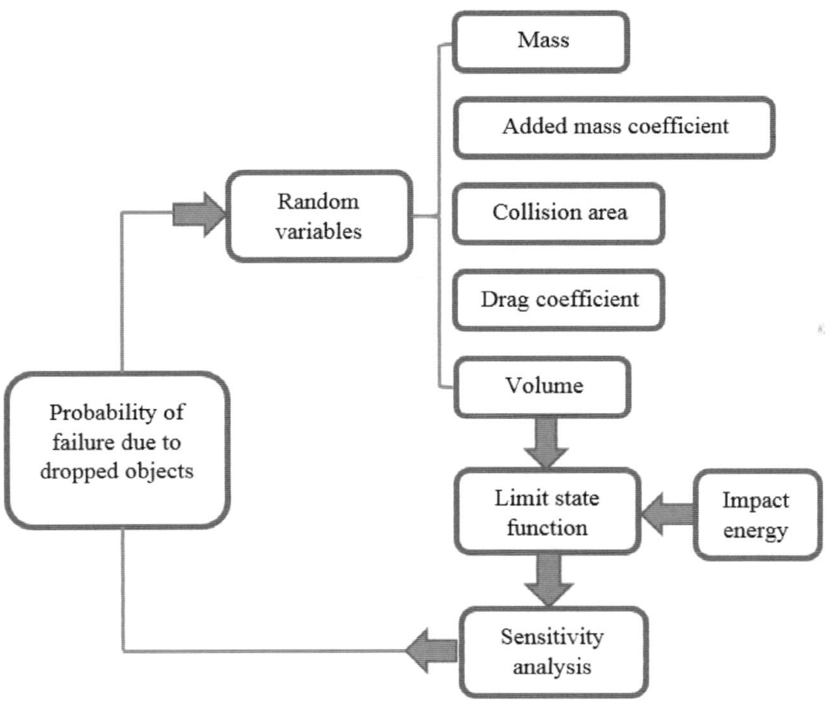

Figure 1: Flowchart for assessing pipeline protection against accidental external loads [17, 18]

As we all know, there are many factors which may affect an dropped object's trajectory an impact energy, such as its size, shape, weight, volume, drop angle, added mass, added damping, contact area, etc. Therefore, a more generic risk-based approach, which is based on the energy criteria, will be developed in this study. A limit state function is firstly proposed to define the probability of failure P_f based on the impact energy absorbed. The Monte Carlo (MC) Method is adopted to perform the reliability sensitivity analysis, in which the mass (M), collision area (A), the volume (V), added mass coefficient (C_a) and drag coefficient (C_d) of the objects are assumed to follow a certain distribution, and how each of them may affect the probability of failure P_f will be investigated. The study case is from "the detailed risk assessment of dropped objects on a 20-inch pipeline at seabed", as described in DNV's recommended practice [1].

2. FORMULAS

2.1 VARIOUS TYPES OF ENERGY

In the risk-based approach for assessing pipeline protection against accidental external loads, various types of energy have been introduced to classify the pipeline's capacity to resist the impact due to the dropped objects. For submarine pipelines, the energies which they are subjected to can be roughly categorized into four types: E_S – dent- absorbed energy for steel pipelines; E_E – the kinetic energy which includes the terminal energy E_T and the energy of added hydrodynamic mass E_A; E_C – concrete coating energy; E_P – polymer coating energy.

While dropped objects finally hit on the subsea pipelines, most impacts are expected to result in a relatively "smooth" dent shape. The dent-absorbed energy relationship for steel pipelines is given as following:

$$E_s = 16\left(\frac{2\pi}{9}\right)^{\frac{1}{2}} \cdot m_p \cdot \left(\frac{D_0}{t}\right)^{\frac{1}{2}} \cdot D \cdot \left(\frac{\delta}{D_0}\right)^{\frac{3}{2}} \tag{1}$$

where, m_p is the plastic moment capacity; δ is the pipe deformation (or dent depth); t is the wall thickness; D_0 is the steel pipe's outer diameter. The plastic moment is defined as the moment at which the entire cross section has reached its yield stress. This is theoretically the maximum bending moment that the section can resist - when this point is reached a plastic hinge is formed and any load beyond this point will result in theoretically infinite plastic deformation. The additional failure of punching through the wall, leading to leakage, can occur for higher velocity impacts or locally small and sharp impact geometry [1]. The possibility of leakage and total rupture is included as a progressive conditional probability, where probability increases with increasing impact energy. The proposed damage classification used for bare steel pipes are shown in Table 1.

Table 1: Different damage level corresponding to different Dent/Diameter ratio [1]

Levels of damage	Dent /Diameter (%)	Damage description
Level 1	< 5	Minor damage
Level 2	5 - 10	Major damage; leakage anticipated
Level 3	10 - 15	Major damage; leakage and rupture anticipated
Level 4	15 - 20	Major damage; leakage and rupture anticipated
Level 5	> 20	Rupture

For the kinetic energy E_E, we calculate the terminal velocity first based on Equation (2):

$$(m - V\rho)g = \frac{1}{2}\rho C_d A v_T^2 \tag{2}$$

Then, we could get the square of terminal velocity as:

$$v_T^2 = \frac{2(m - V\rho)g}{\rho C_d A} \tag{3}$$

where, m is the mass of dropped object; V is the volume of dropped object; ρ is the density of sea water; A is the collision area; C_d is the drag coefficient.

Since kinetic energy includes both terminal energy E_T and the energy of added hydrodynamic mass E_A, it can be written as:

$$E_E = E_T + E_A = \frac{1}{2}(m + m_a)v_T^2 \tag{4}$$

where, m_a can be directly calculated by $C_a\rho V$ and C_a is the added mass coefficient. After substituting m_a and v_T^2 into Equation (4), we could easily obtain kinetic energy as Equation (5) below:

$$E_E = \frac{g}{C_d A}\left(\frac{m^2}{\rho} - V.m + C_a V(m - V\rho)\right) \quad (5)$$

For concrete coating energy, different from DNV rules, E_C in this paper is chosen as the mean of the concrete coating energy for long/flat shaped objects E_{C1} and the concrete coating energy for box/round shaped objects E_{C2}.

$$E_{C1} = YB\frac{4}{3}\sqrt{Dx_0^3} \quad (6)$$

$$E_{C2} = YBhx_0 \quad (7)$$

$$E_C = \frac{E_{C1}+E_{C2}}{2} \quad (8)$$

where, Y is the crushing strength; B is the width of the dropped object; h is the height of the dropped object; x_0 is the penetration depth, assume it is equal to wall thickness.

2.2 LIMIT STATE FUNCTION AND FAILURE ANALYSIS

In order to apply the sensitivity analysis, we define the limit state function as:

$$G = R - L \quad (9)$$

where, R represents the pipeline's capacity to resist the external load L which comes from the dropped objects. Figure 2 illustrates the relationship between R and L.

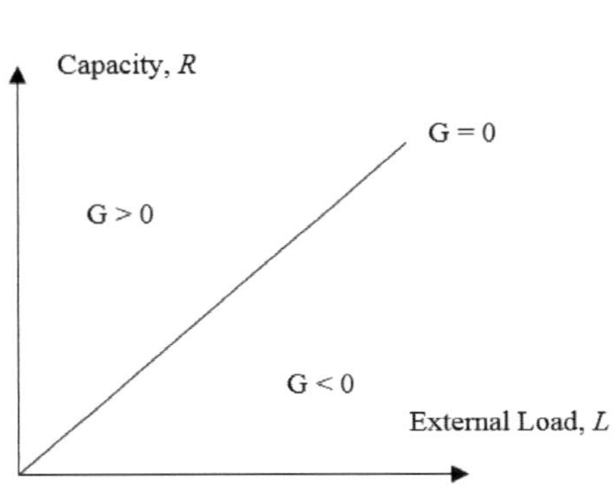

Figure 2: Limit state *G*, pipeline's capacity *R* and external load *L*

The critical line for limit state is described by $G = 0$. When the strength is larger than pipeline load, the failure occurs with $G < 0$, similarly, the survival region is $G > 0$. As a result, the probability of failure can be defined as

$$P_f = Pr(G \le 0) = Pr(R \le L) \quad (10)$$

2.3 SENSITIVITY ANALYSIS BASED ON MC METHOD

According to the limit state function in Equation (9), we can have $R \sim N(\mu_R, \sigma_R^2)$ and $L \sim N(\mu_L, \sigma_L^2)$ if we assume that both R and L follow Normal distribution. Because of the nature of the normal distribution, we should easily obtain $G \sim N(\mu_R - \mu_L, \sigma_R^2 + \sigma_L^2)$. Let:

$$\mu_G = \mu_R - \mu_L \quad (11)$$

$$\sigma_G = \sqrt{\sigma_R^2 + \sigma_L^2} \quad (12)$$

then we can have:

$$\frac{G-\mu_G}{\sigma_G} \sim N(0,1), \text{when } G \le 0 \quad (13)$$

As mentioned in Section 2.1, E_E, E_S, E_C and E_P are four types of energies. Let σ_{EE}, σ_{ES}, σ_{EC} and σ_{EP} be the standard deviation of E_E, E_S, E_C and E_P, respectively. Since E_S, E_C and E_P are assumed to follow the normal distribution, the coefficient of variation (CV) can be used to correlate their standard deviations σ and the mean value μ as following:

$$\mathrm{CV} = \frac{\sigma}{\mu} \tag{14}$$

In this paper, the CV value for E_E, E_S, E_C and E_P are assumed to be 10%, 5%, 20% and 20%, respectively. It should be noted that σ_R^2 and σ_L^2 are calculated by:

$$\sigma_R^2 = {\sigma_{ES}}^2 + {\sigma_{EC}}^2 + {\sigma_{EG}}^2 \tag{15}$$

$$\sigma_L^2 = {\sigma_{EE}}^2 \tag{16}$$

The main impact from dropped object comes from kinetic energy. As most limit state situations are implicit, we apply the response surface method to transform the implicit function to explicit one [20]. Based on the assumption of neglecting the deformation of dropped objects and considering the impact energies absorbed completely by submarine pipelines [1], the limit state function can be rewritten as:

$$G = (E_C + E_P + E_s) - E_E \tag{17}$$

As can be seen from the Equation (17), there are totally five random parameters, including the mass, volume, added mass coefficient, drag coefficient and collision area. In this paper, the sensitivity of all these variables will be performed, and some assumptions are summarized in Table 2.

Table 2: Distribution types of random variables

Parameter	Distribution Type
Mass (m)	Uniform
Volume (V)	Normal
Collision Area (A)	Truncated Normal
Drag coefficient (C_d)	Truncated Normal
Add mass coefficient (C_a)	Truncated Normal

For the sensitivity analysis, we denote $\tilde{P}_f(\theta; \sigma_G)$ as an approximation of the probability of failure as following [19]:

$$\tilde{P}_f(\theta, \sigma_G) = \int_{D(x)} \phi\left(-\frac{G(x;\theta)}{\sigma_G}\right) f_X(\boldsymbol{x})dx \tag{18}$$

where, X is an n-dimensional vector of random variables; $f_X(x)$ is the joint probability density function; $D(x)$ is the set of all n-tuples of real numbers; ϕ is the standard normal cumulative distribution function; θ is the parameter about which sensitivity analysis is performed.

The reason for integration on normal cumulative distribution is that this is a distribution based on the mean value. Taking the derivative of Equation (18) with respect to θ, and according to Leibniz integral rule we can have:

$$\frac{\partial \tilde{P}_f(\theta;\sigma_G)}{\partial \theta} = -\int_{D(x)} \frac{1}{\sigma_G} \phi\left(\frac{G(x;\theta)}{\sigma_G}\right) \frac{\partial G(x;\theta)}{\partial \theta} f_X(\boldsymbol{x})dx \tag{19}$$

It should be noticed that since standard normal distribution is symmetric at the origin, the negative sign in ϕ function could be removed for symmetry of ϕ. Finally, we use Monte Carlo samples $\{\boldsymbol{x_k}, k = 1, \dots\dots, n_s\}$ to simplify and estimate the domain integral in Equation (19), we can get:

$$\frac{\partial \tilde{P}_f(\theta;\sigma_G)}{\partial \theta} \approx \frac{1}{n_s} \sum_{k=1}^{n_s} \left[-\frac{1}{\sigma_G} \varphi\left(\frac{G(\boldsymbol{x_k};\theta)}{\sigma_G}\right) \frac{\partial G(\boldsymbol{x_k};\theta)}{\partial \theta}\right] \tag{20}$$

3. CASE STUDY

3.1 FIELD LAYOUT AND PROPERTIES OF DROPPED OBJECTS AND SUBSEA PIPELINES

In this paper, the "detailed risk assessment of dropped objects on a 20-inch pipeline at seabed", as described in DNV recommended practice rule, is selected as the study case [1]. The field layout is shown in Figure 3. It can be seen that there are some areas of the subsea pipes shielded by the platform legs and bracing, within the range from 0m to 60m.

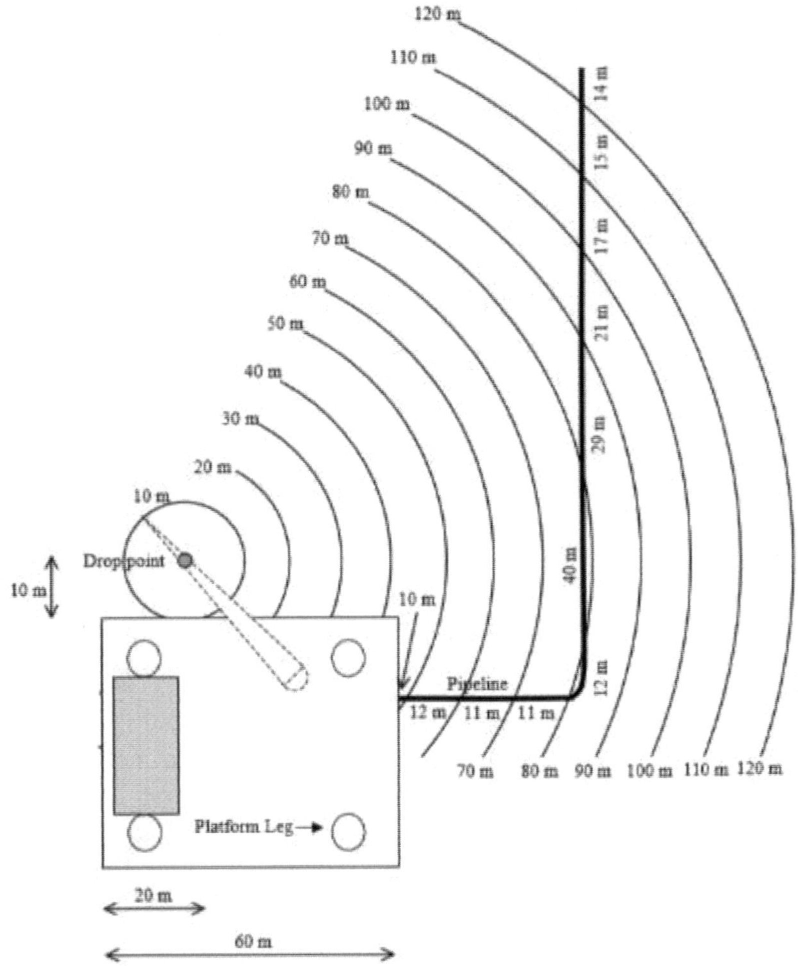

Figure 3: Field layout with 10-meter internal rings [1]

Three different groups of flat/long shaped dropped objects (Cases 1 ~ 3) and three different groups of box/round shaped dropped objects (Cases 4 ~ 6) are used in the model with weights less than 2 tones (Case 1, Case 4), between 2 tones and 8 tones (Case 2, Case 5) and greater than 8 tones (Case 3, Case 6), as summarized in Table 3. The basic properties of the dropped objects and subsea pipelines are shown in the Table 4 and Table 5, respectively.

Table 3: Object classification [1] and case index

Description	Weight (tons)	Typical objects	Case Index
Flat/long shaped	< 2	Drill collar/casing, scaffolding	Case 1
	2 – 8	Drill collar/casing	Case 2
	> 8	Drill riser, crane boom	Case 3
Box/round shaped	< 2	Container (food, space parts), basket, crane block	Case 4
	2 – 8	Container (space parts), basket, crane test block	Case 5
	> 8	Container (equipment), basket	Case 6

Table 4: Properties of dropped objects

Properties	Unit	
Length for long shape object (l)	m	12
Length for box shape object (l)	m	5
Height for impacting object (h)	m	0.3
Breadth for impacting object (b)	m	0.03
Gravity acceleration (g)	m/s^2	9.81

Table 5: Properties of subsea pipelines

Properties	Unit	
Pipeline diameter (D)	m	0.63
Wall thickness (t)	m	0.018
Concrete thickness (c)	m	0.06
Yield stress (s)	N/m^2	4.5e06

4. Results and Discussions

In the reliability sensitivity analysis, the MC method is applied, and the sample size is 10000. Five random variables, including mass, volume, collision area, added mass coefficient and drag coefficient, are selected to be investigated how those parameters may affect the probability of failure P_f. The results are shown from Figure 4 to Figure 8. The horizontal axis represents the parameter about which the sensitivity analysis is performed, and the vertical axis shows the extent to which the relevant parameter affects the probability of damage, that is $\frac{\partial \tilde{P}_f(\theta;\sigma_G)}{\partial \theta}$ in Equation (20). It should be noted that the term $\frac{\partial \tilde{P}_f(\theta;\sigma_G)}{\partial \theta}$ in Equation (20) is dP_f/dm, which is the parameter on the Y axis as shown in Figure 4. It represents the changing rate of probability of failure related to the mass; in Figure 5, the vertical axis represents dP_f/dC_a; in Figure 6, the vertical axis represents dP_f/dA; in Figure 7, the vertical axis represents dP_f/dC_d; in Figure 8, the vertical axis represents dP_f/dV.

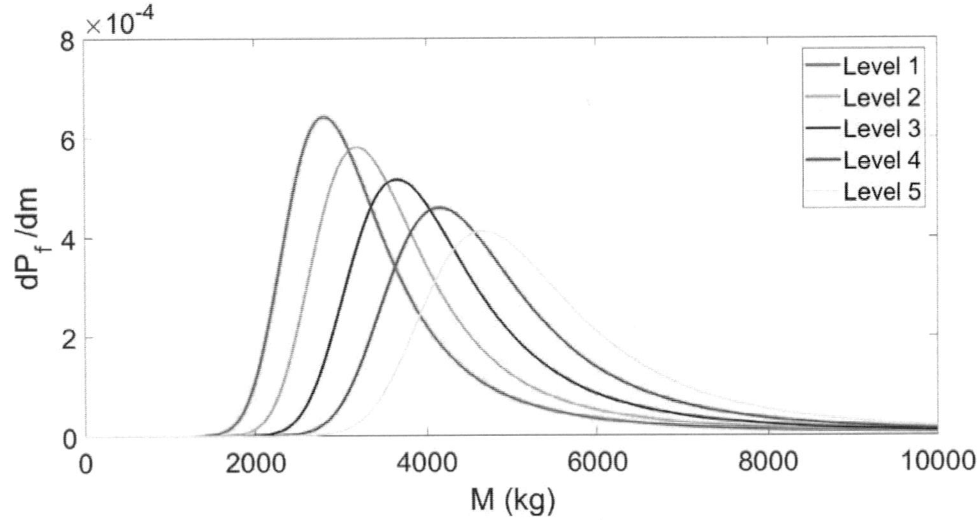

Figure 4: Sensitivity analysis about the mass

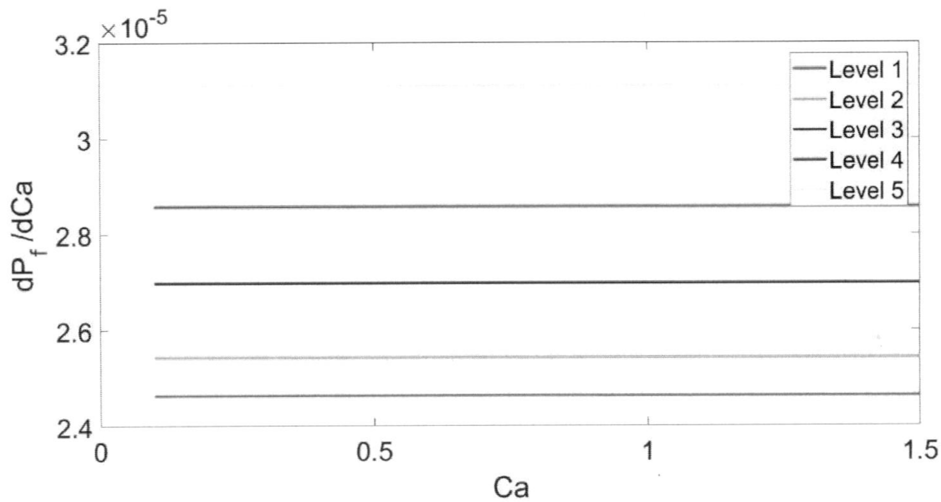

Figure 5: Sensitivity analysis about the added mass coefficient

As indicated in Figure 4, the purple curve (Level 1) corresponds to the lightest object, followed by green (Level 2), blue (Level 3), red (Level 4), and yellow (Level 5), which successively represent heavier objects. It can be observed that the extreme points of these curves are arranged in order from left to right, corresponding to increasing mass, which is consistent with common sense. In addition, each curve clearly shows a normal distribution, which may be due the assumption of normal distribution for the mass.

Further, the effect of the added mass coefficient on the probability of failure is not significant if compared with the mass, since the curves shown in Figure 5 are almost flat lines. The reason may be that the kinematic energy E_E is related to the quadratic power of mass but the first power of added mass coefficient, as shown in Equation (5).

In addition, the collision may have more effects than the added mass, as indicated greater slopes about those curves in Figure 6. In addition, all the curves in Figure 6 show a monotonically increasing trend, indicating that the probability of failure increases with increasing collision area.

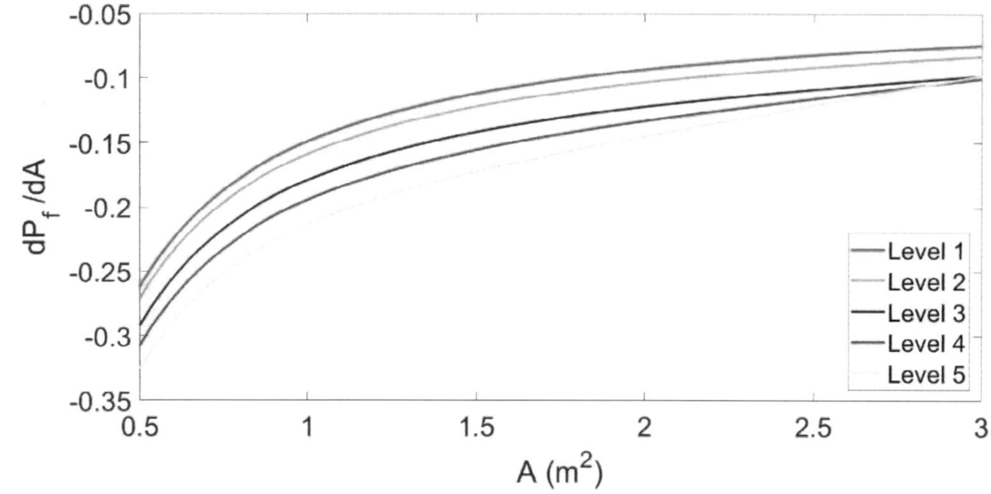

Figure 6: Sensitivity analysis about the collision area

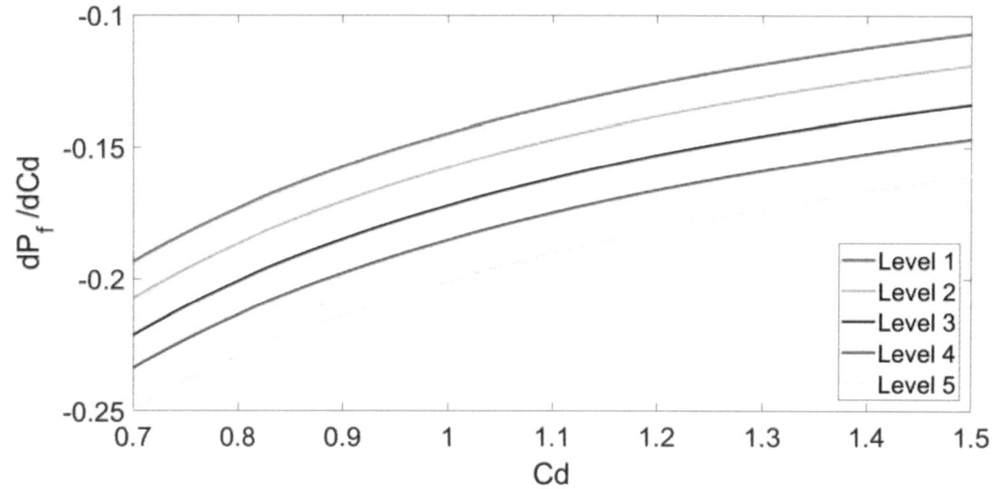

Figure 7: Sensitivity analysis about the drag coefficient

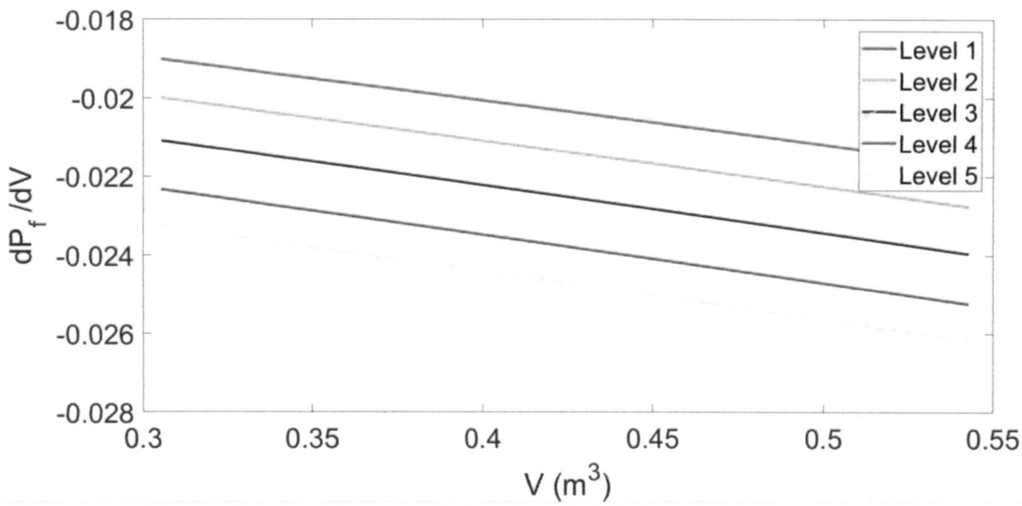

Figure 8: Sensitivity analysis about the volume

The drag coefficient may affect the probability of failure in a very similar way, as shown in Figure 7. This may also be deduced from the mathematical formula in Equation (4), because drag coefficient and collision area are both on the denominator. The curves shown in Figure 6 and Figure 7 do show very similar pattern.

Finally, Figure 8 shows a monotonically decreasing trend, indicating that the probability of failure decreases with increasing volume. This may be because a greater volume intends to provide more cushioning to reduce impact damage.

5. Conclusions

In this paper, a sample case from DNV (2010) is selected as the study case, to investigate how the five random variables, including mass, volume, collision area, added mass coefficient and drag coefficient, affect the defined reliability. With MC method applied in the calculation, the sensitivity analysis is used to study the influence of five random variables, which are mass, collision area, volume, added mass coefficient and drag coefficient.

Overall, the reliability sensitivity analysis shows that the probability of damage is most sensitive to the mass among those five selected parameters. Then it should be the collision area, volume and the drag coefficient. It seems that the probability of damage is not quite sensitive to the selection of added mass coefficient C_a. It could be explained by that the amount of kinetic energy associated with the added mass is always a small amount if compared with the kinetic energy from the dropped object's mass. Finally, these results may be of some help in the analysis of dropped objects in offshore engineering.

References

[1] DNV, Det Norske Veritas (DNV), 2010, "DNV Recommended Practice: Risk Assessment of Pipeline Protection", DNV-RP-F107, Norway.

[2] ABS, American Bureau of Shipping (ABS), 2017. "Dropped Object Prevention on Offshore Units and Installations", Houston, TX, USA

[3] ABS, American Bureau of Shipping (ABS), 2010. "Guidance Notes on Accidental Load Analysis and Design for Offshore Structures", Houston, TX, USA.

[4] Aanesland, V., 1987. "Numerical and Experimental Investigation of Accidentally Falling Drilling Pipes", 19th Annual OTC in Houston, Texas.

[5] Luo, Y., and Davis, J., 1992, "Motion Simulation and Hazard Assessment of Dropped Objects", Proceedings of the 2nd International Offshore and Polar Engineering Conference, Vol. (4), pp.14 - 19.

[6] Colwill, R.D. and Ahilan, R.V., 1992. "Reliability Analysis of the Behavior of Dropped Objects", 24th Annual OTC in Houston, Texas.

[7] Katteland, L. and Øygarden, B., 1995. "Risk Analysis of Dropped Objects for Deep Water Development", The 14th Intl Conf on Offshore Mechanics & Arctic Engineering; 18-22 June 1995; Copenhagen, Denmark.

[8] Majed, A. and Cooper, P., 2013. "High Fidelity Sink Trajectory Nonlinear Simulations for Dropped Subsea Objects", Proceedings of the 23rd International Offshore and Polar Engineering Conference, USA.

[9] Yasseri, S., 2014. "Experiment of Free-falling Cylinders in Water", Underwater Technology, Vol. (32), No.3, pp.177 - 191.

[10] Awotahegn, M., 2015, "Experimental Investigation of Accidental Drops of Drill Pipes and Containers," Master's Thesis, University of Stavanger, Stavanger, Norway.

[11] Xiang, G., Birk, L., Li, L., Yu, X. and Luo, Y., 2016. "Risk Free Zone Study for Cylindrical Objects Dropped into water", Ocean Systems Engineering, Vol. (6), No. 4, pp. 377-400.

[12] Xiang, G., Birk, L., Yu, X., Lu, H., 2017a, "Numerical Study on the Trajectory of Dropped Cylindrical Objects," Ocean Engineering, Vol.130, pp. 1-9.

[13] Xiang, G., Birk, L., Yu, X. and Li, X., 2017b. "Study on the trajectory and landing points of dropped cylindrical object with different longitudinal center of gravity", International Journal of Offshore and Polar Engineering, Vol. (27), No. 3, 274-282.

[14] Xiang, G., Li, X., Yu, X., Luo, Y. and Cao, Y., 2019. "Motion dynamics of dropped cylindrical objects in flows after water entry", Ocean Engineering, Vol. 173, pp. 659 - 671.

[15] Yu, X., Meng, H., Zhen, Y. and Li, L., 2019, "Trajectory Prediction of Cylinders Freely Dropped into Water in Two Dimensions Based on State Space Model," Ocean Engineering 187 (2019) 106154.

[16] BSEE, 2016. Chapter II – Bureau of Safety and Environmental Enforcement (BSEE), Department of the Interior (https://www.gpo.gov/fdsys/granule/CFR-2016-title30-vol2/CFR-2016-title30-vol2-sec250-714).

[17] Edmollaii, S.T., Edalat, P. and Dyanati, M., 2019. "Reliability Sensitivity Analysis of Dropped Object on Submarine Pipelines", Ocean Systems Engineering, Vol. (9), No. 2, pp. 135-155.

[18] Yu, H., 2019. "Reliability Sensitivity Analysis of Dropped Objects Hitting on the Pipeline at Seabed", Master Thesis, University of New Orleans.

[19] Papaioannou I, Breitung K , Straub D ., 2013. "Reliability sensitivity analysis with Monte Carlo methods", In: Ellingwood B, Deodatis G, Frangopol DM, editors. 11th Inter- national Conference on Structural Safety and Reliability (ICOSSAR). New York, NY, USA.

[20] Miranda, J., 2013. "Structural Reliability Analysis with Implicit Limit State Functions", Technical Report, Instituto， Superior Técnico, University of Lisbon: Lisbon, Portugal.

NUMERICAL SIMULATION OF FLUID SLOSHING IN A ROLLING TANK

Wen-Huai Tsao
Postdoctoral Fellow, Ocean Engineering Group, Department of Civil, Architectural and Environmental Engineering, The University of Texas at Austin, Austin, Texas 78712, USA

Spyros A. Kinnas
Professor, Ocean Engineering Group, Department of Civil, Architectural and Environmental Engineering, The University of Texas at Austin, Austin, Texas 78712, USA

Proceedings of the 25th Offshore Symposium, February 27th 2020, Houston, Texas
Texas Section of the Society of Naval Architects and Marine Engineers

ABSTRACT

The dynamic response of a fluid in a rolling tank is one of the most important issues for the integrity of LNG tanks and the overall stability of the vessel. When the tanks oscillate near the sloshing resonance frequencies, sloshing phenomena with strong nonlinearity and large free-surface deformations can be observed. For some liquid depths and external forcing characteristics, hydraulic jumps may form, wave component interactions would occur, and severe impacts on the walls would follow. In this paper, the formations of different types of sloshing waves as well as the impact forces on the walls are simulated by an Eulerian-Lagrangian method, which was developed on the past for the horizontal excitation problem and is extended in this paper to a tank subject to roll motions. This approach allows the use of boundary element method (BEM) and fourth-order Runge-Kutta scheme (RK4) to track the particles on the free surface. In the case of rolling excitation, some highly nonlinear waves and thin jets where the wave impacts on the boundaries of the tank are more likely to occur. Therefore these very thin wave jets would be slightly truncated to alleviate the associated numerical instability. Furthermore, the artificial damping effect, which is linearly proportional to the fluid velocity, is introduced in the dynamic boundary condition on the free surface and the impermeable condition of the walls to take the effects of viscosity into account. This has been proven to be effective in improving the numerical accuracy of the method. Correlations of our results with experimental observations and measurements will be presented.

Keywords: Sloshing, Hydraulic jump, Wave impact, Numerical simulation

Introduction

Sloshing is known as the dynamic motion of the free surface inside a container partially filled with liquid. It is one of the most important issues in ocean engineering for the integrity of LNG tanks and the overall stability of the vessel. The roll-motion of a floating body, such as floating storage regasification unit (FRSU) and floating liquefied natural gas (FLNG) vessel, in an ambient wave-field can be severe. Moreover, the small damping of the ambient flow field is unable to mitigate its resonance effect. Many researchers have put a lot of effort to understand its dynamic characteristics and develop effective damping devices, such as bilge keels or anti-roll tanks (ARTs), for stabilization. Earlier researchers started from the linear wave analysis based on potential theory [1, 2]. Such methods are simple and have been widely used by engineers [3-7]. However, when the tanks subjected to near-resonance excitations, strong nonlinearity and large deformation of the free-surface motion may take place. The nonlinear wave equations have been further solved numerically by potential-based methods [8, 9], finite difference method [10, 11], and finite element method [12]. In recent years, the smoothed particle hydrodynamics (SPH) [13-15] method became popular for simulating the violent sloshing behavior involves turbulence and vortex.

The past research can be categorized into two topics. The first one is the global phenomenon mainly focus on the interaction between liquid and its substructure, such as vehicle stabilization [16], damage analysis of liquid storage tanks [17, 18], seakeeping of membrane tanks [19, 20], tuned liquid damper [21], and flexible-wall effect [22]. The second one is the local phenomenon emphasizes the characteristics of the free surface motion and the dynamic effects it produces. This topic involves the breaking waves [23, 24], free jet [25], contact angle of wetting lines [26, 27], nonlinearity effect [28], impulsive pressure [29], etc. Cariou and Casella [29] used various numerical schemes but found the peak value of the impulsive pressure is difficult to predict. Kim [30, 31] set a buffer zone condition near the top when the free surface reaches the roof to avoid solution divergence and numerical instability. Nasar et. al.'s experiment [32] shown the impact pressure on the roof is about 1.6 times the pressure on the sidewall. Hamlin et. al. [33] found the high impact on walls may come from the hydraulic jump traveling back and forth in the tank. Disimile et. al. [34, 35] discussed through experiments about the condition that hydraulic jump could easily occurs. Wemmenhove et. al. [36] analyzed the wave run-up in a rolling tank by one-phase and two-phase volume of fluid (VOF) method. Actually, the great impacts on the walls and roofs due to some hydraulic jumps have been observed in past experiments [37]. Therefore, to understand the formation of hydraulic jumps inside a rolling tank becomes important.

In this paper, the numerical scheme for the free-surface calculation was an extension of the past work by the authors [8, 25]. It involves two steps, i.e. the Eulerian and then the Lagrangian approach. The Eulerian approach allows the use of BEM to solve the velocity of a free-surface particle. Then the Lagrangian approach uses the fourth-order RK4 scheme to track the particle at the next time step [38]. When the tank oscillates in nearly resonant with the fluid, some violent free surface movements, including highly nonlinear waves and thin jets where the wave impacts on the boundaries of the tank, are more likely to occur. Therefore the nonlinear condition should be applied on the free surface. Some special skills are also applied to obtain better numerical results. The thin wave jet is slightly truncated when certain criterion is met to alleviate the associated numerical instability [25, 39]. The artificial damping condition is also introduced on the free surface (presents in the dynamic boundary condition) [8] and walls (presents in the impermeable condition) [40, 41] to describe the linear damping effect that takes all possible kinds of energy dissipation during fluid sloshing into account. This has been proven effective for improving the numerical accuracy and stability. In addition to BEM, VOF method has also been used for verification. This is implemented with the k-ω SST turbulence model by the commercial software ANSYS Fluent 18.2. Correlations of our numerical results with Disimile et. al.'s experimental observations and measurements [34] are also presented. In this experiment, the formation procedure of a hydraulic jump in a rolling rectangular tank can be observed. The energy dissipation in each case can be visualized by calculating the hydraulic heights and Froude numbers. The mathematical formulae are shown in Section 2; several examples are discussed in Section 3; the conclusions are drawn in Section 4.

METHODOLOGY

1. Governing Equation and Boundary Conditions

The sloshing behavior in a two-dimensional rotating rectangular tank with a unit width is considered. The coordinate system and tank dimensions are shown in Figure 1. Assuming the fluid is potential flow so the flow field can be described by a velocity potential ϕ which satisfies Laplace's equation within the specific domain as [42]:

$$\nabla^2 \phi = 0 \tag{1}$$

The velocity vector of a fluid particle is defined as

$$\boldsymbol{q} = \left(\frac{\partial \phi}{\partial x}, \frac{\partial \phi}{\partial z}\right) \tag{2}$$

On the free surface, the kinematic boundary condition is given as:

$$\frac{D\boldsymbol{R}}{Dt} = \left(\frac{\partial}{\partial t} + \boldsymbol{q} \cdot \boldsymbol{\nabla}\right) \boldsymbol{R} = \boldsymbol{q} \tag{3}$$

where $\boldsymbol{R} = (x, y)$ is the position vector. The dynamic boundary condition is given as

$$\frac{D\phi}{Dt} = \frac{1}{2}|\boldsymbol{\nabla}\phi|^2 - g\eta - \alpha\phi \tag{4}$$

where g is the gravitational acceleration, η is the wave elevation, α is the artificial damping coefficient and its value could be chosen as 0.1 s^{-1} [8]. The atmospheric pressure is assumed to be zero for simplicity. Another boundary condition applied on the walls and bottom can be described as:

$$\frac{\partial \phi}{\partial n} = (\boldsymbol{\omega} \times \boldsymbol{R}) \cdot \boldsymbol{n} - \frac{\alpha}{g}\frac{\partial \phi}{\partial t} \tag{5}$$

where $\boldsymbol{\omega}$ is the angular velocity of the tank, $\boldsymbol{n}$ is the unit outward normal vector on the boundary surface. The first term on the right-hand side of Equation (5) is the impermeable boundary condition, while the second term handles the energy dissipation [40, 41].

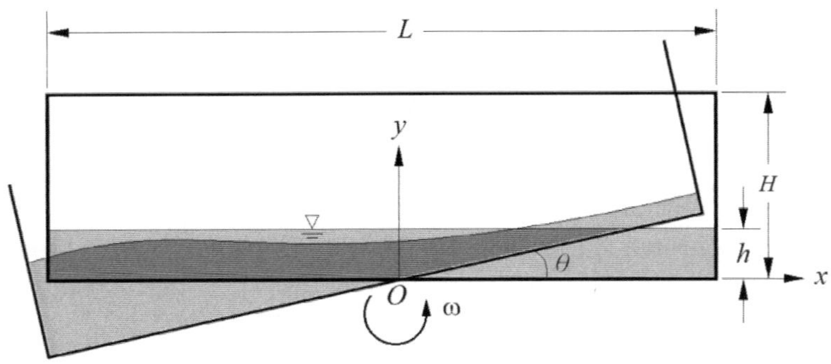

Figure 1: Coordinate system for the oscillating tank

2. Eulerian approach

The fundamental solution of two-dimensional Laplace's equation (1) is given by the Green's function as:

$$G(\boldsymbol{x}_p, \boldsymbol{x}_q) = \frac{1}{2\pi} ln r \tag{6}$$

where r is the distance between a source point $\boldsymbol{x}_p$ and a field point $\boldsymbol{x}_q$ both located on the boundary of the domain. By applying Green's second identity to Green's function and velocity potential, the boundary integral equation (BIE) can be obtained as:

$$c(\boldsymbol{x}_p)\phi(\boldsymbol{x}_p) = \int_S \left[\frac{\partial G(\boldsymbol{x}_p, \boldsymbol{x}_q)}{\partial n}\phi(y\boldsymbol{x}_q) - G(\boldsymbol{x}_p, \boldsymbol{x}_q)\frac{\partial \phi(\boldsymbol{x}_q)}{\partial n}\right] d\boldsymbol{S} \tag{7}$$

The coefficient $c(\boldsymbol{x}_p)$ is referred to as the flux that depends on the location of the source, and its value is 1/2 when $\boldsymbol{x}_p$ is on a smooth boundary and is equal to its solid angle divided by 2π when $\boldsymbol{x}_p$ is at the boundary point between two boundaries. For the sloshing problem, the velocity potential and its normal derivatives are given on the free surface and wetted walls, respectively. This can be regarded as a mixed type boundary value problem by giving a set of Dirichlet and Neumann conditions on the domain boundaries in Equation (7). In this paper, this procedure is implemented by the conventional BEM with linear element approximation. The detail numerical procedure can be referred to some books [43]. After the velocity potential and its normal derivative are solved, the fluid velocity inside the domain can be obtained by the following equation:

$$c(\boldsymbol{x}_p)\frac{\partial \phi(\boldsymbol{x}_p)}{\partial x_i} = \int_S \left[\frac{\partial G(\boldsymbol{x}_p, \boldsymbol{x}_q)}{\partial x_i \partial n}\phi(\boldsymbol{x}_q) - \frac{\partial G(\boldsymbol{x}_p, \boldsymbol{x}_q)}{\partial x_i}\frac{\partial \phi(\boldsymbol{x}_q)}{\partial n}\right] d\boldsymbol{S}, \boldsymbol{i} = \boldsymbol{1}, \boldsymbol{2} \tag{8}$$

3. Lagrangian approach

The Lagrangian method is used in the time-marching procedure to track the fluid particle on the free surface. Since the kinematic and dynamic boundary conditions on the free surface are both first-order differential equation, the fourth-order Runge-Kutta scheme can be used as:

$$\boldsymbol{\sigma[k+1] \cong \sigma[k] + \frac{\Delta t}{6}(k_0 + 2k_1 + 2k_2 + k_3), k = 0, 1, 2 \cdots} \tag{9}$$

where $\dot{\sigma} = f(t, \sigma)$ and σ could be the position vector (x, y) or the velocity potential ϕ, Δt is the time interval. Those four terms in the right-hand side of Equation (9) can be obtained by the following four steps:

$$\boldsymbol{k_0 = \Delta t\, f(t^k, \sigma^k)} \tag{10}$$

$$\boldsymbol{k_1 = \Delta t\, f(t^k + 0.5\Delta t, \sigma^k + 0.5k_0)} \tag{11}$$

$$\boldsymbol{k_2 = \Delta t\, f(t^k + 0.5\Delta t, \sigma^k + 0.5k_1)} \tag{12}$$

$$\boldsymbol{k_3 = \Delta t\, f(t^k + \Delta t, \sigma^k + k_2)} \tag{13}$$

Through this step-by-step algorithm, the position of a fluid particle associated with its potential can be described in the next time step with fourth-order accuracy in time.

The pressure on the walls and bottom can be obtained by the Bernoulli equation as:

$$\boldsymbol{P = -\rho\left(\frac{\partial\phi}{\partial t} + \frac{1}{2}|\nabla\phi|^2 + g\eta\right) = -\rho\left(\frac{D\phi}{Dt} - V_g \cdot \nabla\phi + \frac{1}{2}|\nabla\phi|^2 + g\eta\right)} \tag{14}$$

where $\boldsymbol{V}_g = \left(\frac{Dx}{Dt}, \frac{Dy}{Dt}\right)$ is the local grid velocity of the body surface, and the atmospheric pressure is assumed to be zero for simplicity. The derivative $\frac{D\phi}{Dt}$ and $\boldsymbol{V}_g$ can be also calculated by the fourth-order Runge-Kutta scheme [44] as:

$$\boldsymbol{\frac{D\sigma[k+1]}{Dt} = \frac{1}{4\Delta t}\left\{-2\sigma[k] - 4\sigma^{(1)}\left[k + \frac{1}{2}\right] - 4\sigma^{(2)}\left[k + \frac{1}{2}\right] + 10\sigma[k+1]\right\} - \frac{1}{2}\frac{D\sigma[k]}{Dt}} \tag{15}$$

where the value with superscripts (1) and (2) indicating the value in the first and second mid-steps respectively.

During the time-marching process, the local Courant number C_{ri} is introduced to examine the stability as [45]:

$$\boldsymbol{C_{ri} = \frac{q_i \Delta t}{\Delta x_i}, i = 1, 2} \tag{16}$$

where q_i and Δx_i are the particle velocity and spatial interval along each direction, respectively. The Courant number is always kept less than 1.0, but it is better to be smaller than 0.5 to ensure the numerical stability for nonlinear sloshing problems.

4. Treatment of jets

In the case of large-amplitude waves impacting on the tank walls, some high-speed jets develop along the wetted part. These jets form a critical boundary geometry which needs extremely small elements to capture. Without special treatment, it will lead to inaccurate and unstable solutions. In this paper, it is simply truncated to alleviate the numerical issues. The contact angle β between the free surface and the wetted wall shown in Figure 2(a) is used as the truncation criteria. Once the angle reaches the threshold, which is set to be $\pi/9$, the free-surface element adjacent to the body is truncated [46]. The new free-surface element will be aligned perpendicular to the wall as shown in Figure 2(b), which is a similar treatment to that used by previous researchers [39]. Then the value of the velocity potential at the intersection of the new free-surface element with the tank wall is obtained via cubic-spline interpolation of the potentials on the wetted wall.

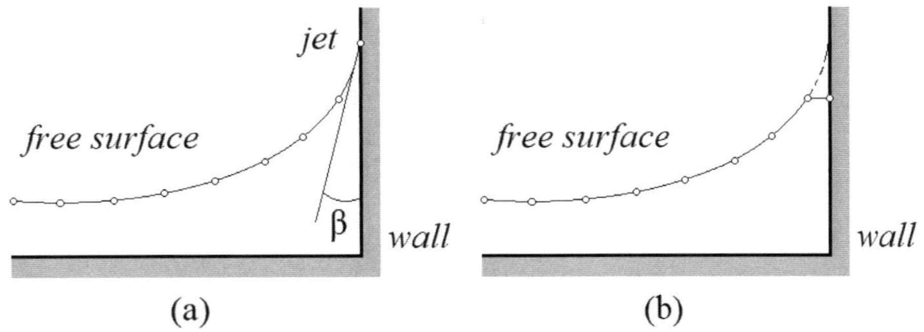

Figure 2: Details of jet truncation model: (a) before truncation, (b) after truncation

5. Froude number

Froude number is a ratio between the depth-averaged velocity $\bar{q}$ and wave celerity c defined as:

$$\boldsymbol{F_r} = \frac{\bar{q}}{c} \tag{17}$$

where $c \cong \sqrt{gd}$ for shallow liquids, d is hydraulic height, $c \cong \sqrt{g\lambda/2\pi}$ for deep liquids, and λ is the wave-length equals $2L$ for the first sloshing mode. When Froude number is greater than unity, which as known as supercritical flow, the water velocity exceeds the wave celerity so that the flows propagate in the same flow direction at a velocity greater than the wave celerity. Otherwise, it is a subcritical flow where the wave may propagate against the fluid flow. When the supercritical flows enter the zone of subcritical flows, hydraulic jump may form. Therefore it can characterize different types of hydraulic jump and different energy dissipation. Larger Froude numbers usually indicate the hydraulic jump with greater energy dissipation. In the case of a hydraulic jump occurs in an upward channel shown in Figure 3, Froude numbers can be determined by the upstream and downstream hydraulic depths as [47]:

$$\frac{d_2}{d_1} = \frac{1}{2}\left(-1 + \sqrt{1 + 8F_{r1}^2}\right) \tag{18}$$

$$F_{r2}^2 = (d_1/d_2)^3 F_{r1}^2 \tag{19}$$

where d_1 is the upstream hydraulic height, d_2 is the downstream hydraulic height, F_{r1} and F_{r2} are the pre- and post-jump Froude numbers, respectively. For $1.0 \leq F_{r1} < 1.7$, it can be classified as the undular jump where the ripple on the free surface will be observed. For $1.7 \leq F_{r1} < 2.5$, it is known as the weak jump where the downstream is quite steady so low energy loss in one jump. For $2.5 \leq F_{r1} < 4.5$, rough water surface will be observed and more energy loss due to obvious turbulence. This is known as the oscillatory (or transition) jump. The hydraulic jumps have its Froude numbers $F_{r1} > 4.5$ will not be considered in this article since they have not been observed during the experiments and the numerical simulation.

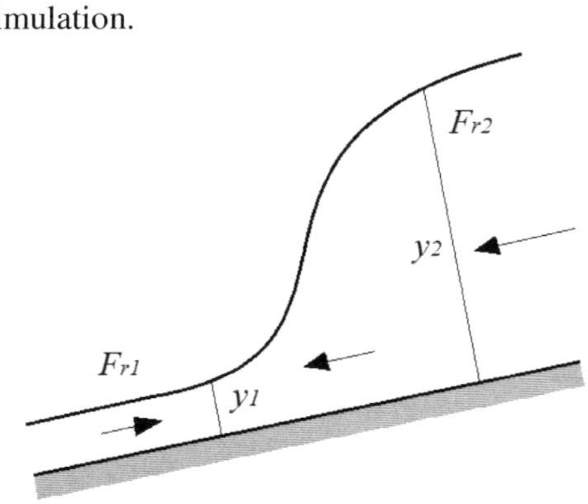

Figure 3: Sketch of a hydraulic jump in an upward channel

RESULTS

The hydraulic jumps and some simple wave and wave interactions in a rolling rectangular tank had been implemented by the past experiments [34, 35]. The tank had its internal length to be 1.9 m, height to be 1.2 m, and width to be 0.94 m. The water depth is 0.265 m so the corresponding first-mode natural frequency is 2.59 rad/s (0.41 Hz). The tank oscillation was started at the tilt platform but the fluid was static before starting the oscillation, and then rotated about the z-axis in a harmonic motion as:

$$\theta = \theta_0 sin(\omega t) \tag{20}$$

where θ_0 is the angular amplitude, ω is the forcing frequency. A complete description of the testing facility and setup can be referred to the past study [34].

In this example, BEM and VOF are used to simulate the sloshing phenomenon. Figure 4 (a) shows BEM applies 100 and 20 linear elements along the x- and y-direction, respectively. The time interval is 0.001 s. to ensure Equation (16). Figure 4 (b) shows VOF applies 23111 nodes (191 and 121 nodes distributed along x- and y-direction, respectively) and 22800 cells (190 and 120 cells distributed along x- and y-direction, respectively) uniformly distributed in the domain. The time interval is 0.01 s. The $k - \omega$ SST turbulence model is used.

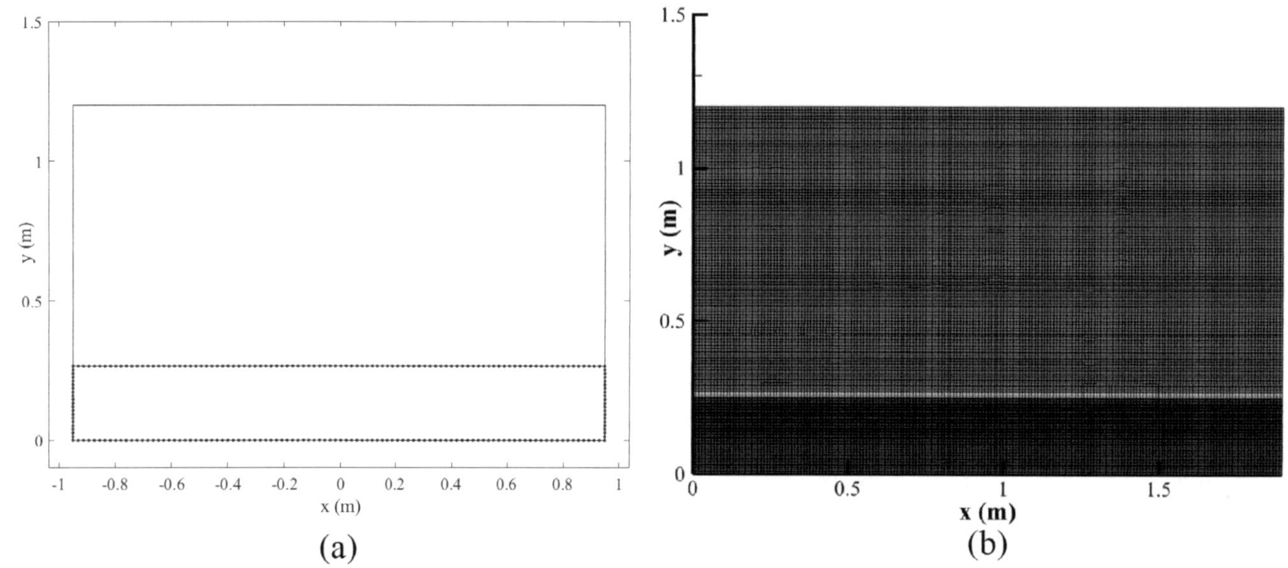

Figure 4: Numerical discretization (a) nodes and elements in BEM, (b) nodes and cells in VOF

1. Example 1: Oscillating hydraulic jump

Since there is no clear video of the free-surface motion provided by Disimile et. al. [34, 35], we can only roughly compare our numerical results with the development of the hydraulic jump in the experiments. However, some phenomenon and characteristics can still be visualized through numerical simulation. As the tank subject to the harmonic oscillation with the forcing frequency ω=1.88 rad/s (0.30 Hz) and angular amplitude θ_0=4.71°, the procedure of the formation of a hydraulic jump in the experiment can be observed as shown in Figure 5. During the first cycle of tank oscillation, the first wave was established owing to sufficient kinetic energy of flowing water. Figure 5(a) shows it reflect from the right-side wall and propagate in the opposite direction with the flow coming from the left side. In Figure 5(b), the high-speed wave on the right enters the low-speed wave zone on the left and makes the upper-part fluid start to moves at a greater speed than the lower part of the incoming fluid. Therefore breaking waves and air entrainment take place. Some turbulence and spray jets are also observed. In Figure 5(c,) the hydraulic jump rapidly forms as the wave propagates to the left. This hydraulic jump quickly fades away since there is no fluid continues flowing from the left, and the kinetic energy is dissipated by the turbulence observed in the hydraulic jump. This type of hydraulic jump is known as the oscillatory jump, which typically occurs under near-resonance condition or large-amplitude excitation, and usually formed on each rotation of the tank.

Figure 5: The development of an oscillatory jump in a rolling tank shown in the experiment [34]: (a) initial formation of upward-propagate wave, (b) breaking wave and air entrainment, (c) fully-developed hydraulic jump

Figures 6 and 7 show the formation of the hydraulic jump solved by BEM and VOF, respectively. In Figure 6(a), BEM simulates the wave reflection very well and provides accurate horizontal velocity. In Figure 6(b), the wave rises up rather than breaks. No air entrainment occurs due to the limitation of kinetic boundary condition on the free surface. In Figure 6(c), the wave profile is in good agreement with the experimental result. However, a significant wave front can be observed while Figure 5(c) shows it will crash into turbulence. Then the BEM algorithm stops after the infinite fluid velocity is solved when the high-speed wave hits the wall right after the first hydraulic jump. The results show the BEM's conventional disadvantage in describing local phenomenon including the turbulence of the hydraulic jump. In this example, only the linear damping mechanism is uniformly applied on the free surface. The fluid velocity will become greater than the actual flow when violent sloshing occurs, and eventually cause numerical instability. The stability issue can be slightly alleviated by truncating the jet, as long as the BEM algorithm stays stable, it can still provide accurate results for the nonlinear sloshing problem. On the other hand, VOF didn't encounter the instability issue. Figure 7(b) shows the free surface start to separate and some breaking wave and air entrainment can be observed. In Figure 7(c), VOF provides a similar flow pattern as the actual flow field where strong turbulence occurs. After this hydraulic jump, the wave hit hard on the right-side wall and a free jet take place. Since BEM crashes due to its infinite velocity at the same moment in this example, their jet formations cannot be compared. However, this will be demonstrated in the next example. The measured Froude numbers of the hydraulic jumps are shown in Table 1. The experiment gave 2.8, BEM gives 2.89 and VOF gives 2.76. BEM has greater value than the experimental result due to lack of local damping mechanism, while VOF is more accurate because the turbulence model can be more useful for the hydraulic jump simulation.

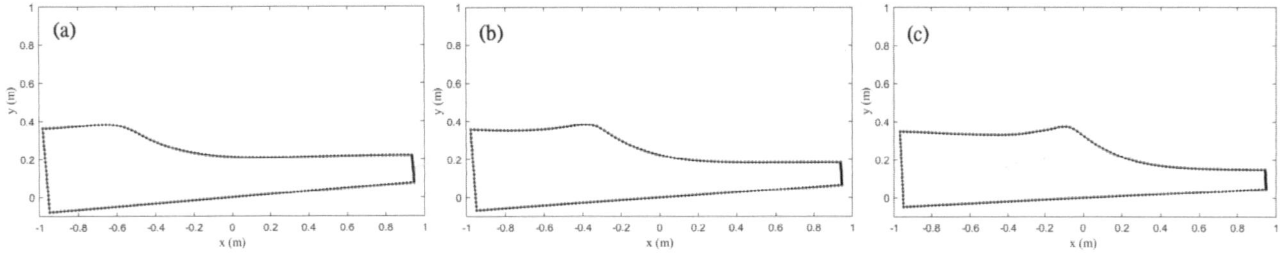

Figure 6: The development of an oscillatory jump in a rolling tank solved by BEM: (a) initial formation of upward-propagate wave, (b) wave confrontation, (c) fully-developed hydraulic jump

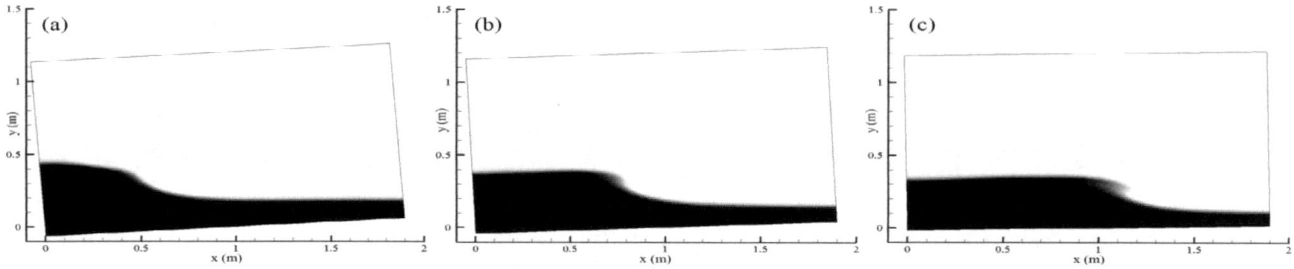

Figure 7: The development of an oscillatory jump in a rolling tank solved by VOF: (a) initial formation of upward-propagate wave, (b) breaking wave and air entrainment, (c) fully-developed hydraulic jump

Table 1: Experimental and numerical hydraulic jump Froude numbers (ω=1.88 rad/s)

	Case 1	Case 2	Case 3
Hydraulic jump type	Oscillatory	Weak	Undular
Forcing amplitude (deg)	4.71	3.50	2.42
Experimental [34]	2.8	2.1	1.6
Numerical by BEM	2.89	2.26	1.37
Numerical by VOF	2.76	2.15	1.60

2. Example 2: Undular and weak hydraulic jump

As the tank subject to the harmonic oscillation with the forcing frequency ω=1.88 rad/s (0.30 Hz) and angular amplitude θ_0=2.42°, Disimile et. al.'s experiment [34] shown that the hydraulic jump can be also observed in the rolling tank. The numerical results by BEM and VOF are shown in Figures 8 and 9, respectively. In Figure 8(a), the first wave bounces back and encounters the second incoming wave. These two waves are similar in amplitude but propagate in the opposite direction. They will constructively merge into one greater wave as shown in Figure 8(b). Afterward, the large wave is separated into two small waves as shown in Figure 8(c). Figure 9 presents the wave and wave interaction very similar to Figure 8. However, a jet on the right-side wall can be clearly observed in Figure 9(a). This may be resulted from the jet truncation treatment carried out in BEM. In this case, the forcing amplitude is very small so the flow can only gain little kinetic energy. Therefore the hydraulic jump is not significant. This type of hydraulic jump is called the undular jump, which typically causes slight ruffles on the free surface. Table 1 shows the experimental hydraulic jump Froude number is 1.6, while BEM gives 1.37 and VOF gives 1.60. Overall, both methods can provide clear wave and wave interaction and satisfied solutions.

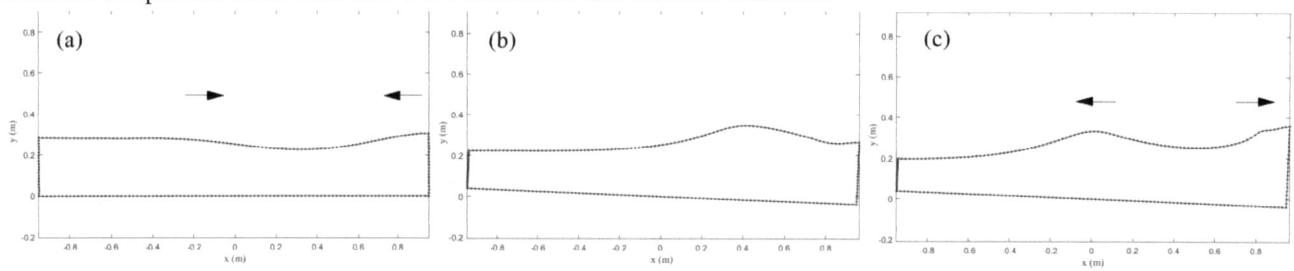

Figure 8: The development of an undular jump in a rolling tank solved by BEM: (a) waves confrontation, (b) fully-developed hydraulic jump, (c) waves separation

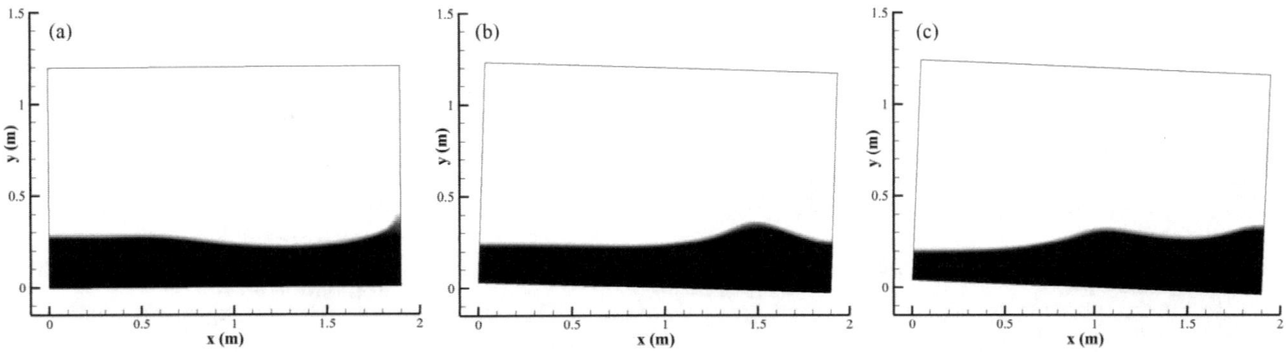

Figure 9: The development of an undular jump in a rolling tank solved by VOF: (a) waves confrontation, (b) fully-developed hydraulic jump, (c) waves separation

As the roll angular amplitude θ_0 is enlarged to 3.50°, a greater hydraulic jump can be observed in a rolling tank. The numerical results by BEM and VOF are shown in Figures 10 and 11, respectively. Figure 10 shows this hydraulic jump is an intermediate phenomenon between the undular and oscillatory jump. Both of their features can be observed in Figure 10(c), which are a small ripple and an obvious wavefront on the hydraulic jump. This type of hydraulic jump is known as the weak jump, which is similar to the undular jump condition. It is noticeable that Figure 10(a) has shown the jet occurs on the left-side wall due to large vertical velocity. The tiny part of this jet will be truncated to maintain the smoothness of the free surface so that BEM could stay stable in this case. In Figure 11(a), VOF has an obvious jet on the left-side wall with breaking waves and air entrainment. This jet immediately drops and vanishes as the wave propagates to the right as shown in Figure 11 (b). Afterward, the hydraulic jump forms and the wavefront breaks as shown in Figure 11(c). Its flow pattern is similar to the first oscillatory jump case. The experimental hydraulic jump Froude number is 2.1, while BEM's is 2.26 and VOF's is 2.15 as shown in Table 1. By BEM, the measured Froude number is apparently greater than the experimental ones because it uses constant artificial damping coefficient which may be insufficient for the violent flow field. In this case, BEM is also unable to describe the local damping effect. Therefore the over-estimated fluid velocity and wave elevation may be obtained. On the other hand, VOF has more reasonable damping mechanism during the hydraulic jumps.

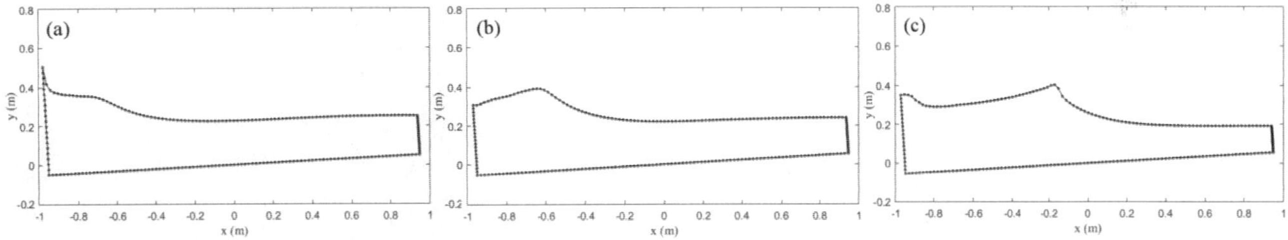

Figure 10: The development of a weak jump in a rolling tank solved by BEM: (a) initial formation of upward-propagate wave, (b) pre-jump, (c) fully-developed hydraulic jump

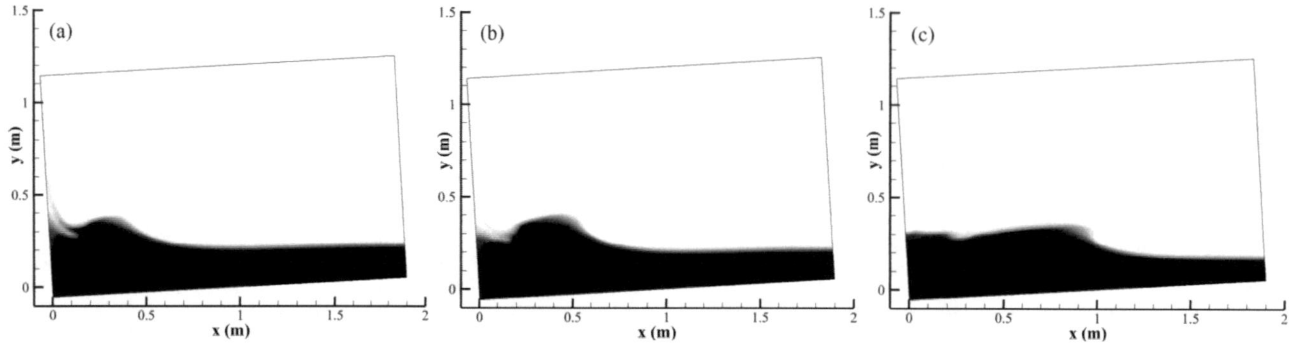

Figure 11: The development of a weak jump in a rolling tank solved by VOF: (a) initial formation of upward-propagate wave, (b) pre-jump, (c) fully-developed hydraulic jump

Figure 12 shows the formation of a jet and its fluid velocity field solved by BEM and VOF. This jet usually occurs under certain oscillatory frequencies which allow the fluid to rise smoothly without breaking the wave. Therefore the vertical velocity on the wall could be extremely high and the fluid may hit the top of the tank. Without breaking the wave, the fluid may keep most of its kinetic energy so that the hydraulic jump could form after the wave bounces back. Both methods provide very similar wave profiles, a sharp spike on the wall in 5 cm from the left-side wall, and the dramatic change in wave elevation rise about 40 cm within the 15 cm near the wall. The rest part of the free surface remains steady. In Figure 12(a), the first two elements aligned next to the wall did not form a smooth free surface. This is because the jet was truncated in the last time step, but the node of intersection (between wall and free surface) rise high again due to great velocity in this time step. Therefore some jagged surface will be obtained which may decrease the numerical accuracy. To avoid the disadvantage, more elements should be applied within this area or special treatment for jet truncation needs to be considered.

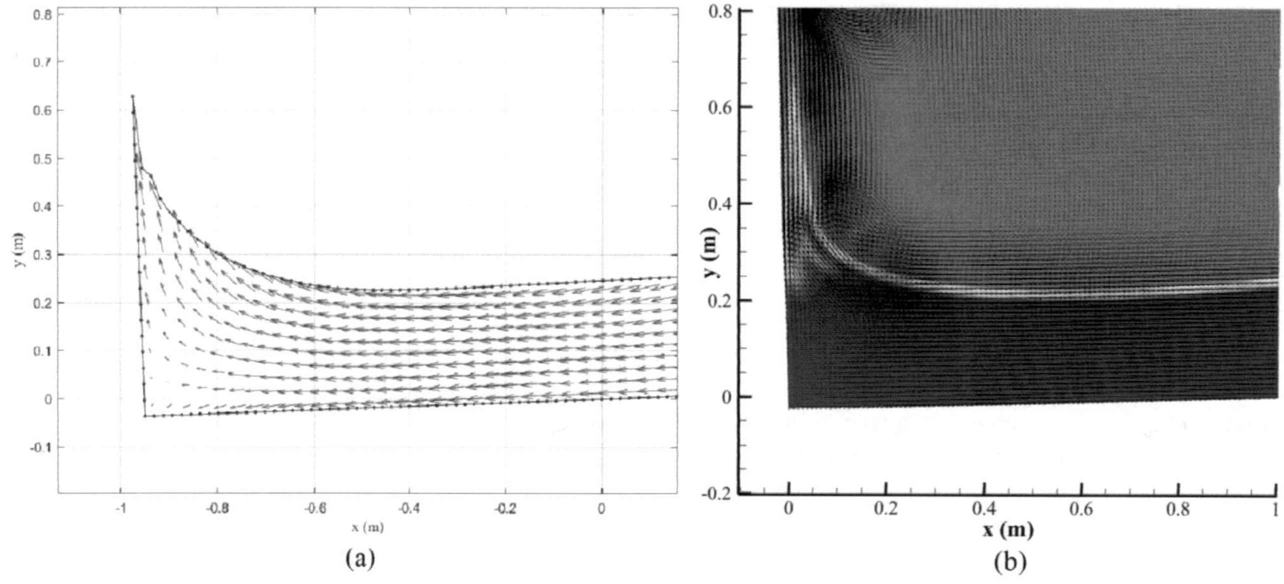

Figure 12: The formation of a jet and its fluid velocity solved by (a) BEM, (b) VOF

Figure 13 shows the pressure distribution solved by BEM and VOF. The contour of the pressure by BEM in Figure 13(a) is calculated by interpolating the pressure of the nodes evenly distributed in the fluid domain from Bernoulli equation. Although its spatial interval is much larger than the VOF's cell, they still show good agreement with each other. They both have stronger pressure near the bottom of the left-side wall. This is because the sloshing pressure mainly comes from the inertia force of fluid. For the jet forms on the top of the left-side wall, it has little effect to the wall since only an extremely small part of mass is impacting on the wall. Therefore, only those jets or splash involve a huge amount of mass hit on the wall and roof of the tank should be considered carefully in practice. By this example, it could be practical to truncate the small jets for the dynamic analysis of the fluid-structure interaction.

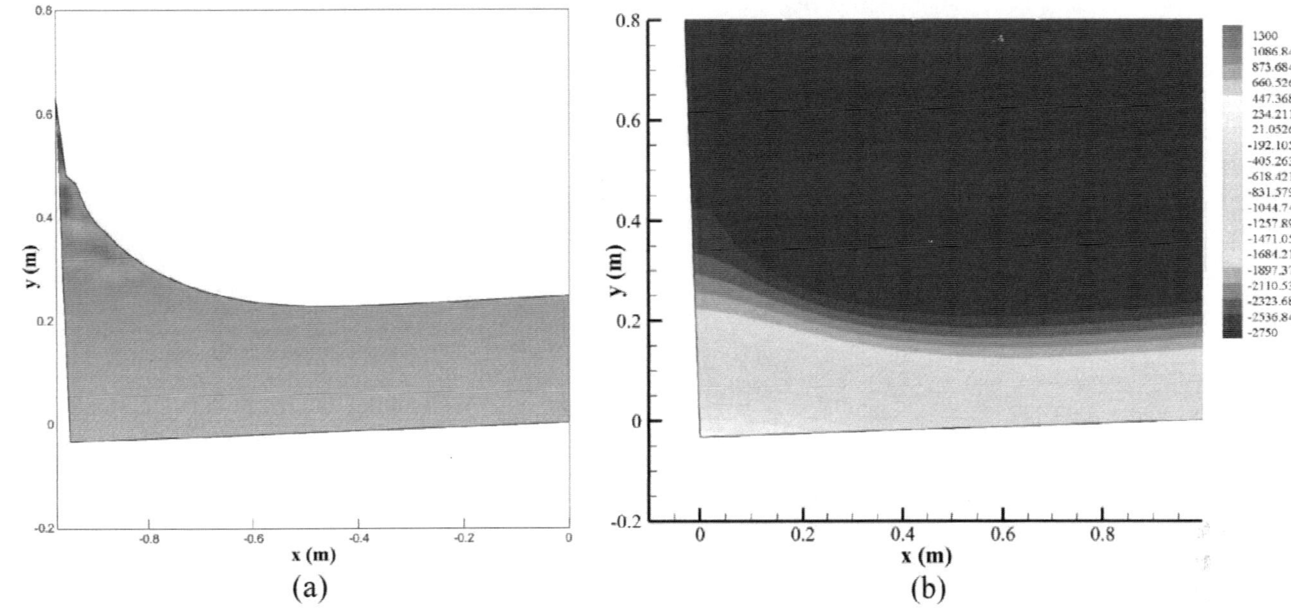

Figure 13: The pressure distribution solved by (a) BEM, (b) VOF

3. Example 3: Parametric study

According to the experimental and numerical results, two facts that are the wave confrontation and enough fluid velocity have a great influence on the occurrence of a hydraulic jump. To take both factors into account, the normalized oscillation condition (NOC) can be defined as:

$$\boldsymbol{NOC} = \frac{y_{max}}{h} \times \frac{\omega}{\omega_0} \quad \textbf{(21)}$$

where y_{max} is the maximum vertical deflection distance of the tank oscillation, and ω_0 is the natural sloshing frequency of the first oscillation mode [42].

Since the hydraulic jump is categorized by Froude number in this paper, Figure 14 shows the experimental and numerical Froude numbers variation for different oscillation condition. The benchmark tests have shown that near-resonance excitation tends to cause wave confrontation condition, and large-amplitude could lead to higher fluid velocity. The frequency ratio (the ratio between forcing frequency and the fluid's first-mode natural frequency) is from 0.6 to 0.85. The rolling amplitude must be large enough to excite the flow become supercritical flow, i.e. to make F_r>1.0. In addition, the higher oscillation condition (because of large amplitude or near-resonance frequency) will lead to the greater Froude number. It stands for the damping effect provided by the hydraulic jump. If applying the linear regression to estimate the experimental results, it will only yield a small correlation coefficient. This means the damping effect is highly nonlinear. Since BEM applied the linear damping effect on the free surface, it can provide accurate estimation in the weak jump region (F_r<2.5). However, in the oscillatory jump region (F_r>2.5), BEM loss its robustness because the nonlinear damping effect was not taken into account in this paper. VOF shows good agreement with the experimental results in the region of weak and strong hydraulic jumps. This has benefited from the turbulence model.

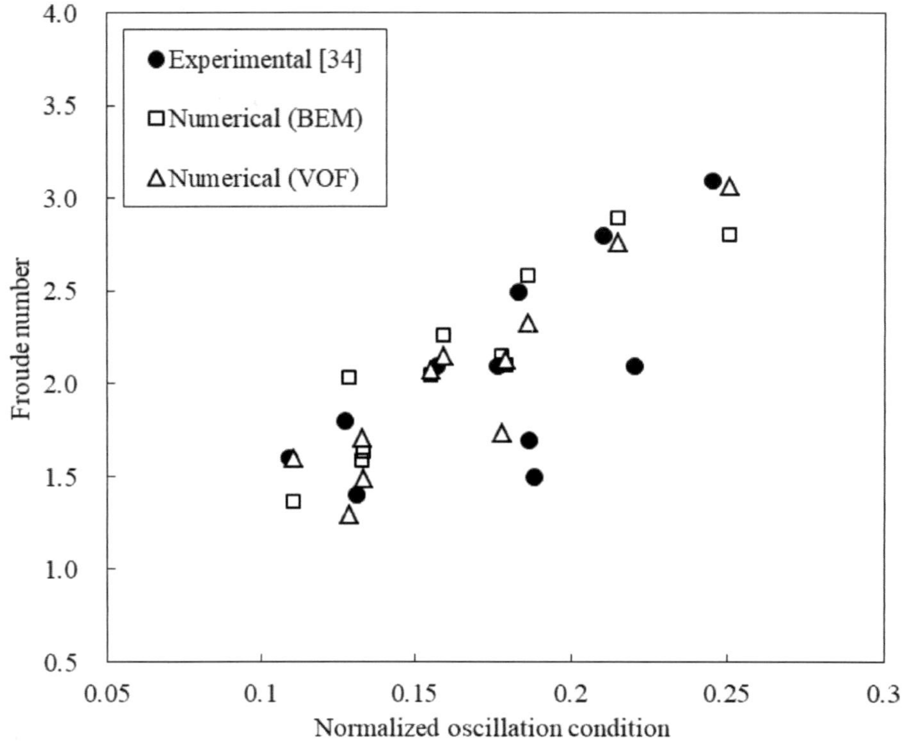

Figure 14: Froude numbers variation for different oscillation condition

Table 2 shows the computational time for a 10-second simulation period by BEM and VOF. The desktop computer with its CPU model intel(R) i7-4790k at 4.0 GHz is used to run both methods. In our case, BEM, which is run with single core, is even much more efficient than VOF run with 4 cores, because it is not required to apply nodes in the entire domain. Although BEM applies fewer nodes, it can still provide very good results in the free-surface shape and the pressure distribution. For the nonlinear sloshing simulation, this advantage can reduce the computational cost and make the extra numerical skills for improving the stability issue remain available in the future.

Table 2: Computational time by BEM and VOF.

	BEM (1 CPU)	VOF (4 CPUs)
Node number	244	23111
Total CPU time (s)	10 minutes	about 2 hours

CONCLUSION

The numerical simulations of the hydraulic jumps occur in a rolling tank are presented in this paper. Some sloshing characteristics and flow patterns can be observed according to the numerical results. There are two main factors that influence the formation of hydraulic jumps. The first one is the large roll-amplitude that can input high kinetic energy to make the flow become supercritical flow. The second one is the forcing frequency, which should be near resonance but not in resonance. This can create two waves to confront each other. In the benchmark tests, the frequency ratio is from 0.6 to 0.85, and the rolling amplitude must be large enough to make F_r>1.0.

Froude number implies the magnitude of the hydraulic jump and its damping mechanism. This depends on not only the oscillatory frequency but also the rolling amplitude. Overall they are positively correlated but strictly speaking, they are nonlinear. In this paper, BEM is accurate enough for the weak jump simulation (F_r<2.5). For the oscillatory jump case (F_r>2.5), it is somewhat difficult to describe local phenomenon such as the breaking waves, and the linear damping mechanism uniformly applied on the free surface is not enough. This may over-estimate the fluid velocity and lead to numerical instability. When breaking waves occur, BEM need more treatment to simulate the local free-surface deformation and its damping effect. When the high-speed jet forms on the walls, it may encounter the numerical instability. In this case, VOF could be another option for numerical simulation.

The sloshing force mainly comes from the inertia force of fluid, so it could be practical to truncate the small jets on the walls since it only has little effect on the dynamic analysis. In addition, this can improve the numerical stability. However, the damping effect of the jets by BEM is not discussed and ignored in this paper.

BEM is very computationally efficient for its simple algorithm and less computational cost, even more nodes may be needed around the region of the jets.

ACKNOWLEDGEMENTS

The authors sincerely appreciate the financial support sponsored by the Ministry of Science and Technology of Taiwan under the Grant 108-2917-I-564-037.

REFERENCES

[1] Graham E.W., Rodriguez A.M. (1951). The characteristics of fuel motion which affect airplane dynamics. Defense Technical Information Center.

[2] Housner G.W. (1957). Dynamic pressure on accelerated containers. *Bulletin of the Seismological Society of America*, 47, 15-35.

[3] Veletsos A.S. (1984). Seismic response and design of liquid storage tanks. *Guidelines for the Seismic Design of Oil and Gas Pipeline Systems*, 255-370.

[4] Peek R., Jennings P.C. (1988). Simplified analysis of unanchored tanks. *Journal of Earthquake Engineering and Structural Dynamics*, 16, 1073-1085.

[5] Ibrahim R.A., Pilipchuk V.N., Ikeda T. (2001). Recent advances in liquid sloshing dynamics. *Applied Mechanics Reviews*, 54, 133-199.

[6] Li Y.C., Wang J.T. (2012). A supplementary exact solution of equivalent mechanical model for sloshing fluid in a rectangular tank. *Journal of Fluids and Structures*, 31, 147-151.

[7] *API 650: welded steel tanks for oil storage*. (2007) American Petroleum Institution.

[8] Chen, Y.H., Hwang, W.S. and Tsao, W.H. (2017). Nonlinear sloshing analysis by regularized boundary integral method. *Journal of Engineering Mechanics*, 143(8), 040170046.

[9] Amano K., Iwano R., Sibata Y. (1993). Three-dimensional analysis method for sloshing behavior and its application to FBRs. *Nuclear Engineering and Design*, 140, 297-308.

[10] Mikelis N.E., Journee J.M.J. (1984). Experimental and numerical simulations of sloshing behaviour in liquid cargo tanks and of its effect on ship motions. *International Conference on Numerical Methods for Transient and Coupled Problems*.

[11] Liu D., Lin P. (2008). A numerical study of three-dimensional liquid sloshing in tanks. *Journal of Computational Physics*, 227, 3921-3939.

[12] Okamoto T., Kawahara M. (1992). Two-dimensional sloshing analysis by the arbitrary Lagrange-Eulerian finite element method. *Structural Engineering and Earthquake Engineering*, 8, 207-216.

[13] Huang J.R., Li H.Q., Liu G.R., Liu M.B. (2012). An improved SPH method for modeling liquid sloshing dynamics. *Computers and Structures*, 100-101, 18-26.

[14] Yang X.F., Peng S.L., Liu M.B., Shao J.R. (2012). Numerical simulation of ballast water by SPH method. *International Journal of Computational Methods*, 9, 1240002-1.

[15] Atif M.M., Chi S.W. (2019). Evaluation of breaking wave effects in liquid sloshing problems: ANCF/SPH comparative study. *Nonlinear Dynamics*, 97, 45-62.

[16] Bauer H.F. (1984). Oscillations of immiscible liquids in a rectangular container: A new damper for excited structures. *Journal of Sound and Vibration*, 93, 117-133.

[17] Hanson R.D. (1973). Behavior of liquid-storage tanks, the Great Alaska Earthquake of 1964. *Proceedings of the National Academy of Science*, 7, 331-339.

[18] Malhotra P.K., Wenk T., Wieland M. (2000). Simple procedure for seismic analysis of liquid-storage tanks. *Structural Engineering International*, 197-201.

[19] Wang X., Arai M. (2011). A Study on Coupling Effect Between Seakeeping and Sloshing for Membrane-Type LNG Carrier. *International Journal of Offshore and Polar Engineering*, 21, 256-263.

[20] Yu Y.M., Ma N., Fan S.M., Gua X.C. (2017). Experimental and numerical studies on sloshing in a membrane-type LNG tank with two floating plates. *Ocean Engineering*, 129, 217-227.

[21] Chen, Y.H., Hwang, W.S. and Tsao, W.H. (2018). Nonlinear Dynamic Characteristics of Rectangular and Cylindrical TLDs, *Journal of Engineering Mechanics*, 144, 06018004-1.

[22] Koh H.M., Kim J.K., Park J.H. (1998). Fluid-structure interaction analysis of 3-D rectangular tanks by a variationally coupled BEM-FEM and comparison with test results. *Earthquake Engineering & Structural Dynamics*, 27, 109-124.

[23] Fujino Y., Sun L.M., Pacheco B.M., Chaiseri P. (1992). Tuned liquid damper (TLD) for suppressing horizontal motion of structures. *Journal of Engineering Mechanics*, 118, 2017-2030.

[24] Sun L.M., Fujino Y. (1994). A Semi-analytical model for tuned liquid damper (TLD) with wave breaking. *Journal of Fluids and Structures*, 8, 471-488.

[25] Vinayan, V., Kinnas, S.A. (2010). Numerical nonlinear analysis of two-dimensional ventilating entry of surface-piercing hydrofoils with effects of gravity. *Journal of Fluid Mechanics*, 658, 383-408.

[26] Baer T.A., Cairncross R.A., Schunk P.R., Rao R.R., Sackinger P.A. (2000). A finite element method for free surface flows of incompressible fluids in three dimensions. Part II. Dynamic wetting lines. *International Journal for Numerical Methods in Fluids*, 33, 405-427.

[27] Walkley M.A., Gaskell P.H., Jimack P.K., Kelmanson M.A., Summers J.L. (2005). Finite Element Simulation of Three-Dimensional Free-Surface Flow Problems. *Journal of Scientific Computing*, 24, 147-162.

[28] Grotle E.L., Bihs H., Pedersen E., Æsøy V. (2016). CFD simulations of non-linear sloshing in a rotating rectangular tank using the level set method. *Proceedings of the ASME 2016 35th International Conference on Ocean, Offshore and Arctic Engineering*.

[29] Cariou A., Casella G. (1999). Liquid sloshing in ship tanks: a comparative study of numerical simulation. *Marine Structures*, 12, 183-198.

[30] Kim Y. (2001). Numerical simulation of sloshing flows with impact load. *Applied Ocean Research*, 23, 53-62.

[31] Kim Y., Shin Y.S., Lee K.H. (2004). Numerical study on slosh-induced impact pressures on three-dimensional prismatic tanks. *Applied Ocean Research*, 26, 213-226.

[32] Nasar T., Sannasiraj S.A., and Sundar V. (2009). Wave-induced sloshing pressure in a liquid tank under irregular waves. *Journal of Engineering for the Maritime Environment*, 223, 145-161.

[33] Hamlin N.A., Lou Y.K., Maclean W.M. (1986). Liquid sloshing in slack ship tanks-Theory, observations, and experiments. *SNAME Transactions*, 94, 159-195.

[34] Disimile P.J., Pyles J.M., Toy N. (2009). Hydraulic jump formation in water sloshing within an oscillating tank. *Journal of Aircraft*, 46, 549-556.

[35] Disimile P.J., Toy N. (2019). The imaging of fluid sloshing within a closed tank undergoing oscillations. *Results in Engineering*, 2, 100014.

[36] Wemmenhove R., Luppes R., Veldman A.E.P., Bunnik T. (2015). Numerical simulation of hydrodynamic wave loading by a compressible two-phase flow method. *Computers & Fluids*, 114, 218-231.

[37] Delorme L., Colagrossi A., Souto-Iglesias A., Zamora-Rodríguez R., Botia-Vera E. (2009). A set of canonical problems in sloshing. Part I: Pressure field in forced roll. Comparison between experimental results and SPH. *Ocean Engineering*, 36, 168-178.

[38] Machane R., Canot E. (1997). High-order schemes in boundary element methods for transient non-linear free surface problems. *International Journal for Numerical Methods in Fluids*, 24, 1049-1072.

[39] Zhao, R., Faltinsen, O. (1993). Water entry of two-dimensional bodies. *Journal of Fluid Mechanics*, 246, 593-612.

[40] Malenica, S., Zalar, M., Chen, X.B., (2003). Dynamic Coupling of Seakeeping and Sloshing. *In 13th International Offshore and Polar Engineering Conference*, 25-30.

[41] Zhao D., Hu Z., Chen G., Lim S., Wang S. (2018). Nonlinear sloshing in rectangular tanks under forced excitation. *International Journal of Naval Architecture and Ocean Engineering*, 10, 545-565.

[42] Currie I.G. (2003) *Fundamental mechanics of fluids*. 3rd edition. CRC Press.

[43] Berker A.A. (1992) *The boundary element method in engineering: a complete course*. 1st edition. McGraw-Hill.

[44] Kim, M.H., Hong, S. Y. (2000). Nonlinear wave forces on a stationary vertical cylinder by HOBEM-NWT. *In Proc. 7th International Offshore and Polar Engineering Conference*, 3, 209-214.

[45] Anderson D.A., Tannehill J.C., Pletcher R.H. (1984). *Computational fluid mechanics and heat transfer*. New York, McGraw-Hill.

[46] Dobrovol'skaya, Z.N. (1969). On some problems of similarity flow of fluid with a free surface. *Journal of Fluid Mechanics*, 36, 805-829.

[47] Defina A. Susin F.M. (2003). Stability of a stationary hydraulic jump in an upward sloping channel. *Physics of Fluids*, 15, 3883-3885.

SNAME
THE INTERNATIONAL COMMUNITY
FOR MARITIME AND OCEAN PROFESSIONALS

THE DESIGN OF THE DEEP WATER JET SLED

CLYDE O. COMBS
RETIRED, CO-OWNER, PROJECT ENGINEER, SMC MCEVER ENGINEERING

Proceedings of the 25th Offshore Symposium, February 27th 2020, Houston, Texas
Texas Section of the Society of Naval Architects and Marine Engineers

ABSTRACT

The deepwater jet sled was first developed to be used in laying and burying pipelines in ocean depths up to 1000 feet in Norwegian waters. Such ocean depths required a paradigm change in jet sled design in order to dig the trenches and lay the pipeline. The BC-500 pipe-bury jet sled was designed in 1972 for Santa Fe International, Inc. and was first operational to transport oil from the Ekofisk Oil Field to a refinery 300 miles away.

This paper presents an overview about the BC-500 jet sled as a new design concept for Santa Fe International, Inc . The design includes an eductor system within the jet sled instead of the air pressure designs which were in use in the waters up to 50 feet in the Gulf of Mexico. The conceptual design drawings were submitted to Texas A&M University to build a scale model to prove the feasibility of the eductor system for trenching. Time was a critical factor, to support Santa Fe International in transportation of oil from the newly found Ekofisk oil field for production in 1973.

Keywords: SNAME, Offshore Symposium, Jet, Sled, BC-500

INTRODUCTION

The discovery of crude oil on the planet Earth in Titusville, Pennsylvania in 1859 was an event that changed the economy of the world, precipitating the Industrial Revolution in the United Stated of America. This was followed shortly, at Spindletop, Texas, by massive amounts of oil gushing uncontrolled from wells near Beaumont, Texas.

Years later, in 1969, and of equal significance for the country of Norway was the discovery of crude oil off the coast of Norway in the Norwegian Sea. Again, the economy of a nation has been directly affected. Up until 1959, Norway had been totally dependent on imported oil, gasoline and other oil related products. All of that was about to change.

The discovery of oil in the Norwegian Sea in the waters off Norway, at a sea depth greater than 500 feet, presented a major challenge in transporting the crude oil to land-based refineries for processing. The Ekofisk Field oil was to be sent by subsea pipeline (30" diameter with 8" of concrete coating) to Scotland, some 300 miles distant. Norwegian regulations specified that the top of the buried line must be 10' feet below the sea bottom, or mud line, due to strong ocean currents. Due to the immensity of waves and weather conditions in the Norwegian Sea, pipe laying could only be done from April to October. This meant that all design testing and fabrication of the jet sled had to be completed and in operation by April 1, 1973.

With no petrol available under the occupation, cars ran on gas generated by wood or carbide burners mounted on the back.

Figure 1: Vehicles in Norway during the German occupation of Norway during WWII. Wood or carbide burners are mounted on the back.

BACKGROUND

The first contract for a pipeline to shore was awarded to Santa Fe International Corporation in early 1972. Subsequently, the contract for the design of a pipe bury sled was awarded to McEver Engineering, Inc., an engineering/construction company The design of the jet sled was dependent on previous experience with design for offshore platforms and underwater pipelines in the Gulf of Mexico,. It was named “BC-500”; the number, “500”, referred to water depth in the area of the Ekofisk oil field. The original design drawings are dated 1972, and can be found below in Figure 2.

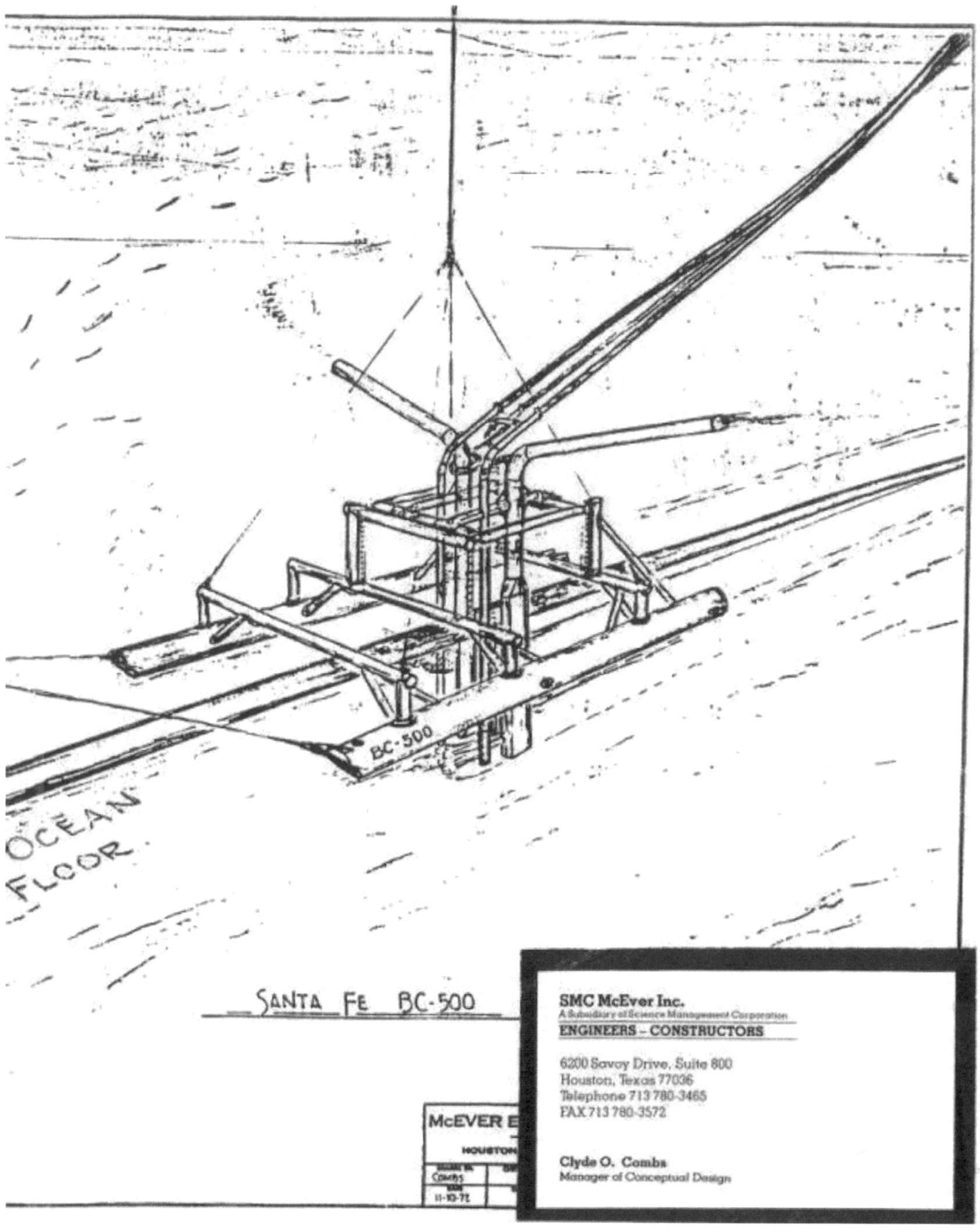

Figure 2: Conceptual drawings of pipe bury sled, BC-500

DESIGN OF THE JET-SLED

The design concept is based on using high pressure water from pumps on a barge on the surface of the sea to the "jet sled" located at the sea floor. Here this water is used for two purposes: 1) The first use is for cutting a trench with water jet nozzles, fourteen feet deep in the sea floor t bury the 36" diameter crude oil pipeline. 2) The second use for the water is to remove these cuttings by means of a vacuum lifting system and dispersing the loosened debris out of the pipeline route. Normal ocean currents would later cover the pipeline to previous water levels. This water eductor system described was designed to work with the other pipelay technology prior to 1972. Ekofisk was the first oil field to be discovered in that region. The original design sketches are figures 3 and 4 below:

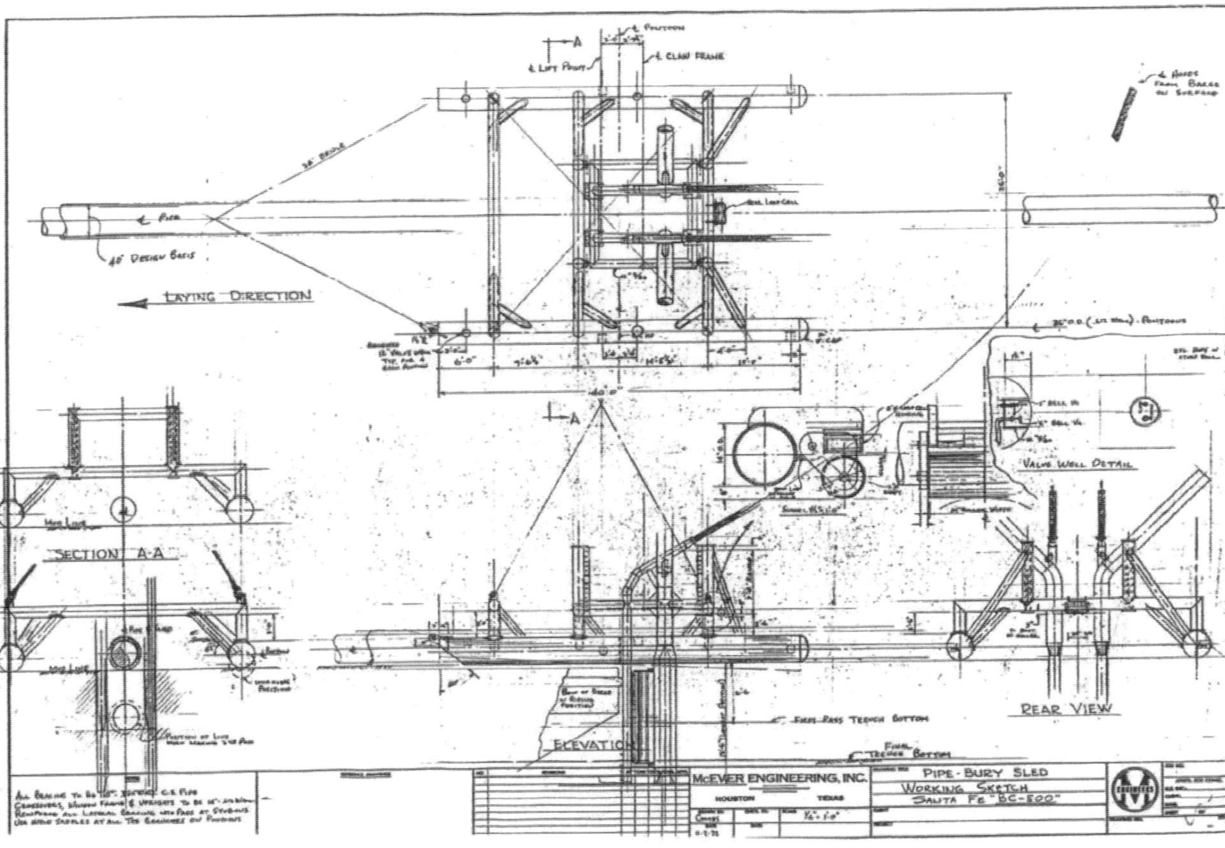

Figure 3: conceptual design drawings of the deep-water Pipe-Bury Sled; Original sketch

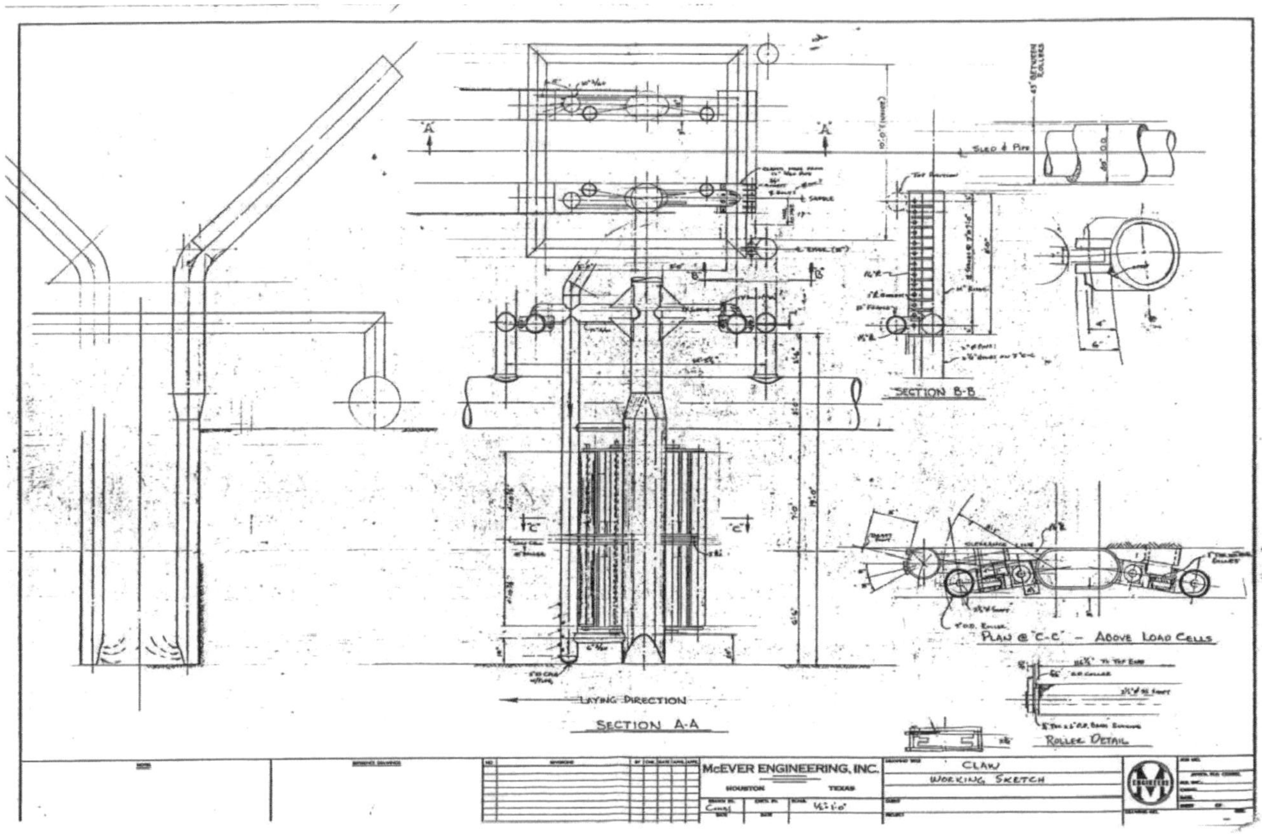

Figure 4: Scanned image from Clyde Combs' copy of the "Claw portion" details of the Pipe-Bury Sled, Sketch.

A 1:10 scaled model of the deep-water jet sled was built and tested at Texas A&M University to prove the design and assure its operation. The model was then sent to Norway with the fabrication drawings to be produced. The original model of the jet sled has been on display in the Norwegian Petroleum Museum up to recent years.

OPERATION OF THE BC-500

The actual pipe laying and pipe burying was done in two phases:

1. The first phase consisted of a 400-foot long "pipe lay barge" on the surface to weld together the 80-foot joints of 30" diameter steel pipe and coat them with an 8" thickness of concrete. The completed pipe line was then lowered to the sea bottom gradually as the barge was towed along its path, and settled on the mud line, or sea bottom.
2. The second phase consisted of another 400-foot long "pipe bury barge" (the "Cherokee"), which was the home for the jet sled, with crane for raising or lowering it to the sea bottom as necessary. Also on this barge deck were 5 large high pressure water pumps for pumping sea water to the jet sled educators which in turn cut the trench in the sea floor for burying the pipeline. See Figure 5.

Deep sea divers were in compression chambers on the deck of the pipe bury barge to be available to go down and position the jet sled over the pipeline initially and to be available for any problems. The jet sled for the BC-500 had to be towed or pulled over the pipeline by the barge on the surface.

Figure 5: Artistic rendering of a pipe-bury barge in conjunction with the deep-water jet sled

CONCLUSION

The jet sled was completed on schedule and successful performance enabled Santa Fe International Corporation to be the first and only company to complete a deep-water pipeline from the Ekofisk Field to a refinery which, in this case, was in Scotland, 300 miles away. The subsequent use of the jet sled model made for the Santa Fe project was not reported to the original design company. Today, Ekofisk oil field is publicly reported to be the first oil field off Norwegian shores.

Today's versions of the jet sled are of similar design, using the 2-stage "high pressure jetting & Airlifting" technology. However, so many improvements have been made due to technology and more, many companies now offer more advanced pipeline burial and excavation to the market.

Figure 6: Photograph of a self-propelled pipe-bury sled

ACKNOWLEDGEMENTS

Figure 1: Norway and the Second World War, by Johs, Andenaes, Olav Riste, Magne Skodvin; 1996, p.65

Figure 2: Image of the conceptual drawing of the piper bury sled, BC-500; designed and drawn by Clyde Combs, 11/10/1972; submitted to Santa Fe International by McEver Engineering, Inc., 1972.

Figure 3: Scanned image of an artistic rendering of a pipe-bury barge at work in conjunction with the deep-water jet sled just 3 years later. Santa Fe International brochure, post 1976

Figure 4: Scanned image from Clyde Combs' copy of conceptual design drawings of the deep-water Pipe-Bury Sled; Working Sketch; 1972.

Figure 5: Scanned image from Clyde Combs' copy of the "Claw portion" details of the Pipe-Bury Sled, Sketch.

Figure 6: Scanned image of photograph of a self-propelled pipe-bury sled displayed in Galveston's Ocean Star Offshore Drilling Rig & Museum.

SNAME
THE INTERNATIONAL COMMUNITY
FOR MARITIME AND OCEAN PROFESSIONALS

COMMERCIAL EXPLOITATION AND DE-RISKING DEEPWATER DEVELOPMENTS WITH EARLY PRODUCTION SYSTEMS: BRIEF HISTORY AND ROADMAP FOR PLATFORM SELECTION

RICHARD D'SOUZA
PRESIDENT – RICHARD B OFFSHORE

SHILADITYA BASU
TECHNICAL ADVISOR – KBR

Proceedings of the 25th Offshore Symposium, February 27th 2020, Houston, Texas
Texas Section of the Society of Naval Architects and Marine Engineers

ABSTRACT

In the early years of the offshore industry, Early Production Systems or EPSs were extensively utilized to develop marginal and/or complex reservoirs in progressively deeper and more remote waters. Their signature attributes were low capital and operating costs, simple designs and accelerated schedules to first oil. Today, Operators are emphasizing capital efficiency, design simplicity, compressed discovery to first oil cycle times and de-risking subsurface uncertainty as deepwater projects compete for capital allocation with onshore shale projects.

Discussing history and applications, the authors show, how an EPS can enable an Operator to sanction developments of marginal reservoirs, mid-size reservoirs with expansion capabilities to capture reservoir upsides and phased developments of giant reservoirs.

This paper addresses major subsurface uncertainties impacting development decisions and strategies to gather relevant dynamic information to mitigate risk. It provides a brief history of EPSs deployed in the North Sea, Brazil and GoM including a database of EPS platforms based on an extensive literature review. A case history in each region to demonstrate the utility of an EPS to de-risk and enable commercial production of marginal, mid-size and large fields.

A discussion is presented for EPS platform selection based on the research of platforms deployed in the three major deepwater regions. This discussion will facilitate to develop a roadmap for the Operators and Development Planners with tools to rapidly deselect or retain options in the early development planning stage while there is a high degree of reservoir uncertainty and pressure to compress cycle time to first oil following a discovery.

Keywords: **Reservoir de-risking, Deepwater, Extended Well Test, Early Production Systems, Phased Field Development, *SNAME, Offshore Symposium,***

INTRODUCTION

All field development decisions are made with the knowledge that there is a probability that the actual return on investment of a sanctioned development may not deliver expected returns. An Operator has to manage and mitigate project uncertainties and maximize the probability of success. Perhaps the single largest risk to the success or failure of a project is the subsurface uncertainty related to geologic complexity, reservoir and well productivity that are the main predictors of the two project critical success factors: expected well rates and ultimate recovery.

The risks associated with subsurface uncertainty are magnified in ultra-deepwater developments by the high cost of drilling and completing wells and the large capital investments required for the subsea infrastructure and floating production platform. There is also the long cycle time from project sanction (when large investments are made) and plateau production (when revenue is realized). The attraction for Operator's to develop deepwater fields is that a well-executed project that performs as well as or exceeds expectations at sanction, provides a very attractive return on investment. The challenge is that with oil prices expected to remain between $50/bbl and $60/bbl for a long while, Operators have low risk tolerance for underperforming deepwater projects. As a consequence, Operators are strongly emphasizing gathering quality subsurface and well performance information before sanctioning a project to reduce major risks and optimize the development.

There are several strategies available to an Operator for gathering requisite subsurface information following a discovery. The most common is to drill and stem test appraisal or delineation wells. But this still leaves a large uncertainty in long term well performance and measures to maximize per well recovery (completions and enhanced recovery). This information is best obtained by Extended Well Testing (EWT) or Early Production Systems (EPS) producing from single or multiple wells for extended durations. However these strategies require significant capital investment and dilute project returns by delaying a go-now decision. This, in short, is the Value of Information (VOI) conundrum that Operators must confront and manage.

In the following section an overview of processes and strategies for managing subsurface uncertainties is presented. The evolution of EPSs performed in the North Sea, Brazil and GOM with selected case histories in each region is presented next with a selected database of EPS platforms. A discussion on EPS platform selection factors are presented to facilitate the development of a roadmap unique for each Operator. The conclusion section summarizes the paper. Acknowledgements, Abbreviations and References section containing the reference list of relevant industry publications that were used to prepare this work round out the paper.

PROCESSES AND STRATEGIES FOR MANAGING SUBSURFACE UNCERTAINTIES

Commercial risks for frontier and offshore ultra-deepwater developments are significantly greater than developments in mature regions. In the past twenty years the E&P industry's Return on Investment (ROI) has averaged below seven percent (Narayanan et al., 2003) despite the fact that individual projects rarely receive committee sanctions unless the expected return is above a "hurdle rate", often 18 percent or more. The discrepancy between expected and actual returns can be traced to decision analysis process that fails to adequately address multiple, interdependent and correlated risks and uncertainties.

There are many risks that impact a successful project outcome (technical, commercial, project execution, operational) and the ones with greatest impacts have to be identified and managed. In frontier and ultra-deepwater developments the greatest risk is related to the uncertainty in reservoir characterization, its effects on production forecasting and ultimately the facility design. Failure of projects to meet their production and economic objectives can be attributed largely to improper or inadequate management of subsurface uncertainty in the field development plan and facility design. Uncertainty leads to risk and opportunity. Managing uncertainty and mitigating risk requires investment. An Operator, in early project phases, is faced with the decision of investing to reduce subsurface uncertainty to help decide whether or not to proceed with the development.

MAJOR SUBSURFACE UNCERTAINTIES

Subsurface uncertainties can be characterized as Static and Dynamic. Reservoir static uncertainties that significantly affect project NPV relate to physical properties of the reservoir, such as reservoir vertical and area limits; faults and transmissibility of the rocks; initial fluid characterization and rock properties (porosity, permeability, and productivity index) etc. On the other hand, reservoir dynamic uncertainties that significantly impact project NPV include: dynamic reservoir fluid behavior and distribution; secondary (or enhanced) recovery performance and wellhead pressure and flow rates.

Static uncertainties can be reduced by getting additional seismic data or drilling appraisal wells. Reducing dynamic uncertainty requires long term well flow.

STRATEGIES TO GATHER CRITICAL SUBSURFACE INFORMATION

Industry has several strategies in its toolkit to gather reservoir static and dynamic information to improve reservoir, geologic and depletion models critical to establish key field development parameters such as: number and location of production and injection wells, well construction (drilling and completions), well production profiles and expected ultimate recovery, impact of enhanced recovery options (water injection, gas injecting, water alternating gas injection (WAG), downhole, mudline or subsea boosting).

In ultra-deepwater one of the biggest cost drivers is the cost of drilling and completing a production well. Reducing uncertainties of key parameters in subsurface models will increase the fidelity of the models and confidence in estimating well rates and ultimate recovery. It also enables optimizing the development plan by increasing recovery factors while reducing well count.

EARLY PRODUCTION SYSTEM (EPS)

Well dynamic information is obtained by EWT and/or EPS (also referred to as Production Pilots in Brazil). Simply stated, an EWT is a production system (generally leased/chartered) that operates a single producing well for a period (typically ranging from 3 to 6 months) to acquire critical reservoir dynamic information, including time varying well rates, downhole and wellhead pressure and temperature, well pressure declines and production behavior.

An EPS or Pilot production system consists of a number of production wells (and occasionally injection wells) tied to a floating production platform, either leased or owned, for durations of 5 to 8 years. Associated gas is either monetized or injected. It is designed to investigate and gather a wider range of reservoir and well performance data than a EWT, with acquired information used to continually update and improve subsurface models. The data enables the subsurface team to optimize the depletion models and improve long term well productivity and ultimate recovery estimates. EPSs are also used to:

- Prove up enabling/enhancing subsea and platform technologies
- Test performance of various enhanced oil recovery strategies (water, gas, WAG injection, downhole ESPs, subsea boosting)
- Test relative efficiencies of various well construction (horizontal wells) and completions (Intelligent Completions)
- Evaluate effectiveness of stimulation treatments

The EPS requires considerable capital investment but is designed so that the expected revenue stream from produced oil and gas will generate an acceptable return for the Operator. Well performance and subsurface information gathered by the EPS is used to determine whether or not to proceed with a Full Field Development (FFD) and to optimize the development if it proceeds to sanction (Final Investment Decision/FID). An EPS can also be regarded as Phase 1 of a phased development strategy.

EVOLUTION AND SELECTED CASE HISTORIES OF EARLY PRODUCTION SYSTEMS

While one of the objectives is to supplement subsurface information obtained from 3D seismic and appraisal wells to optimize the full field development, the other is to ensure that it can meet an Operators' commercial hurdles (NPV, IRR) in its own right, as it represents a major capital investment. The decision is further complicated by the fact that the EPS will dilute the NPV of the full field development versus a go now decision. For an asset team to justify an EPS to upper management on the basis that it could prevent sanctioning a project with a probability of failure is a challenging argument to make.

An EPS decision is closely tied to the size of field to be developed, as defined by a range of expected recoverables. For this paper we have bucketed ultra-deepwater field size into four categories: small, medium, large and giant with expected range of recoverable resources shown in Table 1.

Table 1 Four Categories of Discovery based on Expected Recovery (P50)

Field Size Category	Expected Range of Recoverable Resources (mmboe)
Small	Less than 100
Medium	100 – 250
Large	250 - 500
Giant	500 +

For a small or marginal field in ultra-deepwater, an EPS is the de-facto full field development. In this case the investment in acquiring subsurface information to reduce uncertainty is limited to 3D seismic data and one or two appraisal wells. A strategy to cope with the inherent uncertainty from this somewhat limited data acquisition is to design the EPS with the flexibility to respond to these uncertainties as they arise during development drilling or in the operational phase. Petrobras employs a "Value of Flexibility" strategy to manage and adapt to such uncertainties.

At the other end of the spectrum is the giant field, such as the ones discovered by Petrobras in the ultra-deepwater Campos and Santos basins. In such cases an Operator has significant leeway in investing in subsurface data acquisition to reduce uncertainties and optimize a full field development. A sequential or phased development is justified with a EWT, EPS (or Pilot) and the Definitive Production System (as Petrobras has successfully employed on the SBPSC). Petrobras sometimes fast tracks and overlaps the EWT, EPS and Definitive development phases to improve overall project profitability.

The Medium to Large fields present unique challenges. In the US GOM, Operators have rarely used EWTs and EPSs to manage subsurface uncertainties. The strategy has been to drill more appraisal wells, perform DST and acquire downhole data, use advanced seismic data processing and sophisticated subsurface earth and reservoir models. The commercial performance for many of projects sanctioned on this basis has been disappointing.

In the current low price environment (since 2015) an Operator will require robust justification before committing to the large capital investments required for a full field development for a deepwater development. Operators' have severely curtailed capital spending for upstream deepwater projects since 2015 and major deepwater capital projects have to compete for capital with low risk, low capital projects (such as onshore shale). A deepwater asset team has to make a strong case for investing in data acquisition via an EPS to reduce uncertainty and project risk.

NORTH SEA

Leased, small, fast-track EPSs have been successfully deployed to produce several small or marginal fields in Central North Sea. A sample is summarized in Table 2.

Table 2 Selected EPS in North Sea

Field	Operator	WD (m)	EPS Platform	Owner	Type	Station – keeping	Oil	
							Rat e	Stora ge
							(mb opd)	(mbb l)
Huntington	EON	90	Voyageur Spirit	Teekay	Sevan	Spread	30	240
Chestnut	Centrica	120	Hummingbird Spirit	Teekay	Sevan	Spread	25	300
Varg	Repsol	84	Petrojarl Varg	Teekay	FPSO	Int. Turret	60	470
Ettrick	Nexen	115	Aoka Mizu	Bluewater	FPSO	Disc. Turret	30	600

<u>Chestnut Field Development</u>

The Chestnut field is an example of a successful, innovative development of a marginal field. The field was discovered in 1986. The reservoir had high quality sands, with an estimated 40-60 mmbbls of oil in place. An appraisal well drilled in 1988 confirmed a small complex reservoir. Coupled with weak oil prices at the time the project was shelved. In 2001, a EWT was conducted to reduce uncertainty about connected volumes and flow performance via a horizontal well. The well flowed 1 million barrels to the Crystal Sea EWT platform, over a 125 day period with well rates up to 16,000 bopd. The well was suspended and fully completed as a future producer. The EWT narrowed uncertainty in recoverables to a range of 30-35 million STB and showed the benefit of water injection to enhance recovery.

Centrica (became Spirit Energy in 2017) took over operatorship and with combination of the following factors:

- Improving oil price
- New seismic data and interpretation
- Availability of a leased innovative cylindrical FPSO with flexibility for future expansion

They sanctioned the full field development in late 2005 and first oil was achieved in 2008. The Sevan Hummingbird Spirit platform produced 5.5 mmbbls in the first 20 months with good facility uptime and was decommissioned in 2017 (Figure 1).

The project confirmed the viability of the cylindrical FPSO. A Sevan FPSO was also used as an EPS on the Huntington field in 2013. Petrobras leased the Piranema Spirit Sevan FPSO as an EPS on the Piranema field in 1600 m water depth from 2007 – 2018.

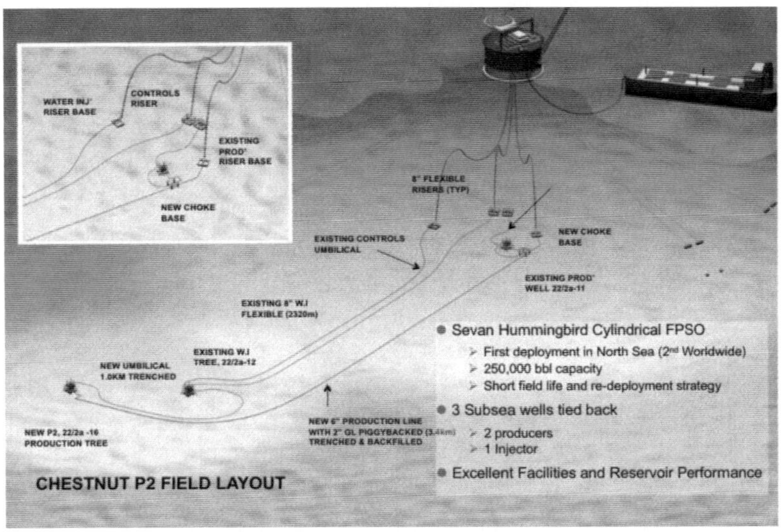

Figure 1 Chestnut Field EPS (Wood and Moore, 2009)

US GOM

Table 3 below presents selected field developments in the US GOM that have used and EPS either to produce a marginal field or a larger field with high subsurface uncertainty.

From published data, project execution and operational performance of Thunderhawk, Gomez and Phoenix EPSs were a success. The commercial performance of these projects is unknown.

Table 3 Selected EPS in GOM Deepwater

Field	Operator	WD (m)	EPS Platform	Owner	Type	Station – keeping	Oil	
							Rate	Storage

							(mbopd)	(mbbl)
Thunderhawk	Murphy	1850	Thunderhawk	SBM	Semisub	Spread	60	-
Gomez	ATP	914	Innovator	ArcLight	Semisub	Spread	30	-
Cascade Chinook	Petrobras	2600	BW Pioneer	BW Offshore	FPSO	Disc. Turret	80	520
Phoenix	Talos	600	Helix Producer	Helix	Shipshape FPU	DP-2	30	-

Cascade and Chinook EPS

The Cascade and Chinook fields are located in the Walker Ridge area in the Central US GOM with Petrobras Americas (PAI) as Operator. Cascade was discovered in 2002 and Chinook in 2003. Two appraisal wells were drilled on Cascade in 2005. They are Lower Tertiary reservoirs in 2600 m of water at a total vertical depth of approx. 7600 m. At the time of the development decision, there were no production analogs for these reservoirs.

Several appraisal and development options were assessed during the conceptual phase including

- An EWT (single well flow test)
- Drilling additional appraisal wells
- Early production system
- Full field development

The first two options were discarded as it was concluded that neither would provide enough information to define mid to long term reservoir and aquifer performance. The risks of proceeding to a full field development were considered too high, which led to the decision of a phased development plan with an EPS as Phase 1.

The initial phase of development was to produce from two subsea wells on Cascade and one on Chinook to an FPSO with a disconnectable riser and mooring turret. Produced oil was transported by shuttle tankers and gas by a pipeline. PAI chose the FPSO option partly because of lack of pipeline infrastructure in the vicinity and partly because of their considerable experience with FPSOs and shuttle tankers in deepwater Brazilian fields and their flexibility for future expansion. The disconnectable turret was chosen to facilitate regulatory approval by the MMS and USCG and provide flexibility for relocation of the FPSO in the event of an underperforming reservoir or replacement with the Phase 2 full field development if warranted. The EPS decision was based on mitigating the reservoir performance uncertainty and optimization of the future development phase. First oil was expected in the first quarter of 2010.

The FPSO contract was signed with BW Offshore in October 2007 for the provision of a converted double hull Aframax tanker with a disconnectable internal turret design to keep the FPSO on station in a 100-yr winter storm but disconnect ahead of an approaching hurricane and sail away under its own power. The disconnected buoy with attached risers and 11-point mooring would submerge to a pre-determined depth of 60 m. The FPSO topsides can process a maximum of 80,000 bopd and 16 mmscfd of gas along with power to drive subsea pumps. The subsea infrastructure was designed to accommodate 8 wells and 25,000 bopd per flowline, connected by two Free Standing Hybrid Risers (FSHR) to the turret. One of the FSHRs collapsed on March 2011 before start of production as a result of a failure in the chain tether connecting the top of the rigid riser to the tensioning buoyancy cans. New chain tethers and dual back up synthetic tethers were installed on the replacement FSHR and all remaining risers.

The initial phase of the development began with Cascade wells starting production in February and December 2012 and one Chinook well in September 2012. The Cascade and Chinook EPS implemented a series of technologies new to the US GOM:

- First FPSO and deepest floater (at the time)
- First use of Jones Act compliant shuttle tankers
- Deepest and highest pressure rated FSHRs

- Deepest subsea boosting system
- First use of single trip multi-zone frack pack system in deep wells.

The EPS (Figure 2) is currently operational. In 2014 PAI began assessing the performance of the FPSO in Phase 1 and deciding on a development option for the next phase based on the information provided by the EPS. After assessing multiple options for Phase 2 – Return to BWO, Extend contract, Replace with full field development – PAI elected to exercise the option to extend the FPSO lease by three years to Q1 2020. Murphy Oil assumed the operatorship of Cascade/Chinook field upon completion of PAI's sale of ownership of their US GOM assets in April, 2019.

Figure 2 Cascade Chinook EPS, BW Pioneer

The implementation of so many new technologies and firsts entailed significant risks. However, PAI in doing so demonstrated the technical/regulatory feasibility of an FPSO in the US GOM and valuable lessons learned for future fast followers.

BRAZIL

Petrobras has pioneered strategies and technologies to develop their giant discoveries in the deep and ultra-deepwater Campos and Santos Basins. Campos Basin developments began with the giant Marlim and Albacora discoveries in 1984-1985 in water depths ranging from 250 m to 1800 m. The Santos Basin developments began with the giant presalt Tupi (now Lula) discovery in 2006. At the time these were frontier developments with very high uncertainties in subsurface characterization, well construction, subsea infrastructure and floating platform technologies – many without analogs or precedents. Spurred by the prospect of becoming a self-sufficient producer of oil and gas, Petrobras committed to a systematic program to develop these fields using Value of Information, Robustness and Flexibility strategies to manage risks and uncertainties while fast tracking the developments. Over time this evolved into a strategy of Phased developments – Pre Pilots (EWT), Pilots (EPS) and Definitive (full field) developments. This section provides a high level overview of the Campos and Santos giant field developments.

Campos Basin Developments: Table 4 summarize the progression of EPS developments on some of the giant discoveries in the Campos Basin that included Marlim, Albacora, Barracuda/Caratinga, Roncador and Jubarte. The progression is well summarized in Fraga, et al. (2003). The main project uncertainties were partially de-risked by using low budget pre Pilots and Pilot systems in the early phases converted from semisubmersible MODUs and tankers. The projects were used as a full scale laboratory for testing enhancing well construction and surface facility technologies. They experienced several failures in the process but used each failure as a learning experience to continually refine and improve the developments. The aggressive risk taking approach was enabled by the scale of the reservoirs. Major technology successes are documented in Fraga, et al. (2003).

Table 4 Selected EPS in Campos Basin

	Caratinga	Albacora Leste	Roncador Module 1A	Marlim Leste	Jubarte	Jabuti	Barracuda
Development Phase	Pilot	Pilot	Pilot	Pilot	Pilot	Pilot	
Discovery Year	1994	1986	1996	1987	2001	2005	1989
First Oil Year	1997	1998	2001	2000	2002	2008	1997
Duration (Months)	72	36	24	60	3	5	72
Water Depth (m)	1000	342	1800	1200	1350	1420	800
Facility Name	*FPU 34*	*FPU 25*	*Seillean*	*FPU 26*	*Seillean*	*Seillean*	*FPU 34*
Facility Type	Conv FPSO	Conv SS	DP FPSO	Conv SS	DP FPSO	DP FPSO	Conv FPSO
# of Producer Wells	3	1	1	1	1	1	8
Peak Production (mbopd)	60	20	25	20	20	20	60

The risk taking and fast tracking approach led to some spectacular failures, the most notable were the March 31, 2001 sinking of the P-36 Pilot semisubmersible in the Roncador field and loss of life, and the near sinking of the P-34 FPSO (ex. PP Moraes) while producing on the Barracuda-Caratinga fields in 2002.

The phased development strategy is one in which capital exposure was minimized in early phases to provide information that reduced risk and allowed cash flow to finance subsequent phases. The Seillean DP FPSO and the turret moored PP Moraes (P-34) as an EWT/EPS to rapidly and cost effectively gather valuable well dynamic data were widely utilized, and set the stage for Pre-Pilot FPSOs deployed for reservoir information gathering on the more challenging SBPSC developments.

Santos Basin Pre-Salt Cluster (SBPSC) Developments: The phased development strategies and technologies and lessons learned on developing the giant Campos Basin were applied and improved upon with great success while developing the giant challenging, frontier presalt reservoirs in the SBPSC. Petrobras was able to install nine production FPSOs, achieving an average oil rate of 700,000 bopd and a cumulative production of 400 mmbbls of oil from 34 production wells, just 8 years from the first discovery well on the giant Lula field (Figure 3).

Fraga and Pinto, (2015) provide an excellent summary of the evolution of SBPSC developments from the 2006 discovery to 2015. It addresses the technical challenges faced in the early years: the heterogeneous nature of the carbonate reservoir, the 2000 m thick salt layers, the variable CO_2 content and compositional gradient of the reservoir fluids, flow assurance issues and special demands on subsea equipment, well construction and topsides. Development strategies were established and institutionalized: staged development based on EWTs, multi-well Pilot Systems (EPSs) and large Definitive Systems prioritizing standardization of wells, subsea infrastructure and production systems.

The use of EWT FPSOs Dynamic Producer and Cidade de Sao Vicente that relocated with relative ease between test locations, allowed gathering of valuable reservoir and well production dynamic data. This allowed the Operators to reduce risk and optimize the Pilot Production Systems or EPSs.

Moczydlower et al., (2012) discusses how the phased development evolved and the information gathering in each phase was used to reduce risks for subsequent phases. It also describes the institutionalization of the VOI, robustness and flexibility strategies employed to cope with critical uncertainties.

The Pilot Production Systems were spread or turret moored FPSOs with oil processing capacities between 100,000 and 120,000 bopd and water alternating gas injection capacities to enhance well rates from 4 or 5 production wells.

The Definitive Production Systems were "standardized" spread or turret moored FPSOs with oil processing capacity of 150,000 bopd. Andrade et al., (2015) describes the standardization of topsides on the Pilot and Definitive systems.

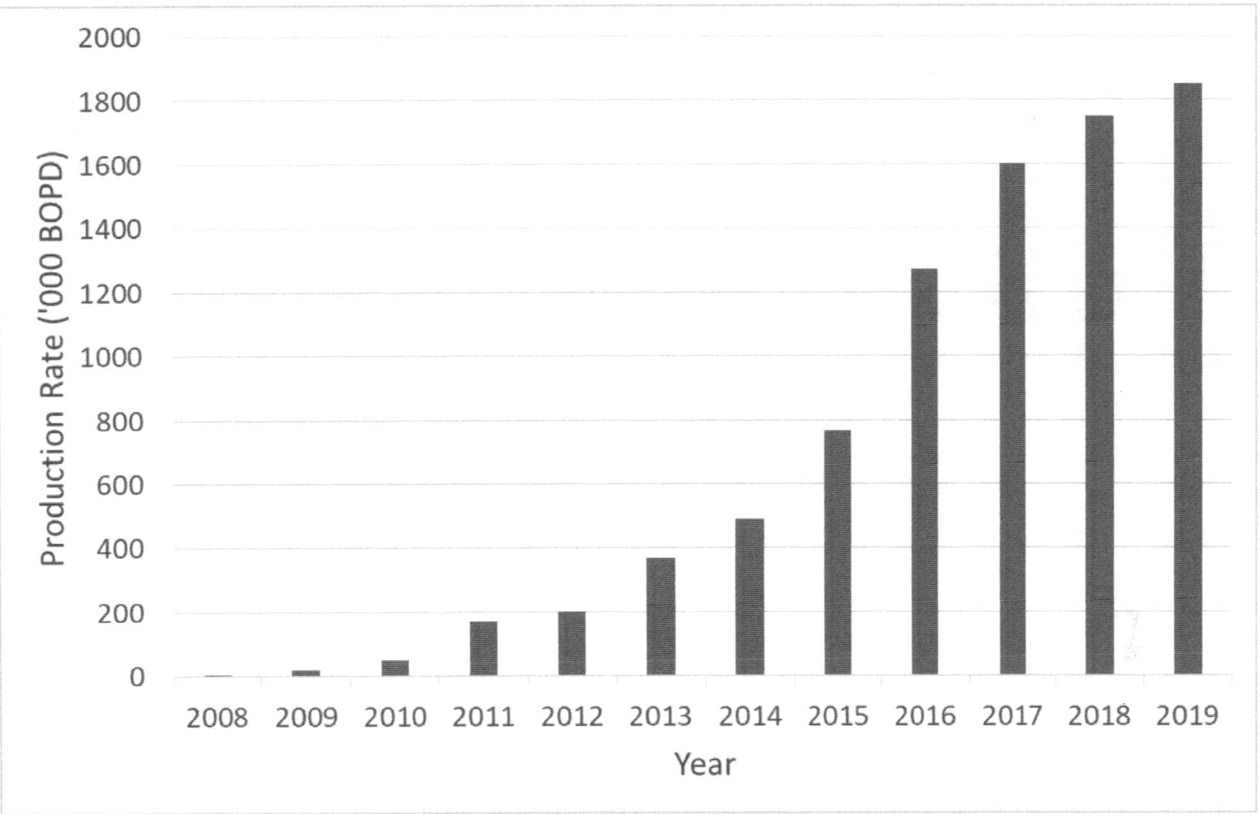

Figure 3 Oil Production in Santos Basin Pre-salt Cluster (Petersohn, 2019)

Cruz and Rosa (2016) and Nakano and Pinio (2009) provide excellent coverage of the progression of the EWT and Production Pilot on the Lula NE development from the planning phase to the operational phase, demonstrating the successful outcomes of the Petrobras' phased development strategy for the SBPSC ultra-deepwater developments.

Atlanta EPS

Situated about 185 km from Rio de Janeiro at 1550 m water depth, Atlanta is a heavy oil field (14° API grade). The reservoir is located about 800 m below seabed with a very low fracture gradient and temperature in the 40°C range. The first vertical appraisal well in Atlanta was drilled in 2001, followed by a horizontal well to test potential of the field in 2006. This test brought to light the challenges to be solved to produce this reservoir. The API density of the crude makes it extremely difficult to characterize the reservoir, designing drilling and completion programs, tackling flow assurance issues from the well head to the FPSO, designing processing scheme as well as offloading stored processed crude to the tankers.

In 2013 and 2014, Operator drilled two wells for EPS production, followed by a third one in 2019. The FPSO Petrojarl-1 (Figure 4) was commissioned and reached first oil in May 2018 with the target production rate expected in Q4, 2019.

Every aspect of the Atlanta field development required multiple innovative solutions. This case history is limited to the ones related to the platform and a couple of examples from other aspects.

To generate and maintain the high production rate, each well in Atlanta EPS has two ESPs, one inside the well and near the reservoir. The other at the mudline, mounted on a skid. Presence of two ESPs provided redundancy and reduced the need for maintenance/workover. These ESPs were also designed to operate on variable frequency range between 40 and 65 Hz. The intentionally designed "inefficient" ESPs would dissipate a significant amount of imputed power as heat thus heating the crude and alleviating some of the flow assurance issues.

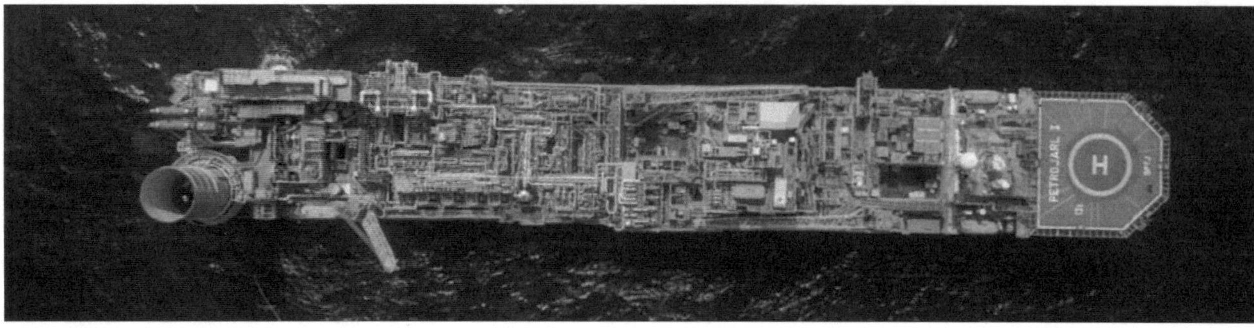

Figure 4 Petrojarl 1 operating in Atlanta (BS-4) Field (courtesy Enauta)

The FPSO also saw the use of the first drag chain system in Brazil for connecting high voltage electrical cables from the turret to the topsides power generation module via variable frequency drives. These were crucial to power the ESPs located in each well.

Converting the FPSO turned out to be quite challenging too. On one hand, the tight schedule and limited life span of the EPS meant maximizing the reuse of systems already installed on the FPSO. On the other hand, those original topsides equipment were not designed to process the heavy oil produced by Atlanta reservoir. The problem mainly arose due very low residence time in the 1st stage separator. To overcome that, a new Degasser vessel was installed with helical path design to enhance the gas/oil separation. The process philosophy was also updated and the crude was heated to help in separation.

The information currently generated by this EPS is being used to design and refine the Definitive Stage FPSO that will deal with a six-fold longer service life and substantially different water cut and fluid profile during that period (Rocha et al, 2019).

EPS Platform Selection Factors

Critical non-commercial selection factors for screening potential EPS platform types during Concept Selection are discussed below.

Sanction to First Oil Cycle Time: A key attribute for EPS selection is short cycle time. As the EPS is expected to last a very short duration compared to the overall life of the field, a concept that provides an earlier start to information gathering and revenue generation is desirable. This usually tilts the decision towards a conversion or an intercept (where typically a tanker already in fabrication is modified) rather than a purpose built platform.

Oil Export: In locations remote from existing infrastructure, such as a gathering pipeline network, platforms with integrated storage and direct offloading to shuttle tankers (FPSOs) will be preferable to semisubmersibles, which will require either an FSO or a pipeline. In the US GOM, depending on the location, it may be less of an issue if there is an accessible pipeline network. In Brazil, some infrastructure exist in the mature part of the Campos Basin and relatively shallower water. For the gigantic pre-salt reservoirs, remoteness and lack of any infrastructure self-selects FPSO systems rather than Semisubmersible and FSO combinations. North Sea, Far East and the Nigeria-Angola corridor also rely heavily on FPSOs. Any frontier basin is almost an automatic FPSO candidate as they lack any infrastructure at all.

Expandability: An EPS is sanctioned on the basis of limited or imperfect reservoir information and needs to be designed with flexibility to react to uncertainties as they are manifest during the operational phase. A concept that has greater adaptability to accommodate future expansion (topsides capacity expansion for additional subsea tiebacks or addition of enhanced recovery equipment) will be preferred.

Operational Risks and Complexities: The concepts selected to go forward should minimize operational risk and system complexities. For example, a disconnectable turret FPSO or DP FPSO has inherently greater operational risk and complexity related to disconnection, sail away and reconnection than a non-disconnectable or permanently moored platform. DP FPSOs also have a higher risk of brown out or drive off compared to a turret or permanently moored option. However, sometimes other considerations can push a development towards a particular option as discussed below.

Regulatory Approvals: This is an issue specific to the US GOM with its limited application of EPSs, particularly for non-disconnectable FPSOs. The Sevan FPSO or permanent turret moored FPSO would fall into this category if they were to be considered. US regulatory requirements for FPSOs stem from the FPSO EIS (USDOI, MMS, 2001) (EIS=Environmental Impact Statement). Though the EIS doesn't rule out a permanently moored FPSO solution, all FPSOs in the US GOM has been disconnectable solutions. As the first operational FPSO in US GOM selected a disconnectable solution, it set a precedence and became the de-facto standard.

Decommissioning: Decommissioning operations and associated costs in ultra-deepwater are not trivial. An EPS may be disconnected earlier than anticipated if the reservoir underperforms in which case it may have to be removed. Similarly a successful EPS will eventually have to be decommissioned and replaced with a larger full field production system. A disconnectable FPSO has some advantages in this regard over a permanent moored FPSO.

CONCLUSION

There is a history of Operators opting for EPSs, beginning in the mid 1980's in the North Sea and Brazil to gather critical subsurface information to increase confidence in making a final investment decision. This was especially the case for marginal, complex reservoirs in the shallow North Sea and giant reservoirs in deep and ultra-deep frontier (at the time of the discoveries) Campos and pre salt Santos Basins in Brazil. The conclusions from published literature researched and presented in this paper is that EPS projects have generally been successful in enabling Operators to make better decisions on whether (or not) to proceed with a full field development.

With the benefit of hindsight a strong case can be made that a phased development beginning with an EPS would have yielded significantly higher returns by minimizing subsurface risk and optimizing the full field development for most cases. A rigorous evaluation of these underperforming projects resulting from the discrepancy between actual and predicted well performance would bolster arguments in favor of conducting a EPS, especially for reservoirs lacking sufficient production analogs or with a high degree of subsurface uncertainty.

ACKNOWLEDGEMENTS

The authors acknowledge the support of KBR/Granherne management team for supporting this work.

LIST OF ABBREVIATIONS

bbl	:	barrels (of oil)
bopd	:	barrels of oil per day
BSSE	:	Bureau of Safety and Environmental Enforcement, previously MMS
DP	:	Dynamically Positioned
DST	:	Drill Stem Test
E&P	:	Exploration & Production
ESP(s)	:	Electric Submersible Pump(s)
FPSO	:	Floating, Production, Storage and Offloading
FPU	:	Floating Production Unit
GOM	:	Gulf of Mexico
IRR	:	Internal Rate of Return (on an investment)
mbopd	:	thousand barrels of oil per day
mmbbls	:	million barrels (of oil)
mmboe	:	million barrels of oil equivalent
MMS	:	Minerals Management Service, currently functions as BOEM and BSEE
mmscfd	:	million standard cubic feet (of gas) per day
MODU	:	Mobile Offshore Drilling Unit

NPV	:	Net Present Value
USCG	:	United States Coast Guard

REFERENCES

Andrade Ana Maria Teixeira., Vaz, Celio Eduardo Martins, Ribeiro, Jonatas, Lopreato, Luiz Guilherme Roquette and do Nascimento, Rafael Fernando Soares, "Offshore Production Units for Pre-Salt Projects," OTC 25691, 2015

Boschee P., "Managing Subsurface Uncertainties in Facilities Design," Oil and Gas Journal, 2013

Coopersmith E. M., Burkholder K., Mendoza L, "Early Production Systems – When do They Make Good Sense for Your Project," OTC – 25440, 2014

Cruz R. O., Rosa M. B., "Lula NE Pilot Project – An Ultradeep Success in Brazilian Pre-Salt," OTC 27297, 2016

Cullick A.S., Narayanan K, Gorell S.B., "Optimal Field Development Planning of Well Locations with Reservoir Uncertainty," SPE 96986, 2005

D'Souza, Basu and Fales, "Selecting the Right Field Development Plan for Global Deepwater Developments", DOT 2012

Duggal A., "Full Field and Early Production in Ultradeep waters: The 'DP-FPSO' and 'DP-EFPSO' Concepts," OTC 16710, 2004

Fraga C.T., Pinto A.C., "Brazilian Pre-Salt: An Impressive Journey from Plans and Challenges to Concrete Results," OTC 25710, 2015

Fraga C.T., et al., "Campos Basin – 25 years of Production and Its Contribution to the Oil Industry," OTC 15219, 2003

Gu, M, Gudmestad O, "Treatment of Uncertainties, Risks and Opportunities in Cost and Schedule Estimates in the Early Phase of Offshore Projects," ISOPE – I-12-662, 2012

Moczydlower B, et al, "Development of the Brazilian Pre-Salt Fields – When to Pay for Information and When to Pay for Flexibility," SPE 152860, 2012

Nakano C., Pinio A., "Pre-Salt Santos Basin – Extended Well Test and Production Pilot in the Tupi Area – The Planning Phase," OTC 19886, 2009

Narayanan et al, "Better field development decisions from multi-scenario, interdependent reservoir, well and facility simulations," SPE 79703, 2003

Palagi C., Ribeiro O., "The Development and Production of Cascade and Chinook Fields in the Gulf of Mexico," OTC 24156, 2013

Palagi C., Ribeiro O., "The Development of Cascade and Chinook Fields", IBP 3522-10, Rio Oil and Gas Expo, 2013

Petersohn, Eliane, "Pre-Salt Super Play: Leading Brazil into the World's Top 5 Oil Suppliers", Adapted from oral presentation given at AAPG Latin America and Caribbean Region Geoscience Technology Workshop, Rio de Janeiro, Brazil, June 13-14, 2019 DOI:10.1306/30625Petersohn2019

Robinson R., "The Economic Impact of Early Production Planning on Offshore Frontier Projects," Offshore Mediterranean Conference – 2009-024, 2009

Rocha, Paulo Sergio, Goulart, Roberto de Oliviera, Kawathekar, Satyajeet and Dotta, Raul, "Atlanta Field Development – Present and Future", OTC-29846, OTC Brasil, Rio de Janeiro, 2019

Saputelli L, et al., "Managing Risk and Uncertainty in the Visualization of Production Scenarios," SPE 107562, 2007

Surotsev et al., "Vaguely Right or Precisely Wrong: Making Probabilistic Cost, Time and Performance Estimates," SPE 181904, 2016

Valenchon C. P., "Early Production Systems in Ultra-deepwater, a Way to Improve Reservoir Management and Field Economics," SPE 6517, 2000

Wood P., Coghlan G., "Chestnut – A successful, innovative development of a marginal North Sea Field," SPE 134312, 2010

Wood Peter, and Moore Rachael, "Chestnut Field: Case Study of an Innovative Development", DEVEX Conference, Aberdeen, 2009

SNAME

THE INTERNATIONAL COMMUNITY FOR MARITIME AND OCEAN PROFESSIONALS

COUPLING OF VORTEX-LATTICE METHOD WITH RANS FOR THE PREDICTION OF OPEN OR DUCTED PROPELLER/AXISYMMETRIC HULL INTERACTION

KYUNGJUNG CHA
GRADUATE STUDENT, OCEAN ENGINEERING GROUP, THE UNIVERSITY OF TEXAS AT AUSTIN

SEUNGNAM KIM
GRADUATE STUDENT, OCEAN ENGINEERING GROUP, THE UNIVERSITY OF TEXAS AT AUSTIN

SPYROS A. KINNAS
PROFESSOR, OCEAN ENGINEERING GROUP, THE UNIVERSITY OF TEXAS AT AUSTIN

Proceedings of the 25th Offshore Symposium, February 27th 2020, Houston, Texas
Texas Section of the Society of Naval Architects and Marine Engineers

ABSTRACT

An iterative method which couples a vortex-lattice method with a Reynolds-Averaged Navier-Stokes (RANS) method is applied to the prediction of propeller/axisymmetric hull interaction. An open propeller and a ducted propeller are considered. In the coupling procedure, the effective wake is evaluated at the propeller control points. In RANS the propeller blades are represented with body forces distributed over the propeller zone. The hull resistance is evaluated in the presence of the propeller, and the propeller RPM is determined by equating the hull resistance with the propeller thrust. The present approach handles the propeller/hull interaction automatically, without the need for assumed values of the wake fraction and of the thrust deduction factor.

Keywords: Vortex-lattice method, Reynolds-Averaged Navier-Stokes solver, effective wake, hull-propeller interaction; ducted propeller

INTRODUCTION

Marine propellers are a basic element of any ocean vehicle. Their successful design is crucial to the overall performance and structural integrity of an ocean vehicle. Thus, prediction of the propeller performance and its interaction with the hull is essential.

Numerical methods for the analysis and design of marine propellers can be largely divided into four representative methods (Kerwin & Hadler, 2010; Kinnas, 2017): (a) Actuator disk and momentum theory in which the propeller is represented via a pressure jump across a disk; (b) Lifting line method in which the propeller blades are represented via their circulation distribution along a radial line; (c) Vortex-lattice method (VLM) which assumes that the blade thickness to chord ratio is small; and (d) Surface panel or boundary element method (BEM) in which the panels are placed on the exact blade surface, and thus makes no assumption on the magnitude of the blade thickness.

This study determines the propeller RPM for given body speed so that the hull resistance in the presence of propeller equals the propeller thrust. An open propeller and a ducted propeller are considered. The propeller/hull/duct interaction is determined automatically through an iterative procedure which couples a vortex-lattice method for the evaluation of the propeller performance with a RANS method, which evaluates the effective wake for the propeller, as well as the hull resistance in the presence of the propeller.

METHODOLOGY

Coupling method

An interface method, Potential Flow to Navier-Stokes (PF2NS), is used for application to axisymmetric body and propeller interaction via coupling the vortex-lattice method (VLM) with the Reynolds Averaged Navier-Stokes (RANS).

Figure 1 indicates the coupling algorithm for the involved. The interface tool includes two parts for coupling which are the body force and the effective wake (Tian *et al.*, 2013, Su & Kinnas, 2017, Du et al., 2018). First, the body force is calculated from the pressure distribution over the blades using the vortex-lattice method MPUF3A (with the effects of viscosity accounted for via a friction coefficient and a viscous pitch correction), and then this force is interpolated to the grid of RANS solver, which is the commercial code ANSYS/Fluent in this work. Furthermore, the effective wake, V_{eff}, is evaluated as:

$$\boldsymbol{V}_{eff} = \boldsymbol{V}_{total} - \boldsymbol{V}_{prop\_induced}$$

where V_{total} is total velocity obtained from the viscous fluid solver, and $V_{prop\_induced}$ is propeller induced velocity from the VLM. The iterative procedure continues till the propeller performance is converged.

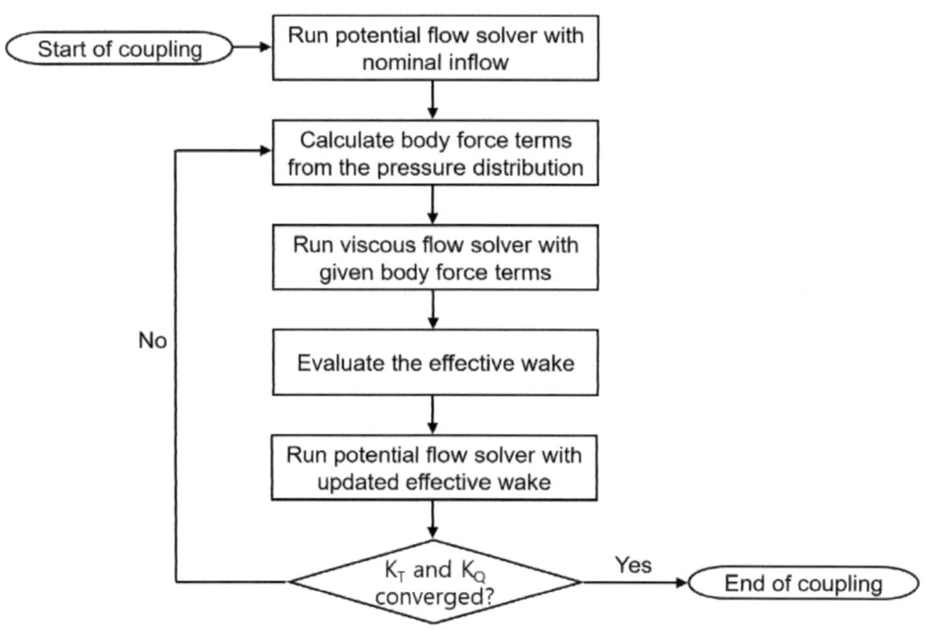

Figure 1: Coupling algorithm

RESULTS

Axisymmetric body and propeller

An axisymmetric streamlined body is considered in our study. The principal particulars of the body, propeller, and duct are laid out in Table 1 and the geometries are shown in Figure 2.

Speed of the body (V_S) is $1m/s$ and the Reynolds number (Re), based on the body length, is 1.42×10^7.

Table 1: Principal particulars of body, propeller, and duct

Body			
Length	L	$[m]$	14.2
Diameter	D	$[m]$	2.2
Propeller			
Diameter	D_P	$[m]$	2.0
Number of blades	Z		3
Pitch-diameter ratio at $0.7R_P$	P/D_P		1.2
Duct			
Diameter	D_D	$[m]$	2.0

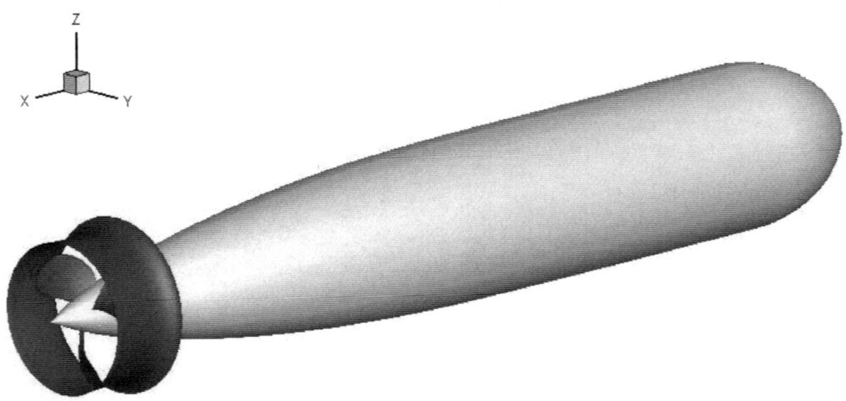

Figure 2: Geometry of body, propeller, and duct

The RANS model

The governing equations are the continuity and the momentum equations in RANS. The turbulence model applied is the SST $k - \omega$ model.

The computational domain and boundary conditions are shown in Figure 3. The size of the computational domain is $-2.5 < x/L < 2.5$ and $0 < y/L < 1.5$, and propeller is located at $x = 0$. Velocity inlet and pressure outlet boundary conditions are applied for the inlet and the outlet boundary plane, respectively. No-slip wall conditions are applied on the body surface and standard wall functions are utilized. Axis condition is applied on the axis of the body.

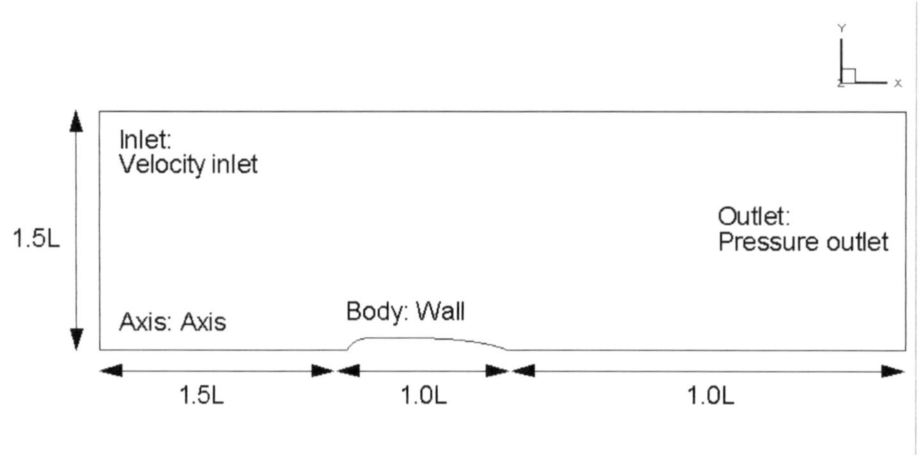

Figure 3: Computational domain and boundary conditions

Under-relaxation factors for pressure and momentum are 0.3 and 0.7, respectively. Convergence is decided based on that the residuals of the continuity equation, the velocity components (u, v, w), and k, and ω are under 10^{-6}.

The total number of grids is 90k for the open propeller case and 100k for the ducted propeller, as shown in Figure 4. Fine grids with $y^+ = 70$ are set near the body.

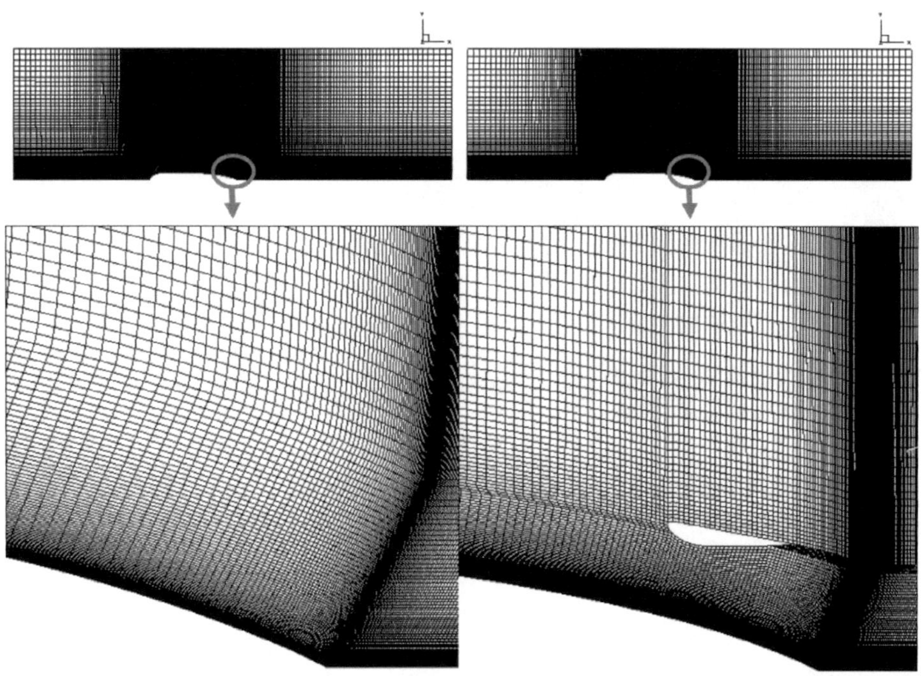

(a) Open propeller **(b) Ducted propeller**

Figure 4: Grid generation in the RANS model

Three grids, with approximately 60k, 90k, and 120k cells, are considered for the grid dependence study, in the case of the body with the open propeller. The study is performed when advance ratio is equal to 1.1. Table 2 offers the result for coarse, medium, and fine grid systems. Given that differences for the predicted hull resistance are under 1%, we perform all the rest of runs with the medium grid.

Table 2: Results for the grid dependence study J_S=1.1.

Grid system	Total grid number	R_P $[N]$
Coarser	60k	235.37
Medium	90k	235.97
Finer	120k	235.56

The coupling of VLM method with the RANS

Figure 5 shows the locations where the effective wake is evaluated in this study are inside the propeller zone, as the propeller control points in MPUF-3A. This approach improves the accuracy of the effective wake and the accuracy of the predicted performance, especially in the case of conical hubs and/or in the presence of a duct, as shown in Tian et al (2014).

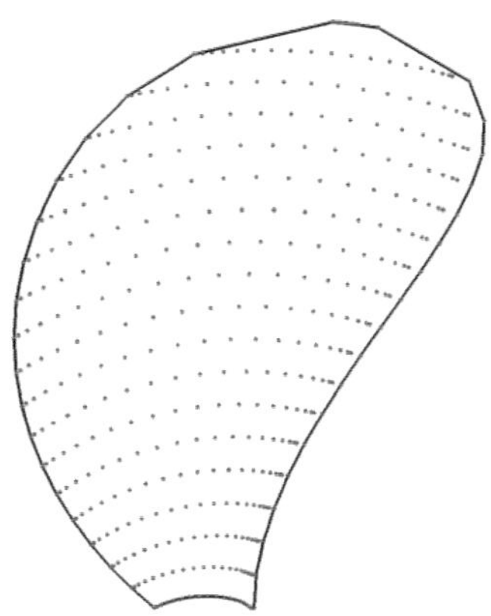

Figure 5: Points where the propeller effective wake is evaluated

Figure 6 displays the predicted hull resistance (R_P) and the propeller thrust (T_P) with number of iterations between MPUF-3A and RANS, for the open and ducted propeller at advance ratio (J_S = 1.1). The coupling procedures converge around the 6th iteration for the open propeller and the 7th iteration for the ducted propeller. It should be noted that the duct drag is also included in the presented hull resistance in the case of the ducted propeller.

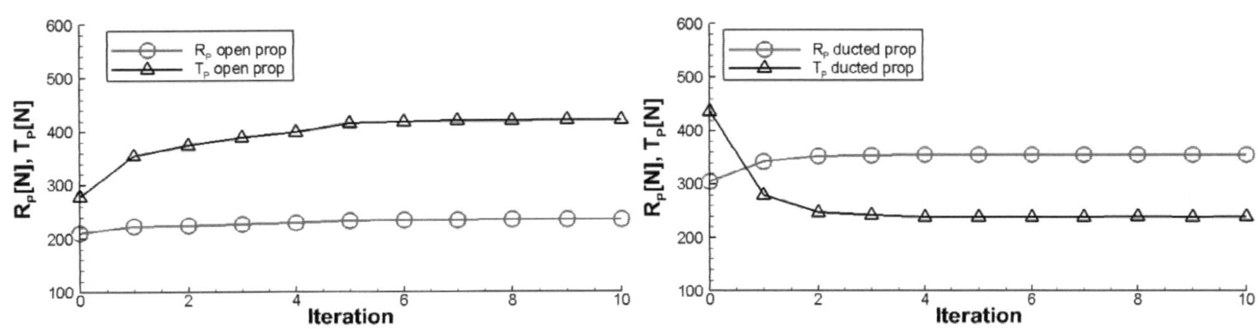

(a) Open propeller **(b) Ducted propeller**

Figure 6: Convergence history for the hull resistance (R_P) and the propeller thrust (T_P) during the coupling procedure

Prediction of propeller performance

Prediction of propeller performance includes factors that are the hull resistance with propeller, the propeller thrust, and the propeller torque. The propeller thrust and the torque are obtained from the thrust coefficient for propeller and the torque coefficient, respectively. The following equations show definitions of the two coefficients.

$$K_T = \frac{T}{\rho n^2 D_P^4}$$

$$K_Q = \frac{Q}{\rho n^2 D_P^5}$$

where ρ and n are fluid density and propeller rotational frequency.

Advance ratio is defined as the following equation.

$$J_S = \frac{V_S}{n D_P}$$

Figure 7 and Table 3 show results for prediction of propeller performance. The resistance for the ducted propeller includes that on the duct. Advance ratio from 0.8 to 1.5 was considered for the open propeller and that from 0.8 to 1.2 was taken into account for the ducted propeller. The resistance, the thrust, and the torque are decreased as increasing advance ratio for the open propeller. The thrust and the torque are more rapidly decreased than the resistance. However, for the ducted propeller, the resistance is increased with the decreased thrust and torque. In addition, it was found that the thrust and the torque of the open propeller is larger than those of the ducted propeller at the same advance ratio. The propeller rotational frequency for the given body speed is 23.4RPM so that the hull resistance equals the propeller thrust at 180.54N. Also, the ducted propeller rotational frequency is 28.5RPM and the forces are equal to 339.65N.

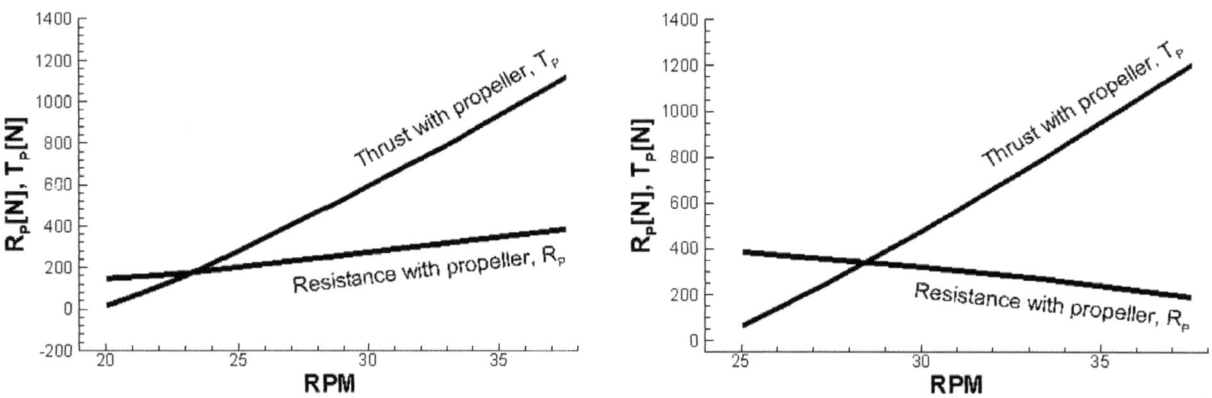

(a) Open propeller **(b) Ducted propeller**

Figure 7: Predicted propeller thrust and hull resistance vs. advance ratio. The intersection of the two curves determines the propeller RPM for which the hull resistance and the propeller thrust equal each other

Table 3: Prediction of propeller performance and hull resistance

Open propeller				
J_S	RPM	R_P $[N]$	T_P $[N]$	Q_P $[Nm]$
0.8	37.5	384.56	1113.28	448.85
0.9	33.3	320.60	811.80	336.72
1.0	30.0	272.85	591.28	254.53
1.1	27.3	235.97	422.73	191.33
1.2	25.0	203.28	278.44	136.86
1.3	23.1	176.10	161.42	92.24
1.4	21.4	157.73	78.08	60.76
1.5	20.0	143.36	11.14	35.64
Ducted propeller				
J_S	RPM	R_P $[N]$	T_P $[N]$	Q_P $[Nm]$
0.8	37.5	190.34	1202.71	489.97
0.9	33.3	270.82	783.27	333.28
1.0	30.0	319.43	472.28	217.92
1.1	27.3	354.88	239.74	132.15
1.2	25.0	383.33	59.46	65.53

The following results are comparisons of the predicted performance of the two propellers when the advance ratio is equal to 1.1 ($J_S = 1.1$). Figure 8 shows that the axial body force for the open propeller is higher than that for the ducted propeller.

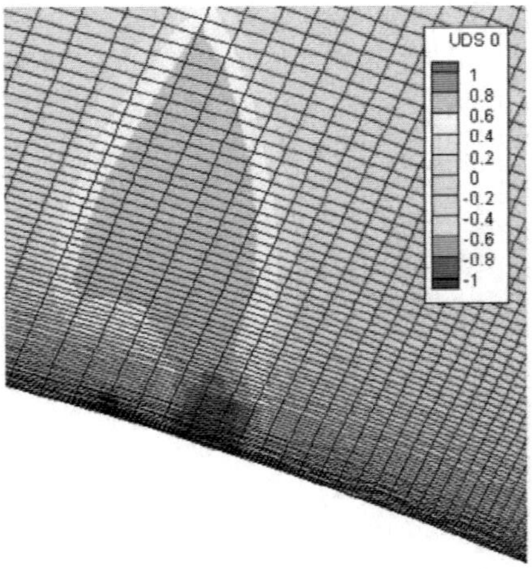

(a) Open propeller

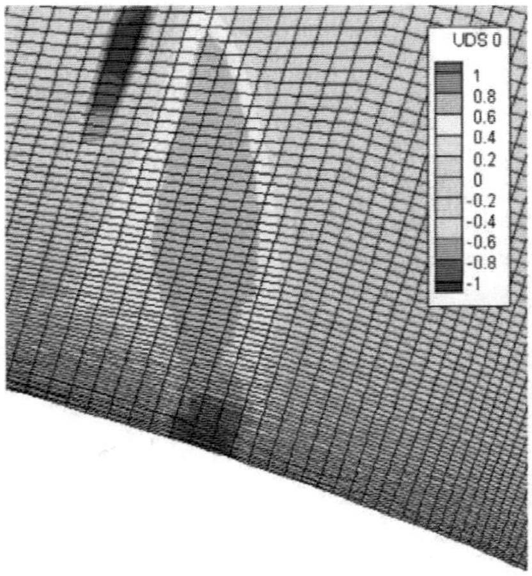

(b) Ducted propeller

Figure 8: Axial body force representing the propeller blade in RANS

The effective wake for the open propeller has lower values (i.e. slower velocities) than that for the ducted propeller, as shown in Figure 9, and this is consistent with the results shown in Figures 8 and 10. Essentially, the chosen duct in this study is an accelerating duct, for the given conditions.

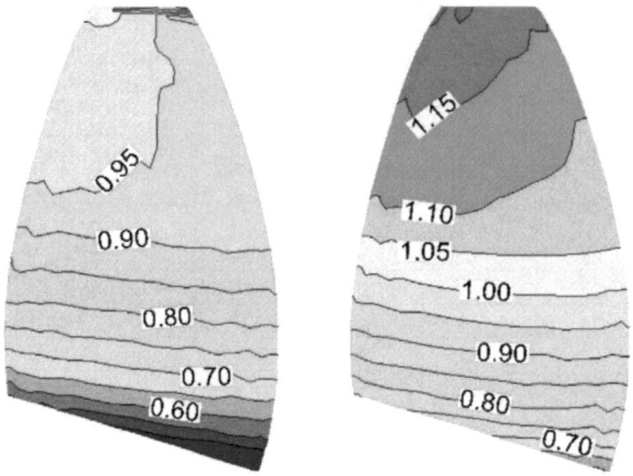

(a) Open propeller **(b) Ducted propeller**

Figure 9: The predicted effective wake shown over the propeller zones

Figure 10 shows the circulation for the propeller blades. The circulation values for the open propeller are higher than the values for the ducted propeller, and consequently the propeller thrust for the open propeller is higher than the thrust for the ducted propeller at the same advance ratio. It should be noted that the final converged circulation distributions have small negative values towards the tip sections, especially in the case of ducted propeller, and that is a consequence of the fact that the chosen propeller has NOT been designed properly to match the chosen hull and duct.

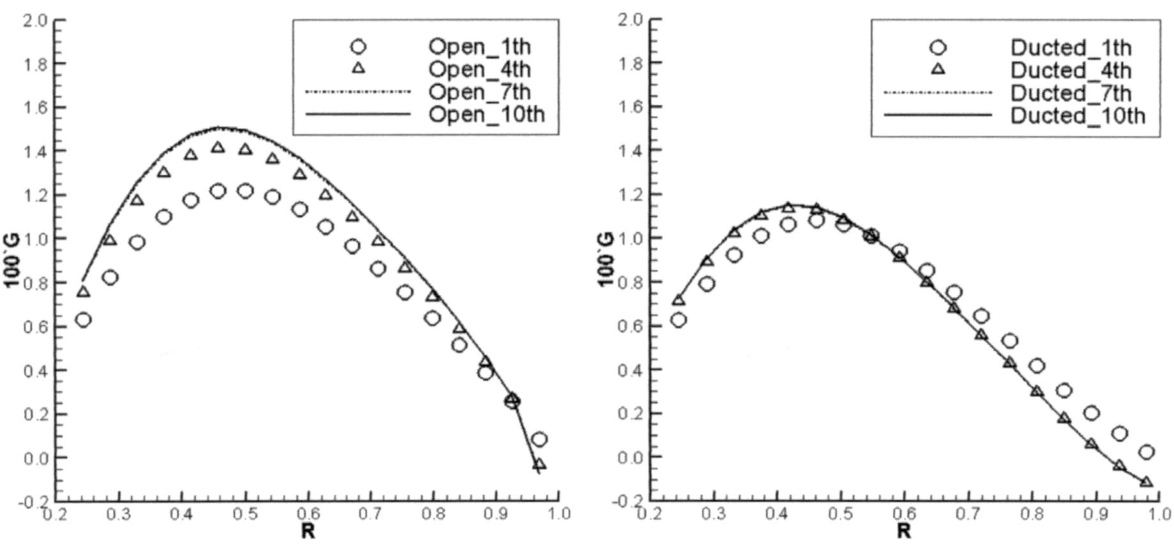

Γ Circulation
G Non-Dimensional Circulation, $G = \Gamma \times 10^2/(2\pi R V_R)$ $V_R = \sqrt{V_S^2 + (0.7 n\pi D)^2}$

(a) Open propeller **(b) Ducted propeller**

Figure 10: Circulation distribution over propeller blades

Figure 11 shows the pressure distributions on the propellers along the chord at 0.7 of the propeller radius. Negative loading is shown at the leading edge of the sections, especially for the ducted propeller, and this is consistent with the circulation distributions shown in Figure 10, and a consequence of the fact that the chosen propeller is far from optimum.

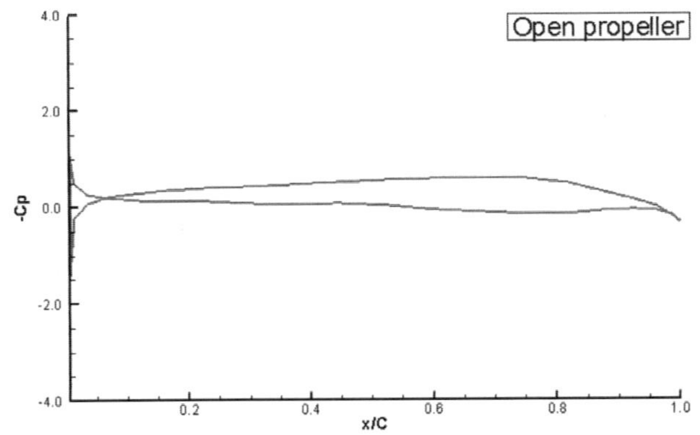

(a) Open propeller

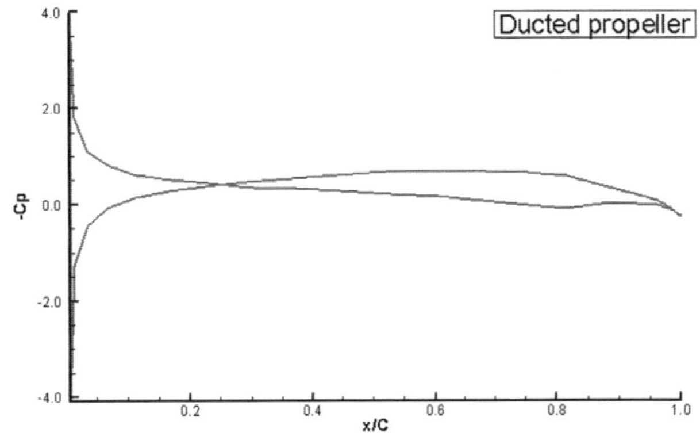

(b) Ducted propeller

Figure 11: Pressure distribution on propellers at $0.7r_P$

Conclusions AND Future Work

An iterative procedure, which couples a vortex-lattice-method with a RANS method, is applied in order to determine the RPM of a given open or ducted propeller in order to move a given axisymmetric hull at a desired speed. The major advantage of the presented work is that it determines the interaction between hull and open or ducted propeller, *automatically and accurately*, WITHOUT assuming any values for the wake fraction or for the thrust deduction factor, both of which would be needed in a traditional approach, which accounts for the effective wake via the wake fraction and for the effect of the propeller on the resistance via the thrust deduction factor (Kerwin and Hadler, 2010).

The authors plan to use the presented approach in order to design the most efficient propeller blades for given hull, hull speed, propeller diameter and RPM, by using the methods presented in Du & Kinnas (2018, 2019). The current approach will also be extended in the case of non-axisymmetric hulls, or hulls of surface ships, in order to design open, ducted, or contra-rotating propellers.

Acknowledgements

Support for this research was provided by the U.S. Office of Naval Research (Grant No. N00014-18-1-2276; Dr. Ki-Han Kim) and by Phase VIII of the "University/Navy/Industry Consortium on Cavitation Performance of High Speed Propulsors".

REFERENCES

Kerwin, J.E., and Hadler, J.B. (2010) Principles of Naval Architecture Series: Propulsion. The Society of Naval Architects and Marine Engineers (SNAME).

Kinnas, S.A. (2017) Hydrodynamic Analysis of Marine Propulsors. Chapter in Encyclopedia of Maritime and Offshore Engineering, Wiley Inc.

Tian, Y., Jeon, C.-H., and Kinnas, S.A. "Effective Wake Calculation/Application to Ducted Propellers", Journal of Ship Research, Vol. 58, No. 2, June 2014, pp. 1–13.

Su, Y. and Kinnas, S.A., 2017 "A Generalized Potential/RANS Interactive Method for the Prediction of Propulsor Performance," Journal of Ship Research, Vol. 61 (No. 4), December 2017, pp. 214-229

Du, A., Kim, S., and Kinnas, S.A. (2018) User Manual and Documentation for MPUF-3A Version R2018, Ocean Engineering Group, Dept. of Civil, Architectural and Environmental Engineering, UT Austin.

Du, W. and Kinnas, S.A., (2018) "Coupling a Numerical Optimization Technique with a Panel Method or a Vortex Lattice Method to Design Cavitating Propellers in Non-Uniform Inflows", 10th International Symposium on Cavitation (CAV2018), Baltimore, Maryland, USA, May 14-16, 2018.

Du, W. and Kinnas, S.A. (2019) "Optimization Design and Analysis of Marine Ducted Propellers by RANS/Potential Flow Coupling Method", 29th International Ocean and Polar Engineering Conference, ISOPE-2019, June 16-21, 2019, Honolulu, Hawaii, USA.

SNAME
THE INTERNATIONAL COMMUNITY
FOR MARITIME AND OCEAN PROFESSIONALS

FLOW PAST A ROTATING CYLINDER PREDICTED BY A VORTICITY METHOD

CHUNLIN WU
PHD CANDIDATE, OCEAN ENGINEERING GROUP, THE UNIVERSITY OF TEXAS AT AUSTIN

SPYROS A. KINNAS
PROFESSOR, OCEAN ENGINEERING GROUP, THE UNIVERSITY OF TEXAS AT AUSTIN

Proceedings of the 25th Offshore Symposium, February 27th 2020, Houston, Texas
Texas Section of the Society of Naval Architects and Marine Engineers

ABSTRACT

A newly developed method that solves the vorticity transport equation in the non-inertial frame of reference is proposed in this paper to simulate the flow past a rotating cylinder. It is a fully viscous method, in which the local phenomenon of vorticity is exploited. The spatial concentration of vorticity leads to a significantly small computational domain and a substantially small number of cells, which results in much less computation time. The present method was applied to simulate the flow past a rotating cylinder in quiescent flow and the flow around a rotating cylinder subject to uniform inflow. The calculated results provided good agreement, which validates the proposed method.

Keywords: Vorticity Equation, Non-inertial Frame, Magnus Flow, Circular Cylinder

INTRODUCTION

The flettner rotor [1] provides an alternative way of vessel propulsion, extracting energy from a renewable source that is available almost everywhere on the oceans - wind. Also, the limits on carbon emissions and the increasing requirements on using fossil fuels in the most efficient and environmentally beneficial way constantly urge engineers and scientists to seek for better substitutions for conventional propulsion systems. The studies in this work provide an effective method for the simulation, analysis and design of a flettner rotor based on the vorticity method. This method takes advantages of the local concentration of vorticity near the wall and along the wake and is proved to be computationally efficient and spatially compact while capable of modeling the complex vortex-vortex and vortex-body interactions.

The VIScous Vorticity Equation (VISVE) method was originally put forward by Tian & Kinnas (2015) aiming at assisting the propeller design subject to high loadings. Li & Kinnas (2017) improved the existing method by modifying and applying it to the unidirectional and oscillatory flows past a 2-D circular cylinder, and a comprehensive study were given by Wu et al. (2019). Later, Xing & Kinnas (2018) implemented the mixture cavitation model in to this method and investigated the cavitating flow past a 2-D hydrofoil. Meanwhile, Wu & Kinnas (2018, 2019) extended the method to model the 3-D flow under the inertial frame with skewness corrections, and it was successfully applied to fully viscous flow past a 3-D hydrofoil of both rectangular and elliptical planform in non-orthogonal grids.

Validations were conducted against other benchmark tests, and reliable results were obtained for all the cases. Eventually, the improved method was applied to flow past a rotating propeller (Wu & Kinnas, 2019). The proposed method was also upgraded with a hybrid parallelism of both Open-MP and MPI to further boost the efficiency (Wu & Kinnas, 2018).

In this work, the present method is extended to consider the non-inertial frame motions. The vorticity transport equation is discretized using the Finite Volume Method (FVM). The velocity field corresponding to a given vorticity distribution is recovered by a direct volume integral method with the acceleration of Fast Multipole Method (FMM) (Lindsay& Krasny, 2001; Ploumhans et al., 2002), and therefore the computational complexity is reduced significantly from N^2 to $N \log N$ for N cells in the grid. Moreover, a vorticity creation scheme is introduced to release the vorticity from the wall into the flow domain by enforcing both the no-slip and non-penetrating boundary conditions. The authors also remark that the present scheme is fully Eulerian without any viscous splitting algorithm.

The remainder is structured as follows. The section METHODOLOGY is devoted to explaining the theories and formulations used in the current method. The results calculated by the present method and discussions are given in the section RESULTS. Eventually, the summaries and future work are detailed in the section CONCLUSION.

METHODOLOGY

1. Velocity Decomposition under Non-inertial Frame

In the non-inertial frame of reference, which translates with velocity $\boldsymbol{U}(t)$ and rotates with angular velocity $\boldsymbol{\Omega}(t)$, the velocity vector $\overline{\boldsymbol{q}}$ under the non-inertial frame reads

$$\overline{\boldsymbol{q}} = \boldsymbol{q} - \boldsymbol{\Omega} \times \boldsymbol{r} - \boldsymbol{U}, \tag{1}$$

where, $\boldsymbol{q}$ is the total velocity under the inertial frame, $\boldsymbol{r}$ represents the location vector.

2. The Vorticity Transport Equation

In the non-inertial frame, the Naiver-Stokes (N-S) equation in the absence of body force reads

$$\frac{\partial \overline{\boldsymbol{q}}}{\partial t} + (\overline{\boldsymbol{q}} \cdot \nabla)\overline{\boldsymbol{q}} = -\nabla p + \nu \nabla^2 \overline{\boldsymbol{q}} - \frac{d\boldsymbol{U}}{dt} - \frac{d\boldsymbol{\Omega}}{dt} \times \boldsymbol{r} - 2\boldsymbol{\Omega} \times \overline{\boldsymbol{q}} - \boldsymbol{\Omega} \times (\boldsymbol{\Omega} \times \boldsymbol{r}), \tag{2}$$

where, p is pressure, and ν is the kinematic viscosity. Apart from the unsteady term, convection term, pressure term, and diffusion term, four additional acceleration terms due to the non-inertial effects appear. They are the translational term, Eulerian term, Coriolis term, and centrifugal term. Taking curl on both sides of the N-S equation, the vorticity equation can be obtained

$$\frac{\partial \boldsymbol{\omega}}{\partial t} + \nabla \times (\boldsymbol{\omega} \times \overline{\boldsymbol{q}}) = -\nu \nabla \times (\nabla \times \boldsymbol{\omega}), \tag{3}$$

where $\boldsymbol{\omega} = \nabla \times \boldsymbol{q}$ is the vorticity. Notably, there is no additional term in the vorticity formulation even in the non-inertial frame of reference. In 2-dimensional case, the vorticity vector $\boldsymbol{\omega}$ degenerates into a scaler ω and so the vorticity equation follows

$$\frac{\partial \omega}{\partial t} + (\overline{\boldsymbol{q}} \cdot \nabla)\omega = \upsilon \nabla^2 \omega. \tag{4}$$

Next, the vorticity equation is discretized using the Finite Volume Method (FVM). The first order upwind and Quadratic Upstream Interpolation for Convective Kinematics (QUICK) scheme (Woodfield, 2004) were hybridly implemented for the vorticity interpolations.

3. Velocity Calculation

To solve the vorticity transport equation (4), the velocity field corresponding to the known vorticity field needs to be determined. This process which has been extensively studied and well established usually involves solving a Poisson's equation.

$$\nabla^2 \psi = -\omega. \tag{5}$$

One can calculate the velocity field by either numerically solving the Poisson's equation or integrating the Green's function. Given a valid vorticity field, the stream-function can be calculated by a direct integral

$$\psi(\boldsymbol{x}_f) = -\int_V \omega(\boldsymbol{x}) G(\boldsymbol{x}, \boldsymbol{x}_f) dV - \int_{\partial V} \gamma(\boldsymbol{x}) G(\boldsymbol{x}, \boldsymbol{x}_f) ds + \psi_b \tag{6}$$

here, $\boldsymbol{x}_f$ is the point of interest where ψ is evaluated, $\boldsymbol{x}$ is the dummy variable running through the whole domain. $G(\boldsymbol{x}, \boldsymbol{x}_f)$ is the Green's function, that reads $(\ln|\boldsymbol{x} - \boldsymbol{x}_f|)/2\pi$ in 2-D. $\gamma(\boldsymbol{x})$ is the wall surface vorticity distribution due to the non-zero net circulation. ψ_b is the stream-function of the uniform inflow $\boldsymbol{U}$.

The direct integral of equation (6) is broken into the summation of the inductions from all the mesh cells with non-zero vorticity in the domain. Obviously, equation (6) is N^2 complexity for N cells in the grid. By employing FMM (Lindsay& Krasny, 2001; Ploumhans et al., 2002), the computational complexity can be reduced to $N \log N$. In 2-D, an analytical expression of equation (6) is obtained (Wu & Kinnas, 2019). By knowing the stream function, the velocity can be achieved

$$\bar{\boldsymbol{q}} = \nabla \times (0,0,\psi). \tag{7}$$

4. Vorticity Creation

The velocity evaluation scheme proposed in the previous subsection will produce a velocity field, $\bar{\boldsymbol{q}}^*$, that has minor normal component, $\bar{q}_n^*$, and tangential component, $\bar{q}_s^*$, on the surface of the wall. Therefore, a vorticity creation scheme is necessary to enforce the non-penetrating and no-slip boundary conditions.

Assuming at time step m, the solution of equation (4) is ω^{m*}, inducing a total velocity filed $\bar{\boldsymbol{q}}^{m*}$. In the current method, the normal component $\bar{q}_n^{m*}$ is used as the input of the BEM solver to solve for the potential ϕ on the wall. Then, the gradient of ϕ on the wall $\nabla_s \phi$ together with the tangential component $\bar{q}_s^{m*}$ gives the created surface vorticity distribution γ_b

$$\gamma_b = -\bar{q}_s^{m*} - \nabla_s \phi, \tag{8}$$

where, ∇_s is the gradient operator along the tangential direction on the wall. In 2-dimensonal case, $\nabla_s \phi = \partial \phi / \partial s$. The surface vorticity is released into the flow domain by solving a diffusion equation. Koumoutsakos and Leonard (1994) proposed a Neumann type vorticity boundary condition that converts the surface vorticity $\boldsymbol{\gamma}_b$ into the normal vorticity flux $\partial \omega_\gamma / \partial n$ across the wall

$$\frac{\partial \omega_\gamma}{\partial n} = -\frac{\gamma_b}{\Delta t}. \tag{9}$$

The surface vorticity is then released into the flow by solving a diffusion equation

$$\frac{\partial \omega_\gamma}{\partial t} = \nu \nabla^2 \omega_\gamma. \tag{10}$$

This diffusion equation is numerically solved also by using FVM. Eventually, the vorticity at time step m reads

$$\omega^m = \omega^{m*} + \omega_\gamma. \tag{11}$$

5. Surface Pressure and Shear Stress

The pressure on the surface of wall can be calculated by knowing the normal vorticity flux across the wall surface

$$\frac{\partial p}{\partial s} = -\mu \frac{\partial \omega}{\partial n}, \tag{12}$$

where, p is pressure, μ is the dynamic viscosity of fluids. n and s are the normal and tangential directions respectively on the wall. As long as the pressure at a reference point on the wall is known, the pressure everywhere on the wall can be obtained via a line integral.

The wall shear stress is related directly to the vorticity on the wall surface, but one needs to keep in mind that the frame rotation also contributes to the stress

$$\tau = -\mu(\omega - 2\Omega). \tag{13}$$

Finally, the body force is comprised of the surface integral of both the pressure and shear stress

$$\boldsymbol{F} = \int_S (-p\boldsymbol{n} + \tau\boldsymbol{s})dl. \tag{14}$$

6. General Solving Procedure

The general solving scheme is shown in Figure 1. The flowchart illustrates the connections between all the solvers.

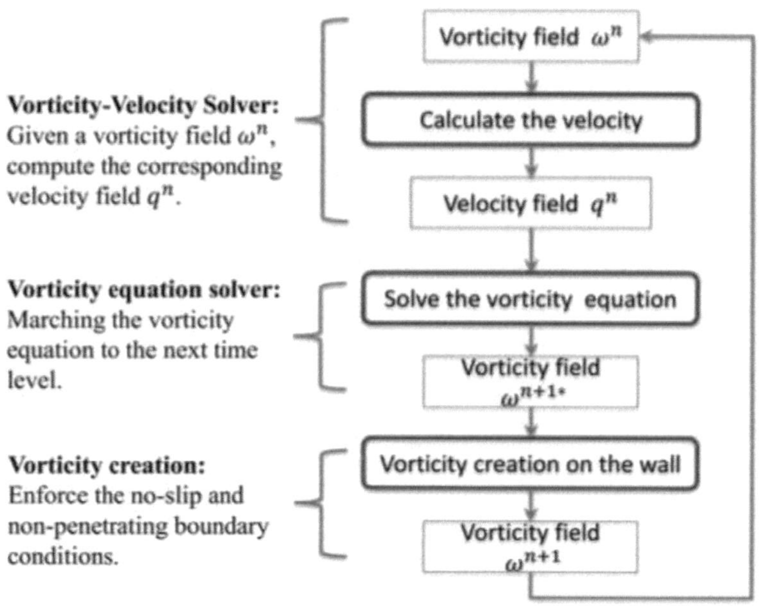

Figure 1: General solving procedures

RESULTS

The proposed method is firstly applied to the flow around a pure rotating cylinder to validate its robustness and accuracy in the rotational frame of reference. Then, it is used to study the magnus flow to provide instructions for the design of the flettner rotor.

The cylinder is characterized by its diameter D. The Reynolds number is defined as

$$Re = \frac{UD}{\nu}, \tag{15}$$

where, ν is the kinematic viscosity of fluids, and U is the velocity of uniform inflow. The non-dimensional time T is

$$T = \frac{tU}{2D}. \tag{16}$$

The rotation rate α is defined to be

$$\alpha = \frac{D\Omega}{2U}, \tag{17}$$

where, Ω is the angular velocity of the rotating cylinder.

1. Rotating Cylinder in Quiescent Flow

The current method is validated by applying to a rotating cylinder in quiescent flow. The radius of the computational domain is set to be $17D$, and the mesh resolution is 200×100 (circumferential direction × normal direction). Notice that there is no uniform inflow in this case. While determining the non-dimensional parameters Re, T, and α, the value of U is set to be 1 (m/s) for convenience, but it does not mean an uniform inflow exists.

The flow at $Re = 200$ and $\alpha = 2.5$ is simulated using the current method. The evolution of radial velocity and vorticity distributions are given in Figure 2. The present results are compared with those predicted by a Navier-Stokes (N-S) method (ANSYS/Fluent). As is observed, the present results overlap with the results produced by N-S.

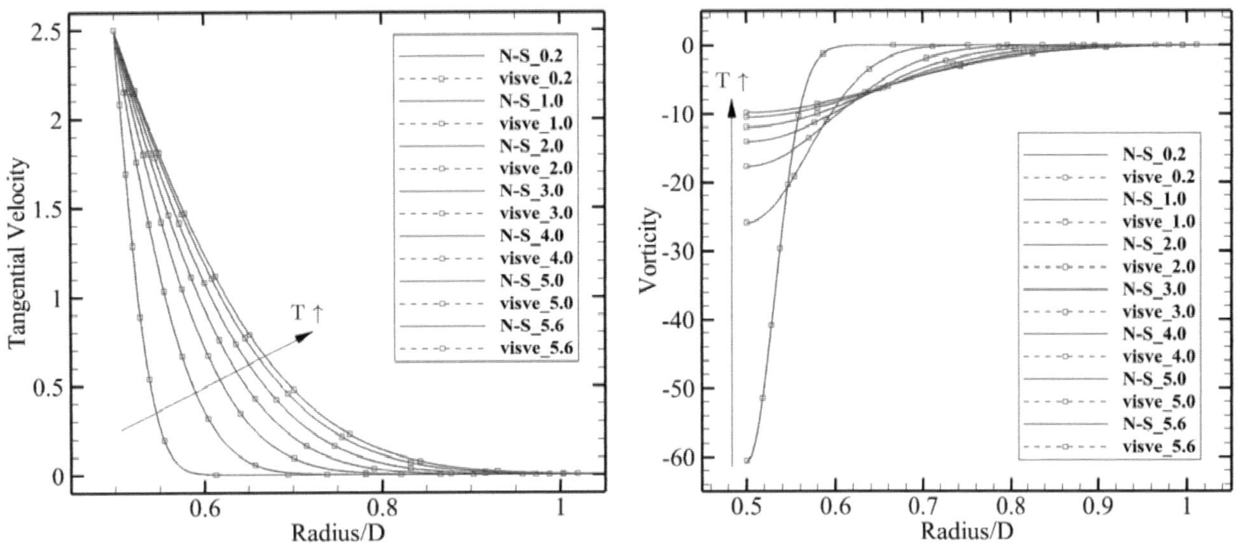

Figure 2: Radial velocity and vorticity distributions compared with predictions of N-S at $Re = 200$ and $\alpha = 2.5$

After the impulsive start of rotation, the flow is initialized with a vorticity field which has a total circulation equal to $\Gamma = 2\pi\Omega R^2$. At the beginning, the vorticity is concentrated near the cylinder and forms a very thin boundary layer. The vorticity gradually diffuses into the domain, and the boundary layer becomes thicker in time with a decreasing maximum magnitude. As the vorticity covers more parts of the domain, the influence of rotating cylinder affects increasing portions of the domain, and the bound of the zero velocity zone is pushed further away from the cylinder. To some point, the non-zero velocity will finally reach the boundary of the computational domain, which has a predefined finite radius. The current method is able to predict the far-field boundary conditions correctly and does not require unperturbed far-field boundary conditions. Therefore, although the radius of the domain is only $17D$, the results calculated by the present method maintain high accuracy.

Eventually, this type of flow will behave like a Rankine vortex when it reaches a steady state as shown in Figure 3. The results calculated by the proposed method is validated against the analytical results of a Rankine vortex. In other words, the steady solutions do not depend on the Reynolds number but are determined only by the rotation speed of the cylinder. The velocity field outside of a Rankine vortex is determined by its radius R and circulation Γ

$$q_{\text{rankine}} = \frac{\Gamma}{2\pi r} = \frac{\Omega R^2}{r}. \tag{18}$$

The steady state of this type of flow is relatively difficult to achieve by the N-S methods since these methods require an extremely large computational domain to ensure the unperturbed outer boundary conditions. Figure 4 displays the results calculated by the N-S method in domains with different sizes at $Re = 10$ and $\alpha = 0.5$. Obviously, the solutions are approaching the velocity distribution of a Rankine vortex as the domain radius increases. Roughly speaking, the radius of the computational domain has to be at least 10^6D to guarantee the residual to drop below 10^{-6} since the velocity of a Rankine vortex scales like $1/r$.

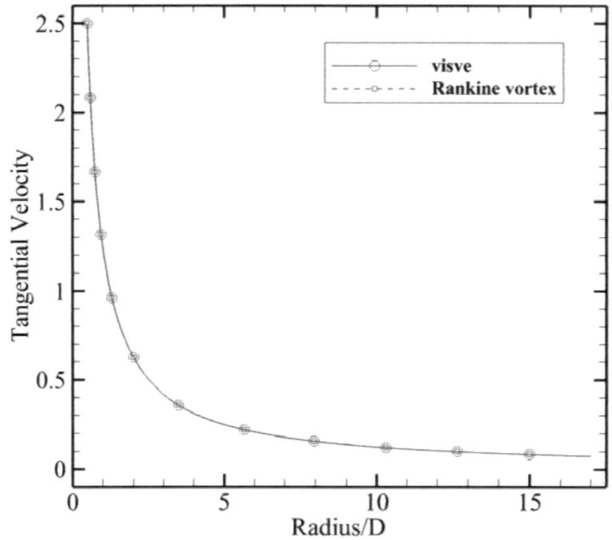

Figure 3: Radial velocity distribution at the converged steady state compared with Rankine vortex at $Re = 200$ and $\alpha = 2.5$

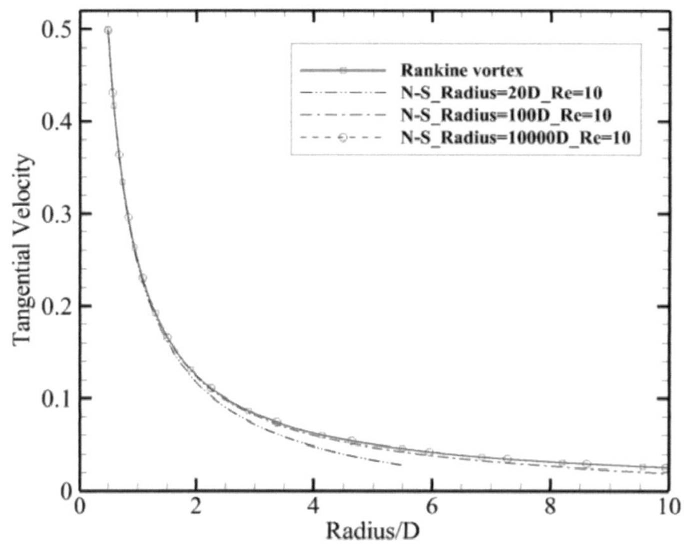

Figure 4: Radial velocity distributions computed by N-S in domains with various radiuses at the steady state at $Re = 10$ and $\alpha = 0.5$

2. <u>Rotating Cylinder Subject to Uniform Flow (Magnus Flow)</u>

In this part, the current method is applied to a rotating cylinder subject to the uniform inflow to study the magus flow. The radius of the computational domain is set to be $150D$, and the mesh resolution is 200×210 (circumferential direction × normal direction). The computational domain is chosen to be relatively large to capture and visualize the vortex structures in the wake. The flow at $Re = 200$ and multiple rotation rates are simulated and the results are given.

In Figure 5, the evolution of lift coefficients is given for rotation rate $\alpha = 0.0, 0.5, 1.0, 1.5, 2.07, 2.5, 3.0, 3.25, 3.5, 4.0, 4.2, 4.4, 4.6, 4.8$, and 5.0, respectively. In addition, the phase diagrams at rotation rate $\alpha = 0.0, 0.5, 1.0, 1.5, 2.0, 2.5, 3.0, 3.25, 3.5, 4.0, 4.2, 4.4, 4.6, 4.8$, and 5.0 are shown in Figure 6. The present results are also compared with those predicted by a N-S method (ANSYS/Fluent). In the N-S method, the radius of the computational domain is 10^3D, and a total number of 640,000 mesh elements, which is approximately 15 times larger than the number used in VISVE, are used.

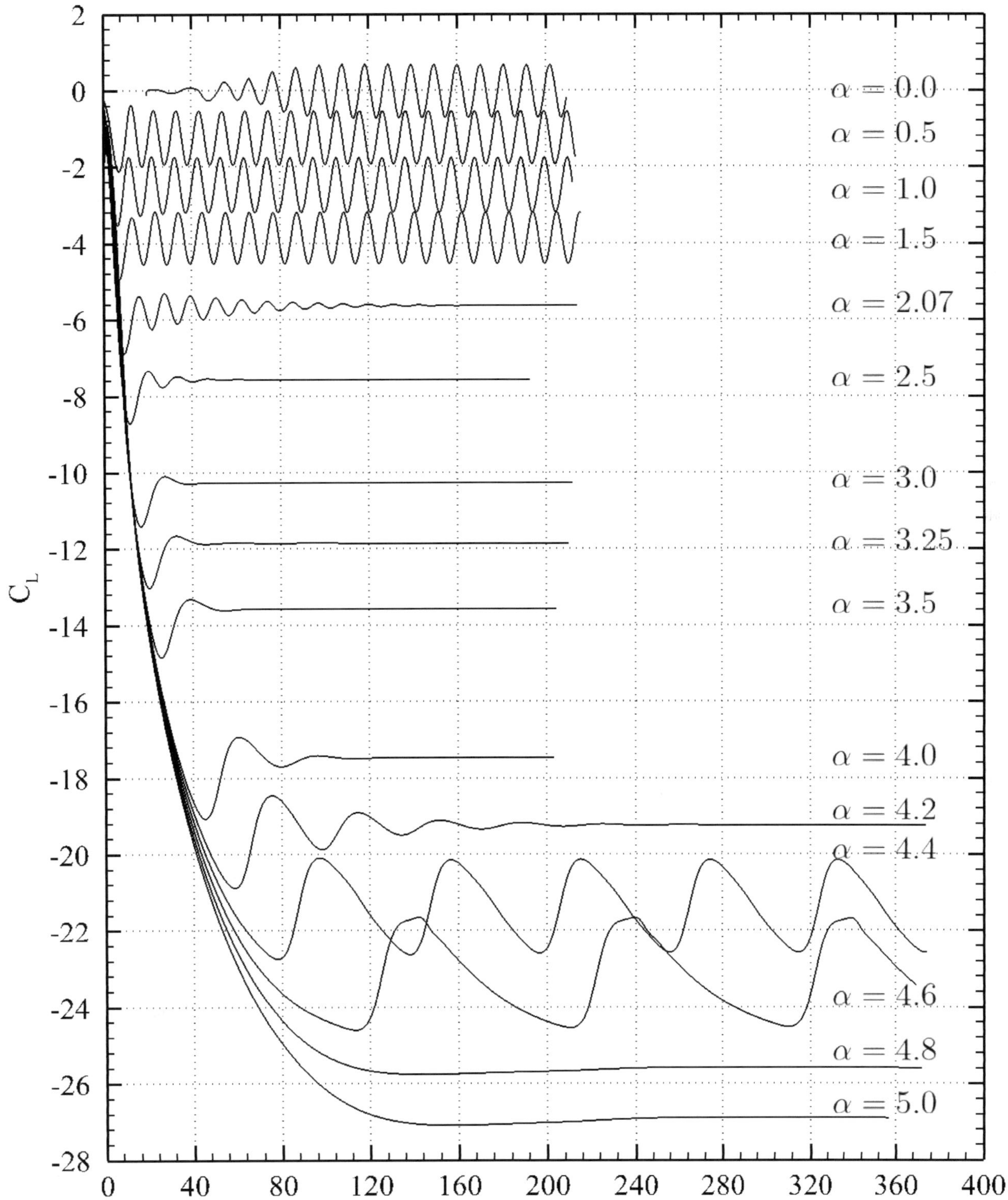

Figure 5: Evolution of lift coefficients in time of the flow past a rotating cylinder at Re = 200 and multiple values of α

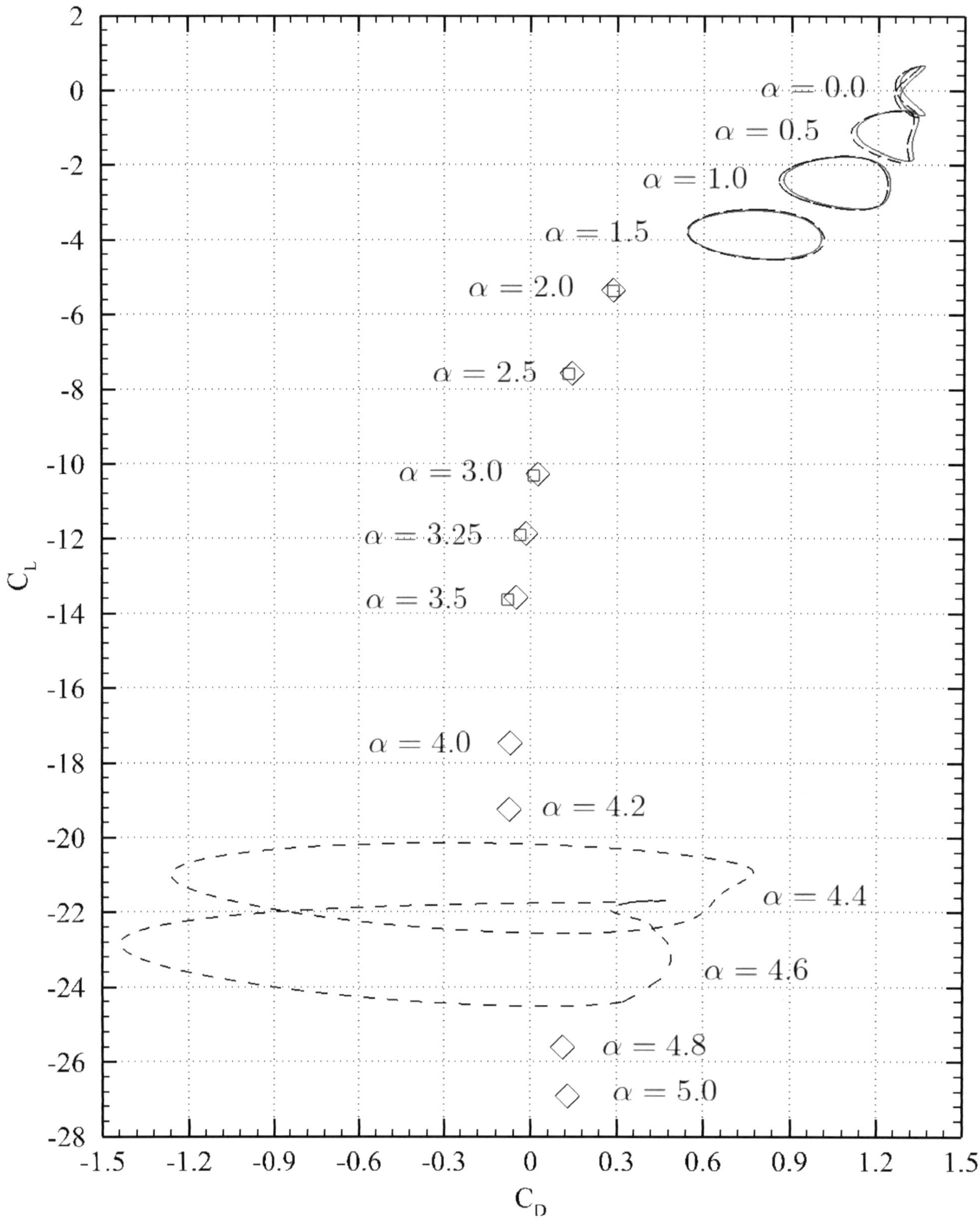

Figure 6: Phase diagrams of the flow past a rotating cylinder at $Re = 200$ and multiple values of α calculated by the current method (-------- and ◇) and the N-S method (——— and □)

From the results, it is observed that the unsteady vortex shedding happens for rotation rate $0 \leq \alpha \leq 1.91$ as shown in Figure 7. This is the first class of periodic vortex shedding regime that happens at lower values of α. The vortex shedding will be suppressed by the cylinder rotation for rotation rate $1.91 \leq \alpha \leq 4.35$. At the beginning, vortex shedding still exits but it fades with time as given in Figure 8. Eventually, the flow can reach a steady state, which is graphically demonstrated in Figure 9. When the rotation rate is further increased to be $4.35 \leq \alpha \leq 4.75$, a much stronger vortex shedding phase is initiated again as illustrated in Figure 10. This is the second class of periodic vortex shedding regime. The cylinder is subject to oscillating body forces with a much larger amplitude and period than it is at smaller rotation rates. For rotation rate $4.75 \leq \alpha \leq 5.0$, the flow develops directly to the steady state without intermediate vortex shedding phase.

Generally, the average lift on the cylinder increases with the rotation rate α. In other words, raising the rotation rate of the cylinder will directly lead to a rise in the thrust provided by the flettner rotor. However, one needs to avoid the possible vibrations resulting from the unsteady vortex shedding for certain range of rotation rates. Especially, the second class of vortex shedding must be accurately detected and prevented since the triggering rotation rates are hidden in between the steady cases, and the periodic body forces have much larger magnitudes.

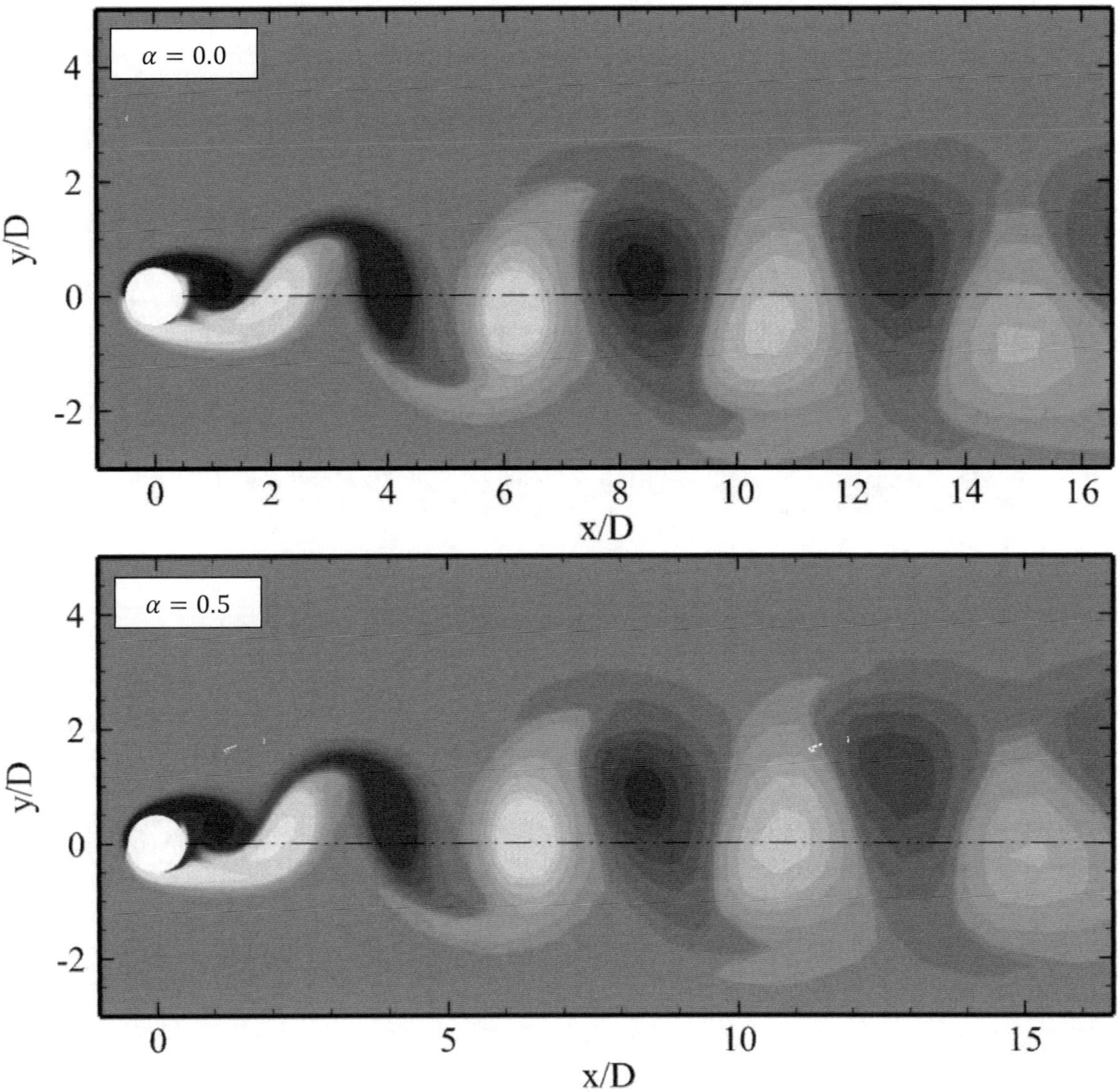

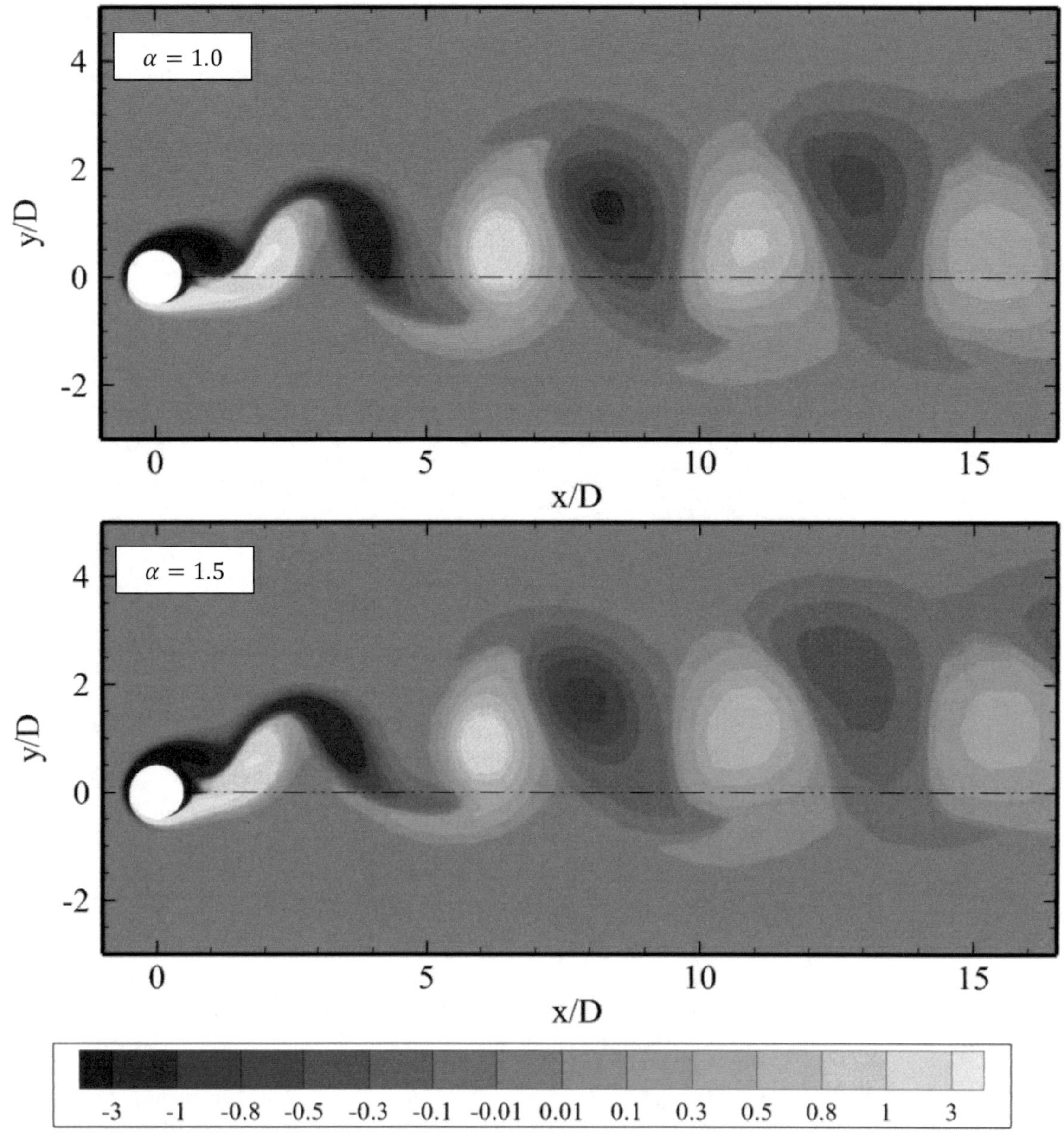

Figure 7: Vorticity field during fully established unsteady vortex shedding phase of the flow past a rotating cylinder at $Re = 200$ and multiple values of α between 0.0 and 1.5

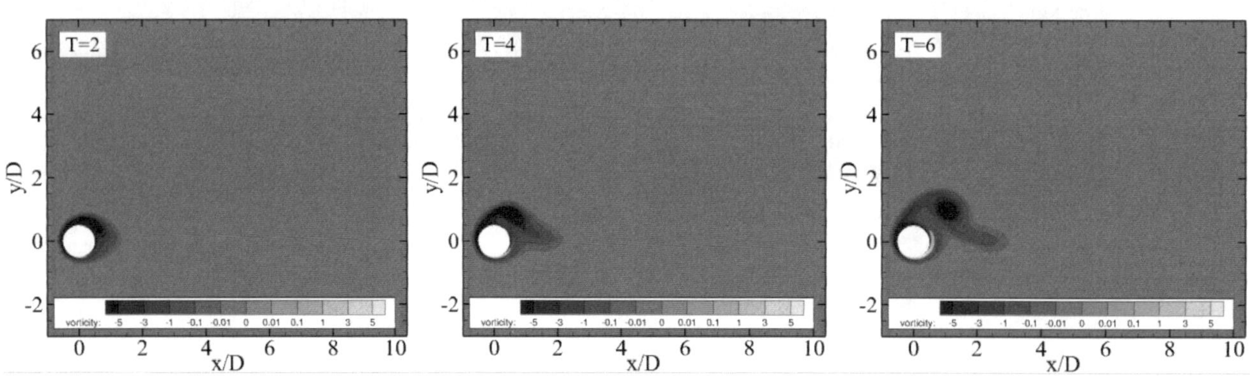

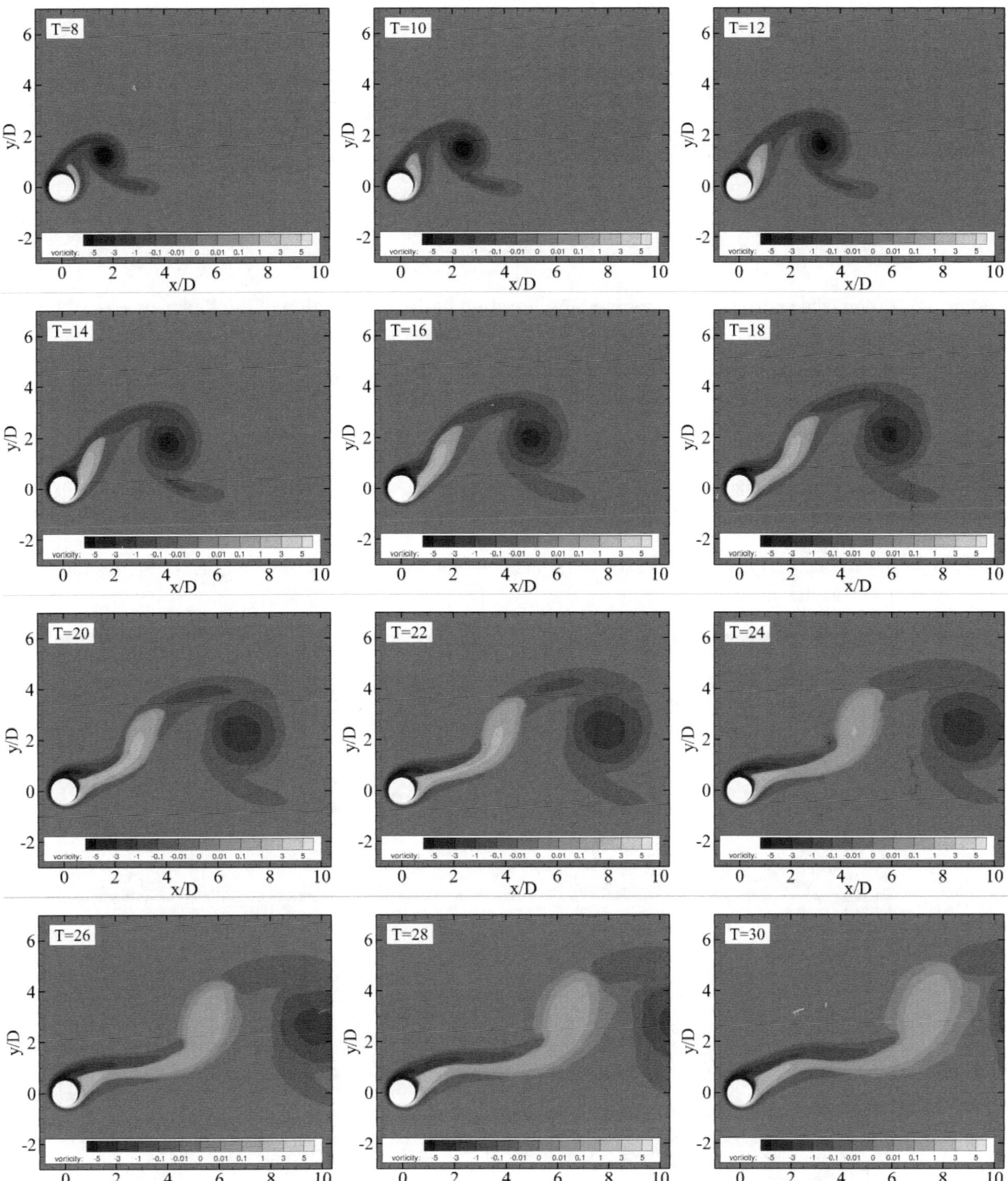
T=8
T=10
T=12
T=14
T=16
T=18
T=20
T=22
T=24
T=26
T=28
T=30
y/D
x/D
vorticity: -5 -3 -1 -0.1 -0.01 0 0.01 0.1 1 3 5

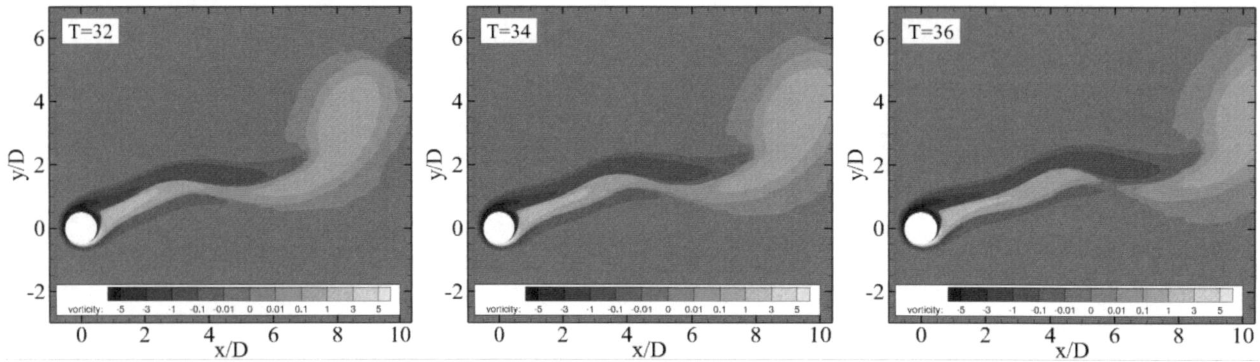

Figure 8: Initial stage vortex development and separation in the wake region of the flow past a rotating cylinder at $Re = 200$ and $\alpha = 2.5$

Also, the rotation of cylinder can effectively suppress the vortex shedding and reduce the drag on the body for proper choices of rotation rate α. At Reynolds number $Re = 200$, the rotation of cylinder can effectively suppress the vortex shedding for $1.91 \leq \alpha \leq 4.35$ and $4.75 \leq \alpha \leq 5.0$. From the evolution of lift coefficients in Figure 5, it is observed that the faster the cylinder rotates, the quicker the vortex shedding disappears. When α reaches over 4.75, no obvious vortex shedding happens, and the flow directly goes to a steady state. The drag on the cylinder keeps dropping as α decreases for $0 \leq \alpha \leq 3.1$. Then, the further decrease of α renders the drag negative but still with a small magnitude. When α falls in the range of the second class of vortex shedding ($4.35 \leq \alpha \leq 4.75$), the drag on the cylinder becomes periodic with an extremely large magnitude, which should definitely be avoided. For $4.75 \leq \alpha \leq 5.0$, the drag turns to be positive again and trends to increase with α.

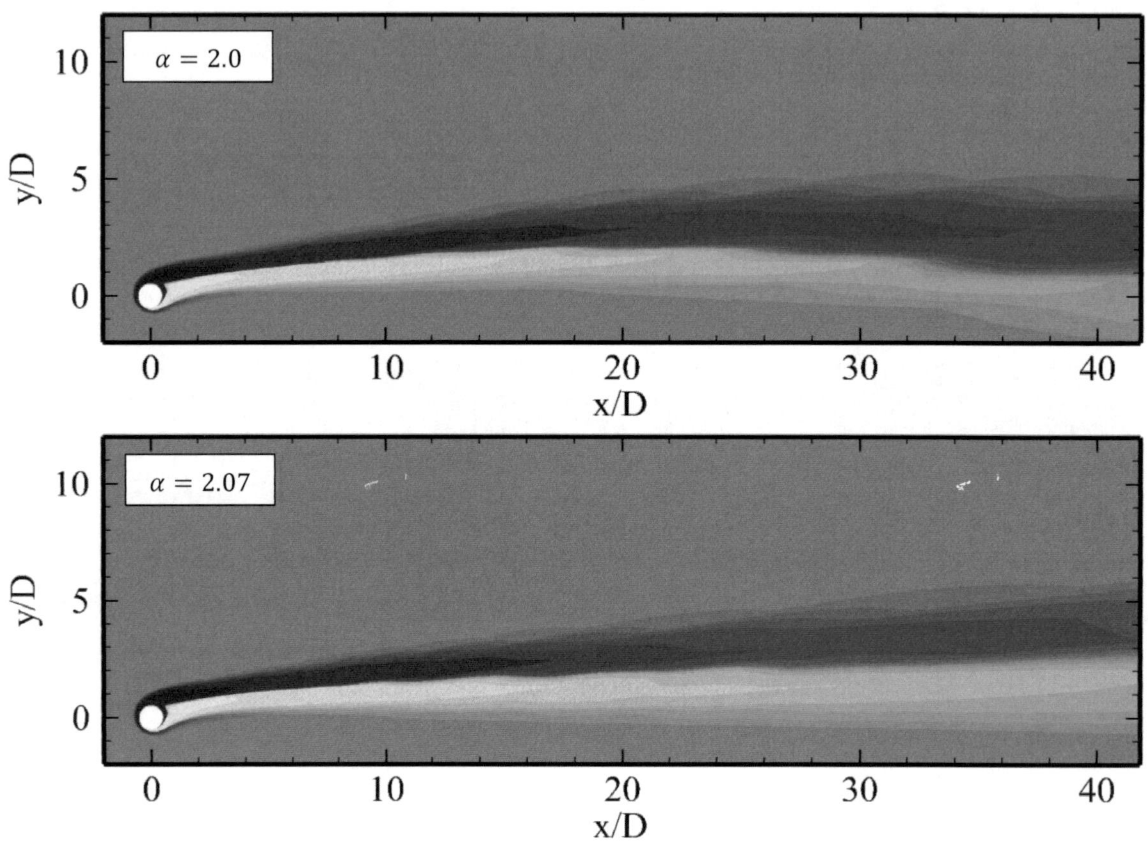

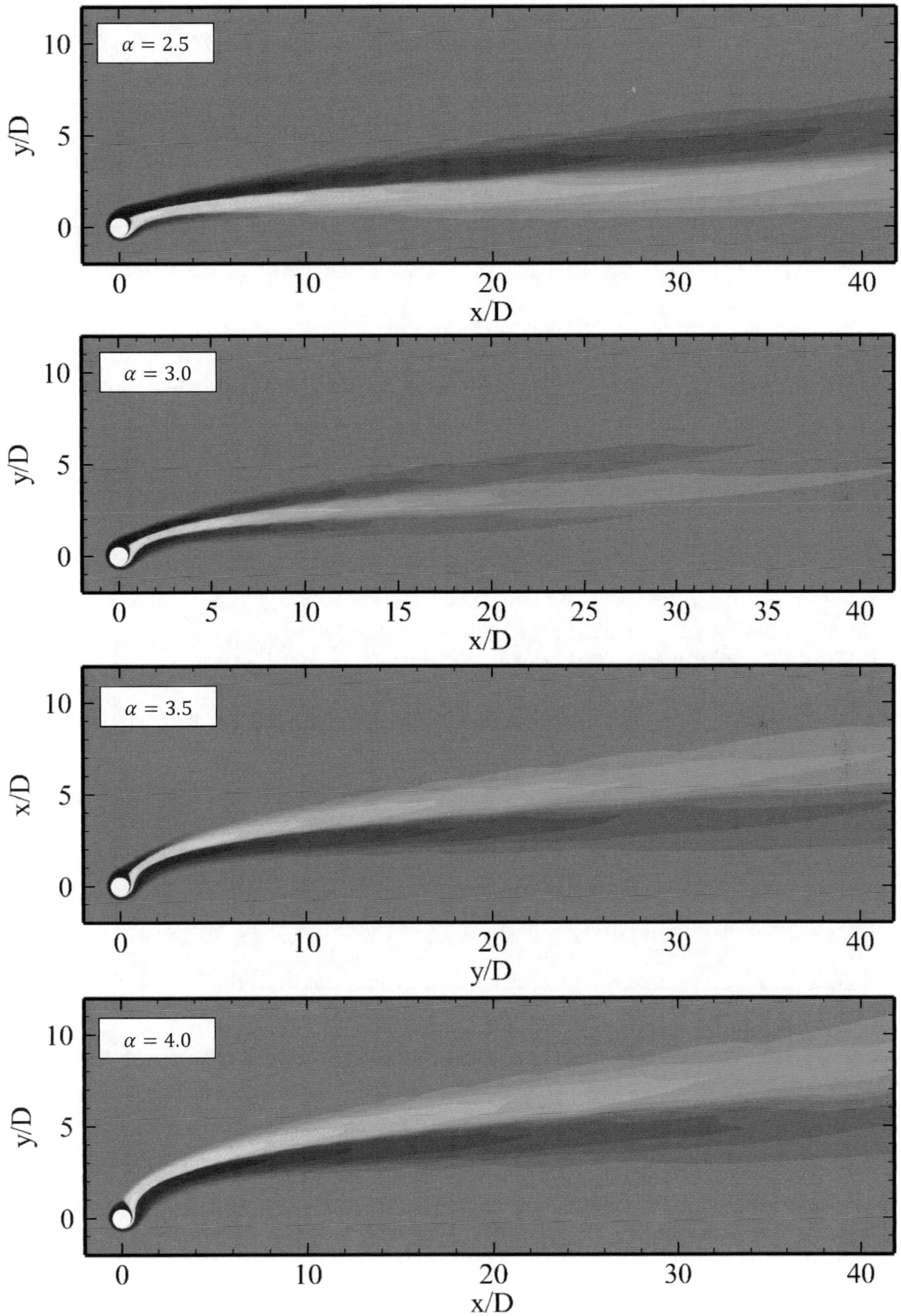
$\alpha = 2.5$
y/D
x/D
$\alpha = 3.0$
y/D
x/D
$\alpha = 3.5$
x/D
y/D
$\alpha = 4.0$
y/D
x/D

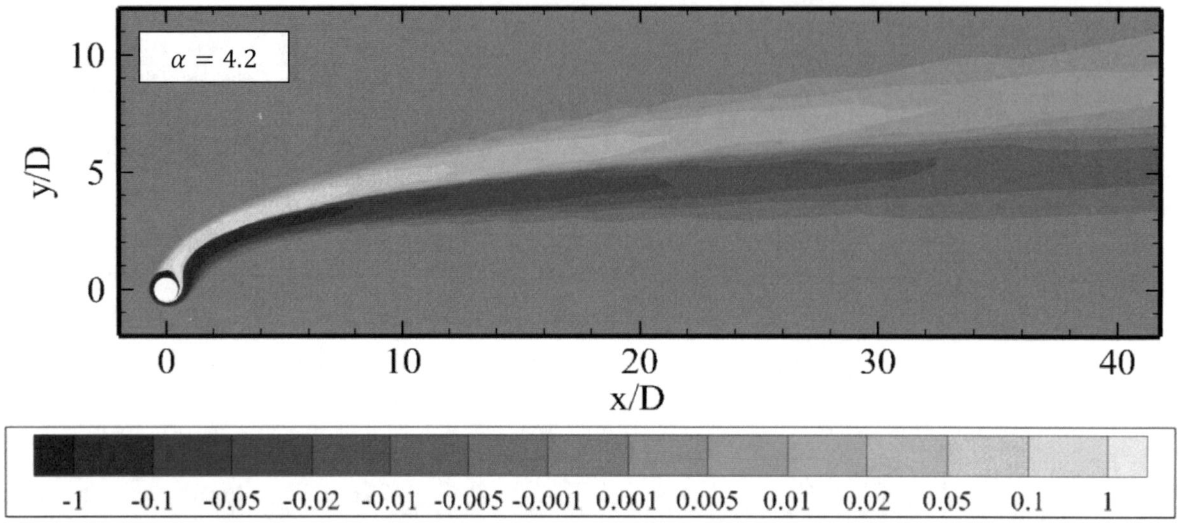

Figure 9: Fully established vorticity field of the flow past a rotating cylinder at steady sate with $Re = 200$ and multiple values of α between 2.0 and 4.2

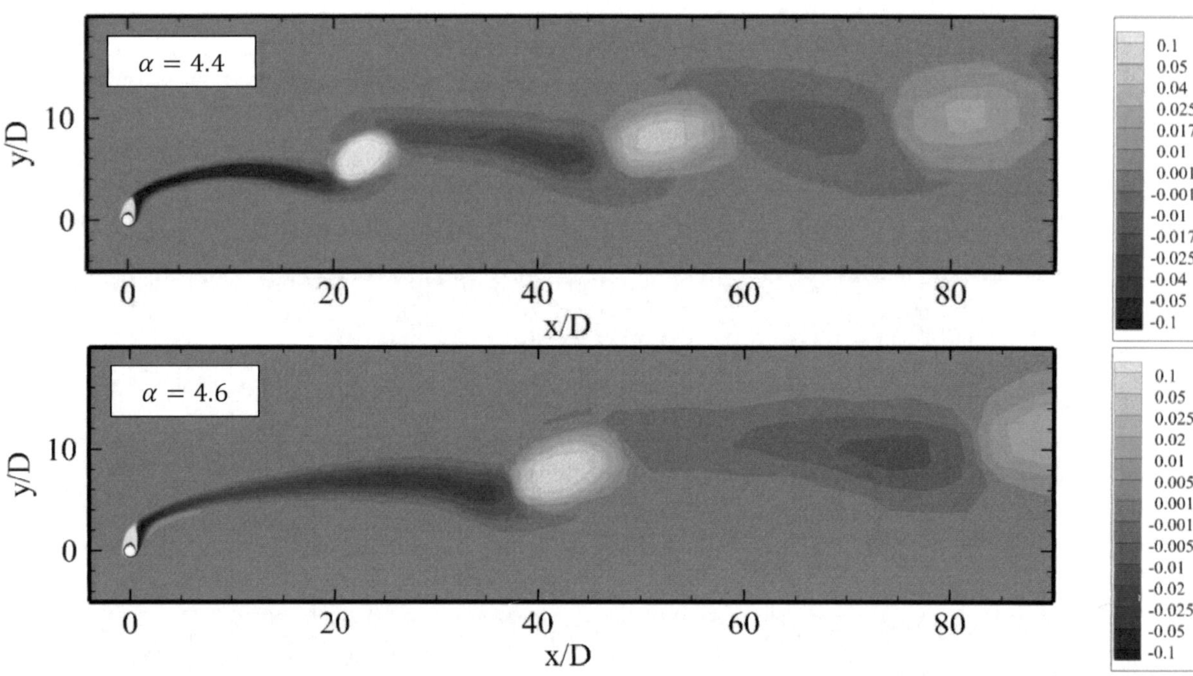

Figure 10: Fully established vorticity field of the unsteady flow past a rotating cylinder with $Re = 200$ and $\alpha = 4.4$ & $\alpha = 4.6$

Last but not lease, the computational performance is compared between the current method and N-S method while simulating the uniform flow past a rotating cylinder. The number of mesh elements in the current method is 40,000, and the N-S method employs a number of 2,560,000 cells. The tests were conducted on the Stampede2 super computer at the Texas Advanced Computing Center (TACC) with 48 CPUs. The performance comparison is given in Figure 12. As is observed, the present method much faster than the N-S method, which is a direct consequence of allowing for a smaller computational domain with a small number of mesh cells.

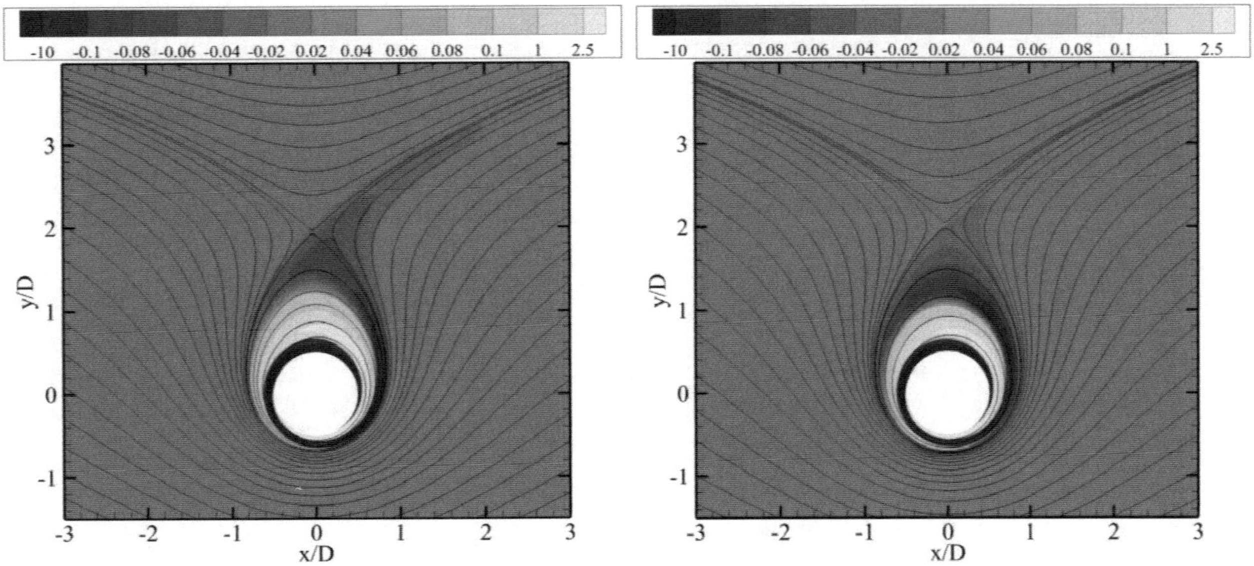

Figure 11: Fully established vorticity field of the flow past a rotating cylinder at steady sate with $Re = 200$ and $\alpha = 4.8$ (left) & $\alpha = 5.0$ (right)

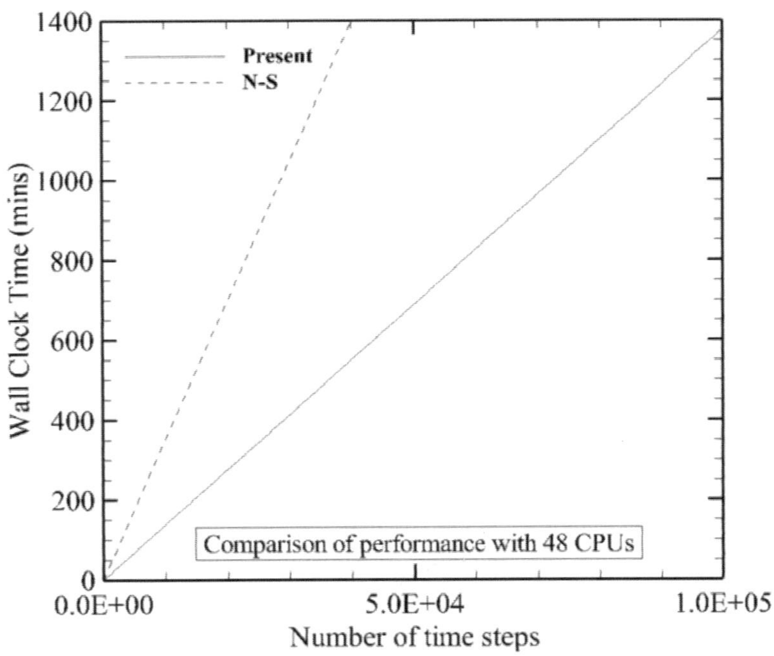

Figure 12: Comparison of computational performance of modeling the flow past a rotating cylinder at $Re = 200$ and $\alpha = 0.0$. Notably, the present method uses 40,000 cells in the grid, and the N-S method has 2,560,000 cells in the grid.

CONCLUSION

In this work, a VIScous Vorticity Equation (VISVE) method is proposed to solve the incompressible laminar flow past a rotating circular cylinder in the non-inertial frame of reference. The current method is designed to be spatially compact and numerically efficient, meanwhile capable of modeling the complicated vortex-body and vortex-vortex interactions. Therefore, the biggest advantage is that it allows for a small as well as compact computational domain, which leads to much less computational time.

The proposed method has been applied to an impulsively started 2-D rotating circular cylinder in the quiescent flow to validate the method in the rotational frame of reference. In addition, this method has also been applied to the uniform inflow past a rotating circular cylinder to study the influence of rotation rate on the vortex behaviors and body forces. The results predicted by the current method were compared and validated with those calculated by N-S methods. The comparisons show good agreement and prove that VISVE produces reliable results.

The authors investigated the magnus flow at $Re = 200$ and $0 \leq \alpha \leq 5.0$. The first class of unsteady vortex shedding regime happens at rotation rate $0 \leq \alpha \leq 1.91$. For rotation rate $1.91 \leq \alpha \leq 4.35$, the vortex shedding will be suppressed by the cylinder rotation gradually, and the flow will reach a steady state eventually. The second class of unsteady vortex shedding regime appears at rotation rates $4.35 \leq \alpha \leq 4.75$, and a stronger vortex shedding phase with much a larger amplitude and period is initiated again. The cylinder is also subject to greater oscillating body forces than the first class of vortex shedding. When the rotation rate reaches $4.75 \leq \alpha \leq 5.0$, the flow develops directly to the steady state without intermediate vortex shedding phase.

Finally, the computational performance is compared between the present method and the N-S method. The present method is proved to be more efficient.

ACKNOWLEDGEMENTS

Support for this research was provided by the U.S. Office of Naval Research (Grant Nos. N00014-14-1-0303 and N00014-18-1-2276; Dr. Ki-Han Kim) partly by Phases VII and VIII of the "Consortium on Cavitation Performance of High Speed Propulsors". The authors are grateful for the inspiring ideas provided by Dr. Ye Tian to complete this method and finish this work. The authors also acknowledge the Texas Advanced Computing Center (TACC) at The University of Texas at Austin for providing HPC resources that have contributed to the research results reported within this work.

REFERENCES

Koumoutsakos, P., Leonard, A., & Pepin, F. (1994). Boundary conditions for viscous vortex methods. Journal of Computational Physics, 113(1), 52-61.

Li, Z., Kinnas, S.A. (2017). VISVE, a Vorticity Based Model Applied to Unidirectional and Alternating Flow around a Cylinder. 22nd SNAME Offshore Symposium, Houston, Texas, USA.

Lindsay, K., & Krasny, R. (2001). A particle method and adaptive treecode for vortex sheet motion in three-dimensional flow. Journal of Computational Physics, 172(2), 879-907.

Ploumhans, P., Winckelmans, G. S., Salmon, J. K., Leonard, A., & Warren, M. S. (2002). Vortex methods for direct numerical simulation of three-dimensional bluff body flows: application to the sphere at Re= 300, 500, and 1000. Journal of Computational Physics, 178(2), 427-463.

Traut, M., Gilbert, P., Walsh, C., Bows, A., Filippone, A., Stansby, P., & Wood, R. (2014). Propulsive power contribution of a kite and a Flettner rotor on selected shipping routes. Applied Energy, 113, 362-372.

Tian, Y., Kinnas, S.A. (2015). A Viscous Vorticity Method for Propeller Tip Flows and Leading Edge Vortex. 4th International Symposium on Marine Propulsors, Austin, Texas, USA.

Wu, C., Kinnas, S.A. (2018). A VIScous Vorticity Equation (VISVE) method applied to 2-D and 3-D hydrofoil in both forward and backing conditions. 23rd SNAME Offshore Symposium, Houston, Texas, USA.

Wu, C., Kinnas, S. A. (2019). A 3-D VIScous Vorticity Equation (VISVE) Method Applied to Flow Past a Hydrofoil of Elliptical Planform and a Propeller. In Sixth international symposium on marine propulsors (smp'19), Rome, Italy.

Wu, C., Kinnas, S. A. (2019). A Distributed Vorticity Model for Flow Past a 3-D Hydrofoil. In SNAME 24th Offshore Symposium. The Society of Naval Architects and Marine Engineers, Houston Texas, USA.

Wu, C., Li, Z., Wu, Y., Kinnas, S.A. (2019). A Conservative Viscous Vorticity Method for Unsteady Unidirectional and Oscillatory Flow Past a Circular Cylinder. Ocean Engineering, 191: 106504.

Xing, L., Wu, C., and Kinnas, S.A. (2018). VISVE, a Vorticity Based Model Applied to Cavitating Flow around a 2-D Hydrofoil. 10th International Symposium on Cavitation, Baltimore, Maryland, USA.

SNAME
THE INTERNATIONAL COMMUNITY
FOR MARITIME AND OCEAN PROFESSIONALS

APPLICATION OF AMR MESHING FOR EXPLOSION ANALYSIS ON OFFSHORE STRUCTURES

J. KEITH CLUTTER
SCIRISQ, INC.

ANDREW STASZAK
SCIRISQ, INC.

Proceedings of the 25th Offshore Symposium, February 27th 2020, Houston, Texas
Texas Section of the Society of Naval Architects and Marine Engineers

ABSTRACT

Computational fluid dynamics (CFD) modeling is routinely used to analyze various potential events on offshore structures. One example is the analysis of vapor cloud explosions (VCEs). Meshing is a critical step in any CFD problem and can present a significant bottleneck. In a traditional approach an experienced user would follow the basic process of: 1) Review the problem, model, and critical questions intended to be answered to identify a base meshing strategy. This step relies heavily on user experience and knowledge of tools being employed.; 2) Construct a starting mesh which provides refinement in proper areas, abides by all meshing rules specific the solvers being used (interface transitions, geometry alignments, stretching, etc.).; and 3) Conduct critical mesh sensitivity analyses where refinement and mesh rules are tested and verified. This process is highly dependent on the type of mesh used and the CFD code being used. As a result of these requirements, there always exists a significant opportunity in the modeling industry to improving meshing strategies and approaches. Here a review of the various methods is presented and their utility for offshore VCE analysis is noted. The application of an automatic mesh refinement (AMR) based approach is highlighted. The fundamentals of the algorithms are reviewed and example applications are presented. The advantages of the AMR approach over other strategies are noted.

Keywords: SNAME, Offshore Symposium, CFD, meshing, explosions

INTRODUCTION

Computational modeling (CFD) approaches and capabilities have made great strides in combating the stigma that this type of modeling is too expensive and time intensive. This has mainly been overcome by the continued exponential growth in computing power. Yet, as our computational capabilities has grown, as has the demand for the number and complexity of problems that are simulated in a CFD study. Thus, hardware efficiencies have created a need for an increase in the number and type of models simulated and, in many cases, this has resulted in a net zero effect. In an effort to improve both the image of CFD as a cost effective and efficient tool and the value of our solutions, we must continue to look to new ways to improve our methods. One area where efficiency can be gained is in the construction of the computational mesh needed for the analysis. Meshing is such a significant step in the modeling process that there exists an entire set of stand alone software packages and applications, outside of the physics models, dedicated to mesh construction and development.

Despite these improvements, current approaches to vapor cloud explosion (VCE) modeling with CFD still rely on mesh techniques in which a "knowledgeable" user is required to build suitable meshes. Meshing is a critical step in any CFD problem and can present a significant bottleneck. In a traditional approach an experienced user would follow the basic process of: 1) Review the problem, model, and critical questions intended to be answered to identify a base meshing strategy. This step relies heavily on user experience and knowledge of tools being employed.; 2) Construct a starting mesh which provides refinement in proper areas, abides by all meshing rules specific the solvers being used (interface transitions, geometry alignments, stretching, etc.).; and 3) Conduct critical mesh sensitivity analyses where refinement and mesh rules are tested and verified.

This meshing process may have to be conducted again and again, for various vapor cloud locations and shapes, when changes to geometry are made, new ignition locations, new cloud locations or sizes, etc. In contrast to this process, there are alternative approaches, one being the use of adaptive meshing. Automatic mesh refinement (AMR) routines can be used to auto-refine meshes around geometry and critical characteristics of the solution. Much like Artificial Intelligence, AMR based techniques can be guided by simple rule sets and allow the model to make live mesh changes as necessary during the simulation. The end result is the ability to reduce meshing requirements, simplify and reduce setup time, all while maintaining and in many cases improving the solution accuracy. In some applications AMR techniques can be used to remove or reduce the "knowledgeable" user requirement of mesh construction.

The goal of this paper will be to provide a look into what adaptive meshing is, how it works, and show how it can be used to increase our modeling efficiency (i.e. reducing costs and time). A sample facility will be used to demonstrate the capability and provide visuals of the AMR effect.

Mesh Types

At the foundation of any computational model is the grid, or mesh, on which the problem is solved. There exist a variety of meshing frameworks for use in computational modeling. The main classes are (1) body-fitted curvilinear or unstructured tetrahedral meshing, or (2) Cartesian meshing. Body-fitted meshing is routinely used in applications such as aerodynamics, such as the example in Figure 1. This is necessary to resolve key characteristics of the problem such as boundary layers. However, they are not easily applied to scenarios such as offshore structures where there may be thousands or tens of thousands of individual elements within the modeling domain.

The Cartesian based meshing is more applicable to offshore structures that are composed of a variety of elements, ranging in size from small piping to large buildings. To reduce computational requirements for these types of problems, sub-domains are often employed to reduce the number of geometric elements represented by the mesh. One popular approach is the use of a porosity representation of all structural elements [1]. An example of this approach is shown in Figure 2. The geometry of interest is imposed on a Cartesian mesh. Then each mesh cell is interrogated to determine how much of the cell is filled. This provides a porosity level that is then used in the solution of the governing equations. This approach is very robust and addresses many of the challenges of the complex nature of offshore structures. However, there are issues in that sometimes items like walls can be skewed since the representation of the objects is directly dependent on the selected mesh spacing. This could have direct impact on the analysis results since it will affect blast focusing and shielding. Thus, additional care is required to ensure proper mesh alignment to certain surfaces, structures and geometric elements included in the domain.

[1]. Chille, F, et al, "Use of CFD tools for prediction of explosion loads", Handling Exceptions in Structural Engineering, Universita degli Studi di Roma La Sapienze, 13 Nov 2008, DOI: 10.3267/HE2008.

Figure 1. Example of body-fitted unstructured tetrahedral meshing applied to an aircraft by the authors.

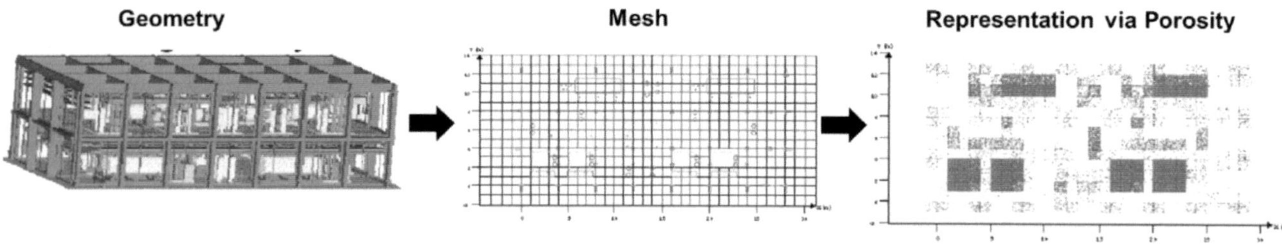

Figure 2. Cartesian meshing using porosity representation of geometry [1].

Here the Cartesian meshing approach is used along with adaptation to facilitate the representation of structures with less user input and effort for development. It also increases efficiency because the mesh can be allowed to dynamically adapt such that there is only refinement were necessary for the solution. The baseline mesh is refined only where there exist surfaces of the geometry. The method employed here uses the basic principles found in previous approaches [2,3,4].

METHODOLOGY

The modeling code discussed here is CAMBER which uses Cartesian Adaptive Meshing for Blast, Explosion and Release scenarios. It is used for a variety of problems to include simulation of high explosives and gas-phase combustion. Therefore, it can be considered either a hydrocode or CFD code. CAMBER was developed by SciRisq, is currently used inhouse but will be commercially available soon. The user interface that allows for easy model construction and data post processing is in a Beta form and currently being finalized.

The code uses the finite-volume approach and employs an object-oriented structure to allow for the integration of various physics models. Here the focus is on the adaptive meshing that improves the efficiency and accuracy of explosion simulations.

Using the Cartesian approach, the domain of interest is defined as any rectangular shape. Also defined is the size of the coarsest grid desired which is referred to as the Level 0 cells. The cells need not be cubic but larger aspect ratio cells can introduce numerical error depending on the particular flow solver used. Also defined is the number of adaptation levels desired, referred to here as n-levels. The two parameters, Level 0 cell size and n-levels will influence both solution accuracy and efficiency.

[2]. M. J. Berger and J. Olier, Adaptive mesh refinement for hyperbolic partial differential equations, J. Comput. Phys. 53, 484–512 (1984).

[3]. M. J. Berger and P. Colella, Local adaptive mesh refinement for shock hydrodynamics, J. Comput. Phys. 82, 64–84 (1989).

[4]. M. J. Berger and R.J. LeVeque, An Adaptive Cartesian Mesh Algorithm for the Euler Equations in Arbitrary Geometries, AIAA Paper 89–1930-CP (1989).

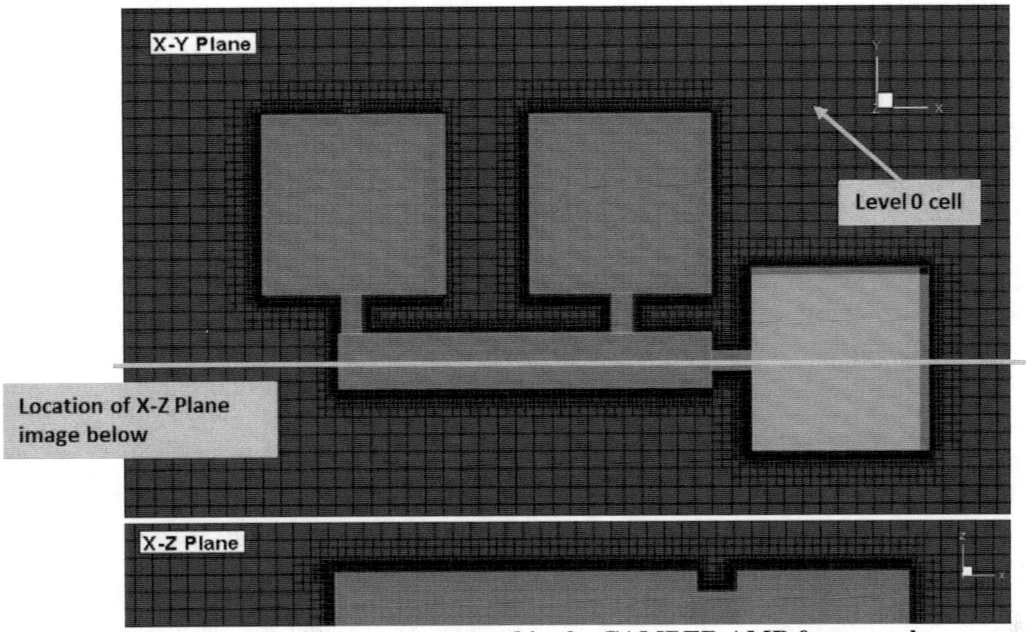

Figure 3. Example of an onshore facility as represented in the CAMBER AMR framework.

The domain of interest is decomposed into a set of Level 0 cells as shown in Figure 3 for a simple onshore structure. Then the mesh is refined to define the components of the structure. If increased resolution is needed these cells are systematically divided into 8 children. The requirement for refinement is dependent on a local feature of interest in the geometry or solution. In the current implementation, the discontinuity between neighboring cells is never greater than a 4:1 ratio. This restriction does require that some additional cells be refined than just those meeting specified criteria. However, it simplifies the flow solver and difference equation solution algorithms and more than adequately accounts for any additional cost [[5]].

To facilitate the management of the computational mesh and the solution process, there are some basic grid accounting parameters that are assigned to each cell. These include the following:

1. level = level of the cell
2. idiv = 0 if cell is not divided, = 1 if cell is divided
3. Parent = pointer to a parent cell
4. C2[i][j][k] = pointer to children cells: i = 0,1; j = 0,1; k = 0,1
5. xN = pointer to a neighboring cell: x = L(eft), R(ight), B(ottom), T(op), U(nder), O(ver)

A cell at level n is divided into 8 children at level n + 1. The first three parameters are self-evident. The parameter C2 defines the children cells, if the cell is divided.

The geometry of the structure is typically done in a CAD package and can be exported as a collection of triangular plates, typically in STL format. Most of the common CAD programs allow such output. The adaptation to the geometry uses the plate information long with the defined grid domain, Level 0 cell size, and the parameter n-levels to construct a representation of the geometry within the Cartesian mesh.

[5]. Clutter, JK and Staszak, A, "Adaptive Meshing – The Future of Explosion Modeling", 16th Global Congress on Process Safety, AICHE, Houston, 2020

Adaptation to Geometry

A method derived from the separating axis theorem (SAT) is used for the adaptation to the structure [6]. The model cycles through the set of plates, refining only the cells which it intersects. The algorithm begins with the level 0 cells and whether the particular level 0 cell has been divided or not, it does not search lower if the plate does not intersect that cell. If the intersection check returns a true then the algorithm steps to the finer levels. If the cell has not been divided earlier, it is divided at that time. This recursive process continues until the volume the plate occupies has been divided to the finest level permitted. At that point, the finest of cells is defined to contain a geometric surface. In Figure 3, these cells are colored "red". The presence of these solid surfaces is represented in the computation by applying the correct boundary conditions.

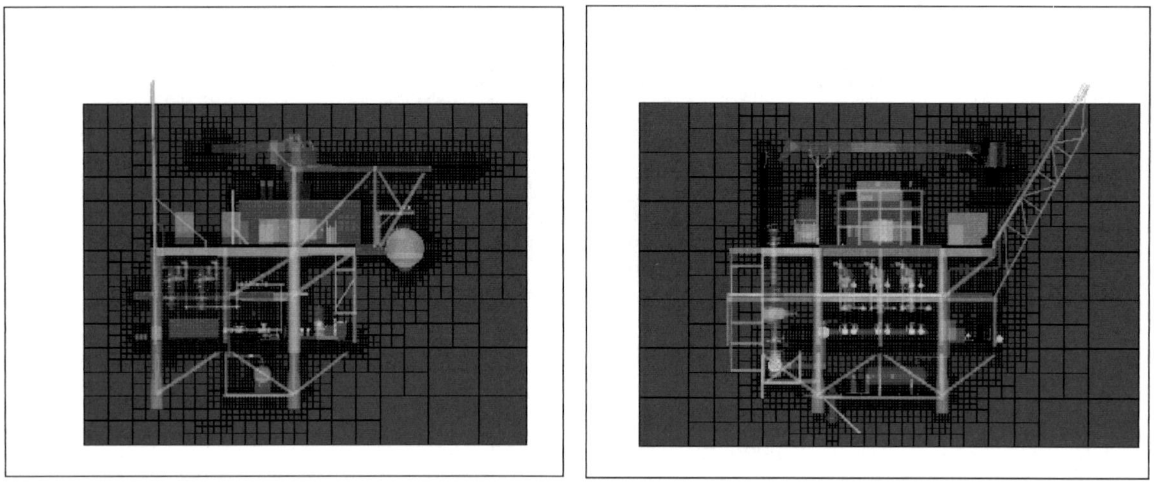

Figure 4. Example of mesh refinement to the model of an offshore platform.

Figure 4 shows the adaptation to a complete offshore platform. Objects can be defined to be solid or hollow. Figure 5 shows cut planes that highlight the adaptation to the various pieces of equipment on the offshore platform. This approach ensures elements like decks and building walls are clearly defined to be obstructions to the flow. Also evident is how large vessels and other objects that can affect the progression of flow or blast are defined. The porosity approach can still be used by importing an object with a defined porosity. For example, a run of piping can either be imported with each pipe defined, or a box encapsulating the piping can be imported. This box would have the appropriate area and volume blockage defined. This is used for items such as grated decking.

The use of the AMR approach dramatically reduces the preparation time for a simulation. If the CAD model is constructed appropriately, it literally only takes minutes to import the model and have a model ready for analysis. The finest mesh spacing is set based on the scale of the smallest items that need to be resolved. Rule sets can be developed for specific problem classes to ensure appropriate resolution and the capturing of critical physics elements.

[6]. T. Akenine-Moller, Fast 3D Triangle-box Overlap Testing, http://www.acm.org/jgt/AkenineMoller01 (2001).

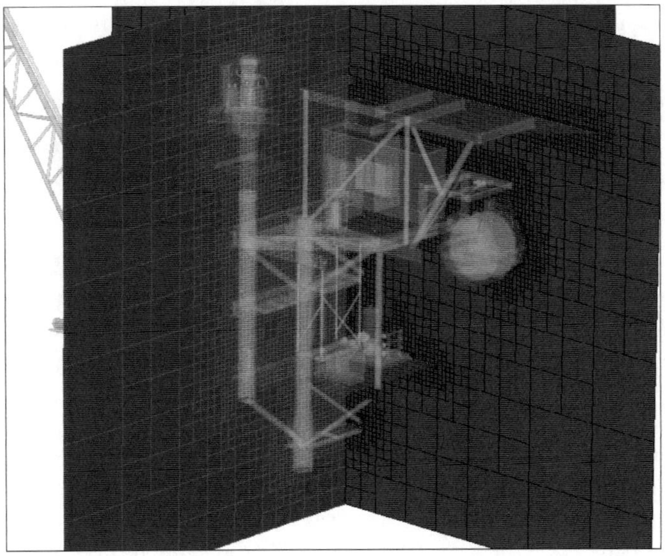

Figure 5. View showing cut planes through the structure showing cells that contain structural components.

Adaptation to Solution

Grid adaptation is also performed during the simulation. For the nature of the problems simulated with CAMBER, refinement is needed in areas of flow property gradients to ensure accuracy while maintaining efficiency. The two primary approaches to setting the criteria for adaptation are (1) using a measure of convergence of a solution or (2) using the local gradient information. Here the second method is used. An adaptation criteria parameter, ε, is calculated and used to determine if refinement is warranted. The parameter assigned to the cell and used in the adaptation decision will be the largest of the set corresponding to the gradient between the cell and its six neighbors,

$$\varepsilon = (\varepsilon^L, \varepsilon^R, \varepsilon^B, \varepsilon^T, \varepsilon^U, \varepsilon^O) \tag{1}$$

where the parameter ε^N corresponds to the largest value as determined by each flow variable of interest and *N* denotes the neighbor. The user can define which variables to refine on with the typical set being pressure, temperature, density, material and velocity. For example, in the case of pressure, the local pressure gradient is determined and used to set the adaptation parameter using

$$\varepsilon^N = \left\{ 1, if \ \frac{|P^N - P|}{min(P^N, P)} > \beta \ 0, \qquad otherwise \right\} \tag{2}$$

where P is pressure in the cell being evaluated and P^N is the pressure in the neighboring cell. The variable β is the criterion that must be met for refinement to occur. The user can set a criterion for each variable such as pressure, velocity or material. These are then used to automatically determine if refinement is warranted.

Mesh refinement for steady-state problems is rather straight forward in that the areas needing refinement are stationary. In that case many codes will refine at every iteration. This is not the case in the class of problems of interest here where there are numerous unsteady wave fronts that need refinement. For the cases involving moving fronts, if refinement is restricted only to the immediate area where the front sits then refinement will have to occur frequently to ensure adequate adaptation is always present. If not, critical results such as the maximum blast pressure produced during the explosion will be under predicted. The requirement may be that adaptation be performed essentially at every time step. This can introduce excessive computational time and counter the goal of the adaptation in the first place. Here an alternative approach is suggested [7].

A method is used here that maps out the current location where key fronts are located and projects their motion. These additional area are refined along with the current location of the front. Such an approach will require that the grid be refined only at discrete times during the time integration. By balancing the mapping and the interval between adaptation then both increased accuracy and efficiency can be achieved.

To develop an algorithm to map out areas in the vicinity of the fronts for refinement, the adaptation criteria already discussed is used. However, it is treated as a domain property similar to temperature and a conduction-diffusion equation is solved to perpetuate the property through the domain. At the intervals when grid refinement occurs the conduction-diffusion equation is integrated in pseudo time, off-line from the time integration of the governing equations of the flow problem.

This is similar to the method used in earlier work [8]. However, unlike the earlier work that used a reaction-diffusion equation, here a convection-diffusion equation was employed. This was done because unlike the reaction-diffusion equation, the c-d equation converges to a steady-state solution. This makes the mapping independent of the time over which the integration is performed. Therefore, the perpetuation of the adaptation property will depend solely on the nature of the flow field and the constants used in the equations.

The approach to map out regions around fronts in unsteady problems that need adaptation begins with the same process used in the steady-state situation. Those regions where the gradient warrants that the adaptation parameter, ε, be set to a value of 1 are defined. These cells are maintained at a $\varepsilon = 1$ state. This parameter is then treated as a property and perpetuated through the domain using the governing equation

$$\frac{\partial \varepsilon}{\partial t} = K\nabla^2\varepsilon - h\varepsilon \tag{3}$$

patterned after the heat conduction-diffusion equation. To show how this works, a simple discontinuity in pressure within a straight duct is used and the results for two sets of parameters are shown in Figure 6. The size of the region around fronts over which ε perpetuated can be controlled with the parameters K and h since those cells where ε is greater than some specified value will be refined. All others will be coarsened. Examples of adaptation to solutions are shown in the results section of this paper.

---

[7]. Clutter, JK, "Application of Computational Modeling for Explosive Hazard Assessments", *International Journal of Protective Structures,* Vol 4, No 3, Sept 2013.

[8]. A.M. Khokhlov, Fully Threaded Tree Algorithms for Adaptive Refinement Fluid Dynamics Simulations, J. Comput. Phys. 143, 519–543 (1998).

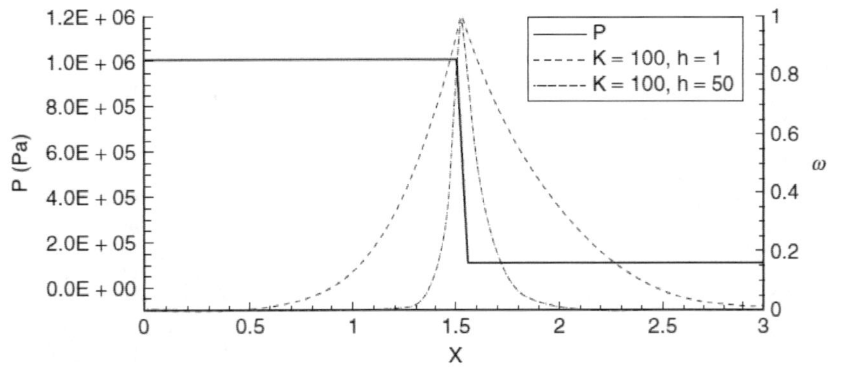

Figure 6. Mapping of the parameter ε for different constants.

RESULTS

For the offshore platform scenario, the adaptation to the structure is evident in Figure 4 and Figure 5. Here adaptation to the solution of the explosion is shown. The vapor cloud is shown in Figure 7 is located on the middle deck. The right image is a cut plane through the center of the cloud. An ignition was assumed to be at the center of the cloud.

Figure 8 shows a sequence of images highlighting how the mesh is automatically refined to the blast front during the explosion. The initial mesh, after the definition of the geometry is shown in Figure 8 Image 1. Image 2 shows the refinement in the vicinity of the ignition shortly after the explosion has begun. Image 3 shows how the AMR logic refines at the blast front and through the high-pressure cloud. Evident in Image 4 is how the volume within the high-pressure cloud, near the ignition is unrefined. When the gradients in the area go away and refinement is no longer needed, the mesh coarsens. This ensures efficiency in the simulation by controlling the mesh to limit computational requirements.

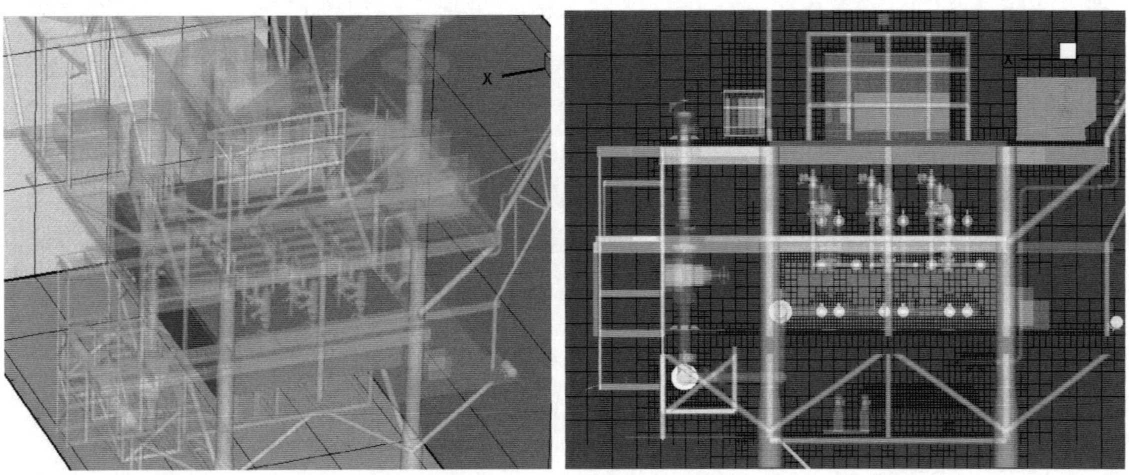

Figure 7. 3D (left image) and 2D (right image) of the vapor cloud used in the analysis.

The refinement to the solution versus the geometry is difficult to see in the images of Figure 8 so Figure 9 show an example of a blast against a simplified onshore structure. The refinement of the mesh in the early stages where there is high pressure in the blast field is evident. Then as the overexpansion occurs behind the front and the gradient in pressure drops, the coarsening of the mesh is seen. In the later stages of the event, it is clear that grid refinement only occurs at the front of the blast.

Using the AMR technique eliminates the need to do numerous simulations to determine what grid spacing provides a grid independent solution for every problem attempted. Rules are set that refine the mesh based on the magnitude of gradients. The finest mesh spacing is set based on a knowledge base as to what is needed for a converged solution. The analyst does not have to determine this for each simulation project.

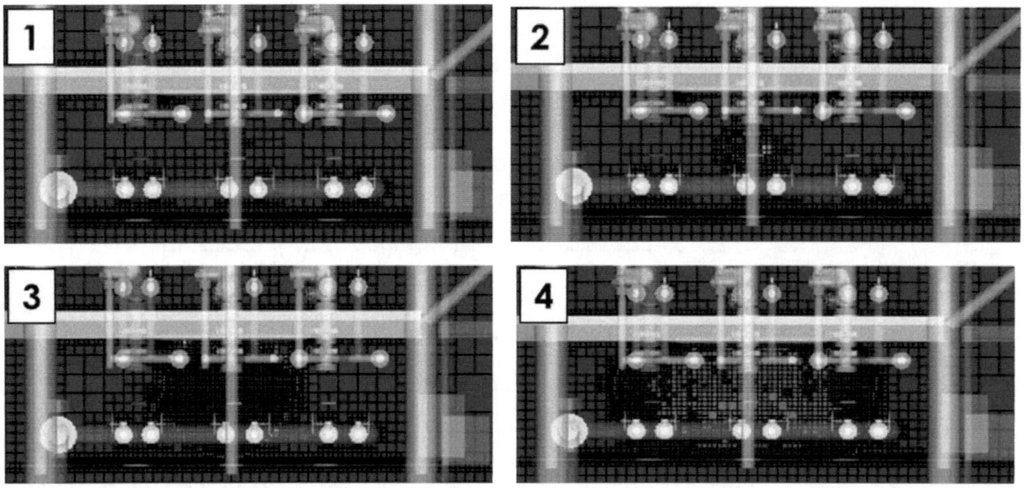

Figure 8. Sequence of images showing pressure highlighting the automatic refinement of the mesh to the blast front.

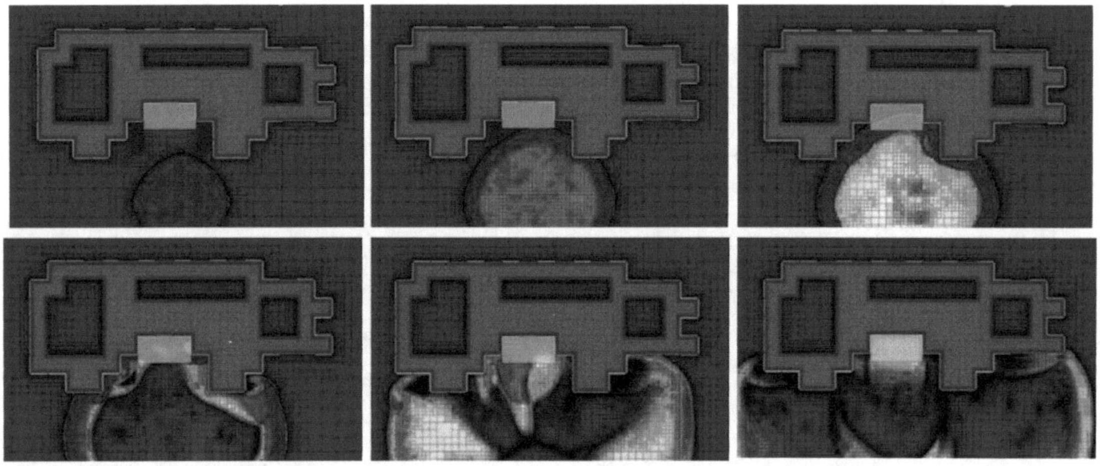

Figure 9. Example of refinement and coarsening of mesh to a blast front.

CONCLUSION

Computational modeling is routinely used for VCE analysis for on and offshore facilities. Much of the labor is related to building computational meshes for the analyses. The accuracy of the results is also highly dependent on the mesh. Applying an AMR approach can address both of these issues. The logic allows for the automatic construction of a mesh around the defined geometry. This tremendously reduces the model construction time. The model can be constructed and simulations can be initiated in literally minutes.

Using predetermined criteria for mesh spacing allows for the accuracy issue to be automatically addressed. The analyst does not have to run numerous simulations to determine the results dependence to the mesh spacing. A fine spacing is used from the beginning. But the AMR logic ensures this spacing is only used where required. This allows the analyst to focus more on the overall objectives of the project and not waste time performing countless benchmark cases.

Rule sets and guidance parameters can be developed for various problems sets. These can be used to simply guide a user, or as default settings, when conducting certain problems. This can eliminate or reduce the requirements of the "knowledgeable" user when setting up simulations. The rules sets can be easily adapted to fit various stages of a project or various solution requirements. For example, in early design, simulations could be conducted with little effort, setting up problems in a matter of minutes. This would allow the design team to gain preliminary understanding of physics effects, at stages were modeling would traditionally be avoided due to costs and efforts required. At later stages, for refined rules sets can be applied to developed refined solutions and explicit design requirements.

SNAME
THE INTERNATIONAL COMMUNITY
FOR MARITIME AND OCEAN PROFESSIONALS

A NOVEL APPROACH FOR SUBSEA INSULATION -NANO COMPOSITE FOR CHALLENGING ENVIRONMENTS

Wen-Tsuen (Vincent) Wang
International Paint

Neil Wheat
International Paint

Chin-Yuan George Ma
Schlumberger

Jay Yun
Schlumberger

Proceedings of the 25th Offshore Symposium, February 27th 2020, Houston, Texas
Texas Section of the Society of Naval Architects and Marine Engineers

ABSTRACT

As we move into deeper, hotter and higher-pressure subsea areas to discover oil; the need for better performance and cost-effective materials to protect subsea equipment in these harsh environments is more prevalent than before. The traditionally used materials have presented challenges in balancing performance, ease of application and cost effectiveness.

This paper describes the materials that have been identified as candidates for the service for subsea thermal insulation and outlines the design and testing philosophy that is being evolved to test and qualify these materials with confidence for very long periods of service. Preliminary test data is presented, along with predictions of what this new class of products will feature. Lessons learned during the project will be of interest to all users of subsea thermal insulation materials and will be valuable to those involved in subsea engineering

Keywords: SNAME, Offshore Symposium

Introduction

As drilling and production operations expand into ever deeper water and new environments are being explored, the temperature of produced oil and gas is either becoming hotter or arctic climates are placing increasing demands on current insulation materials.

Current market trends include enhanced environment and safety performance with a greater focus on insulation qualification for subsea projects. Increasing water depths such as 14,000ft (>4,000m) are being targeted. We are seeing an increase in innovative engineering solutions with the drive to reduce surface facilities to increase well recovery efficiency. More remote fields and subsea tieback to existing facilities are under development. Engineers are ultimately trying to improve recovery rates and focus on the need for increased flow assurance.

Currently within the temperature range of 100-150°C there is no easy hand-applied insulation product. It is understood that a material which is easier to apply by hand and with a maximum temperature rating of 125-150C would prove popular with applicators if the price point was below silicone insulation systems.

Working together with Schlumberger's R&D 3MT group; we have identified several key properties that customers will like to have:

- Apply by hand to provide great flexibility for application.
- No molds needed --reduce the cost and installation time.
- Sag resistance up to 2'' thickness per coat –reduced layers.
- Low k value– less material.
- Low exotherm, no volume limit per pour – reduced installation time.

The project was divided into five phases:

Milestone	Task
Phase 1	Initial Material Screen
Phase 2	Material Characterization
Phase 3	Scale-up testing and process definition
Phase 4	Third-party Simulated Service Test (SST)
Phase 5	Third party qualification test (long-term aging test) and additional tests

Design Philosophy

Project objective was to develop a cost effective, flexible, durable, low thermal conductivity and easy to install insulation coating suitable for use on subsea equipment such as trees, manifolds, PLET, connectors, etc. The product to be suitable for maximum service temperature up to 150°C, tested to ExxonMobil GP-65-08-01 and assessed in accordance with the principles of the ISO 12736 standard.

The product was required to have better flexibility and cost effectiveness compared to the current and traditional 5-6-layer pack-in-place systems in the market. This meant better crack resistance, improved durability, faster handling and improved mechanical resistance.

Simplified product application was a key requirement and a 1-2-layer application was identified in the project brief.

Material Selection

Many different kinds of materials can be used for subsea insulation, including epoxy, hydrocarbon resins, polyurethane, silicone, polypropylene, and various rubbers. Each material has distinct advantages and disadvantages. Extensive testing has shown that thermosetting epoxy resins have superior properties in most hot, wet applications, especially in ultradeep service. An epoxy syntactic normally consists of at least four components: the basic epoxy, a hardening agent (or curative), one or more fillers, and additives, such as diluents, catalysts, modifiers, and coupling agents. Desirable properties include ease of processing, long life under service conditions, high efficiency, and acceptable economics.

Most epoxy resins are rigid, especially those with good high temperature performance. Because they are so rigid, epoxies may be difficult to cast directly onto pipe without cracks. One solution is to add toughening agents or flexibilizers. The most commonly used flexibilizers are CTBN or castor oil, the purpose being to create a two-phase matrix. Unfortunately, the effect of added flexibilizer is to reduce hot-wet performance. Another disadvantage is aging. Most polybutadiene materials, for example, perform well in the initial stages, but deteriorate after thermal aging. As a result, it is important to choose flexibilizing agents carefully and to balance long-term performance with flexibility.

Another way to avoid cracking is to use so-called" soft "materials, such as naturally flexible epoxy, or an elastomer such as polyurethane. Unfortunately, most flexible materials do not provide good high temperature performance, often showing high water absorption and cracks. There is much debate regarding polyurethane as a high temperature insulation material. A concern is the possibility of hydrolysis after the material is exposed to hot-wet conditions for a long period. Extensive testing has shown that most polyurethanes perform well at temperatures below 70 °C, but, above this temperature, the materials lose their water resistance and some even show reversion.

Working with some specialty material suppliers; we have identified an epoxy resin with Nano technology to address the rigidity issue for traditional syntactic epoxy foam. Performance can be optimized by tailoring the structure, chemistry, particle size and distribution. The Nano technology provides:

Enhanced Mechanical Properties: Fracture, Fatigue, Abrasion, Crack, Peel, Scratch, Abrasion

Improved Chemical Properties: Corrosion, Acid Resistance, Shrinkage Reduction

Figure 1. Epoxy with Nano Filler

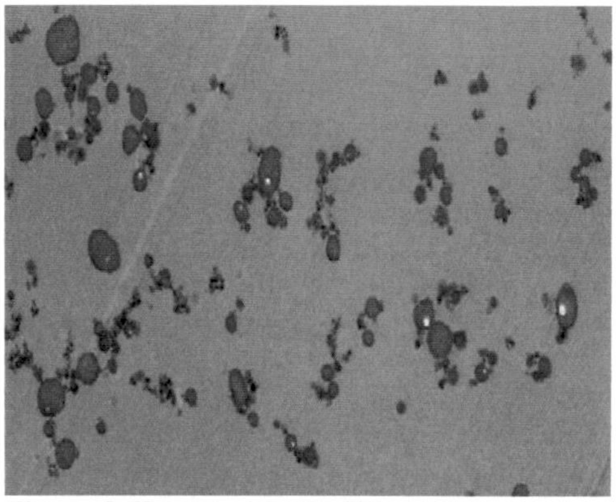

Figure 2. Epoxy with Traditional filler

DEVELOPMENT

Phase 1& Phase 2 Initial Material Screen and Material Characterization

During these phases various formulations were made and tested at different temperatures to compare their mechanical properties (density, hardness, compression, tensile, creep, shear and impact), thermal properties (thermal conductivity, glass transition temperature, specific heat), hydrostatic properties (water absorption at different temperatures and pressures, crush pressures) and aging properties (thermal and humid aging) so we can get more understanding on material performances.

			Average			Standard Deviation	
Formulations	Test temperature	Elastic modulus (MPa)	Tensile strength (MPa)	Strain to break (%)	Elastic modulus (MPa)	Tensile strength (MPa)	Strain to break (%)
A	21°C	50.27	2.8	8.01	7.32	0.5	0.64
B	21°C	54.25	3.23	8.44	9.23	0.42	1.08
C	21°C	28.87	1.62	7.84	2.73	0.04	0.96
D	21°C	43.83	2.38	7.26	6.87	0.33	0.79
A	125°C	43.86	0.82	2.08	1.54	0.27	0.31
B	125°C	51.41	1.36	2.78	1.83	0.13	0.21
C	125°C	23.51	1.36	2.78	0.77	0.11	0.63
D	125°C	39.08	1.09	3.2	1.37	0.19	0.01
A	-35°C	2723	24.11	0.99	104	0.19	0.12
B	-35°C	3049	28.25	1.07	286	1.63	0.17
C	-35°C	2372.5	15.69	0.72	94.05	1.09	0.12
D	-35°C	2644	28.27	1.28	1.41	2.57	0.18

Table 1 Tensile property Comparison

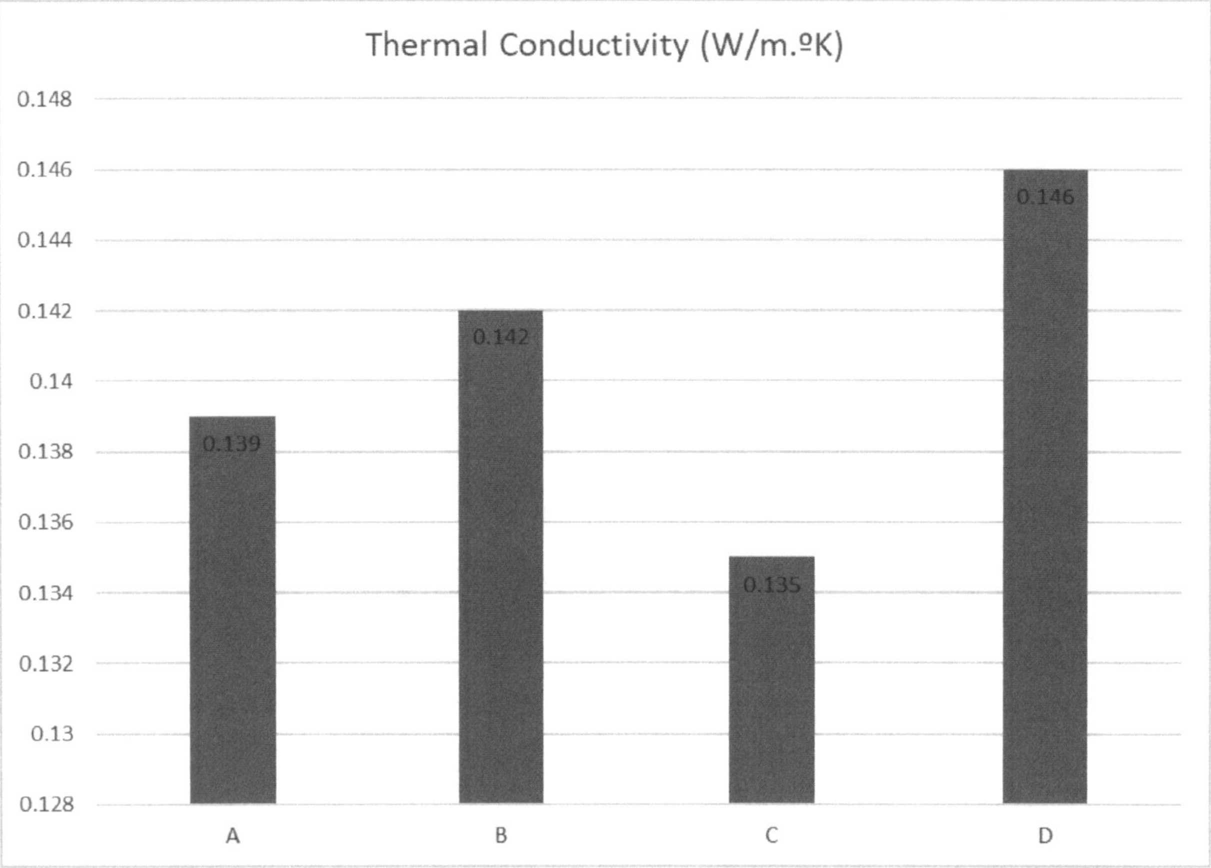

Figure 3. Thermal Conductivity Comparison

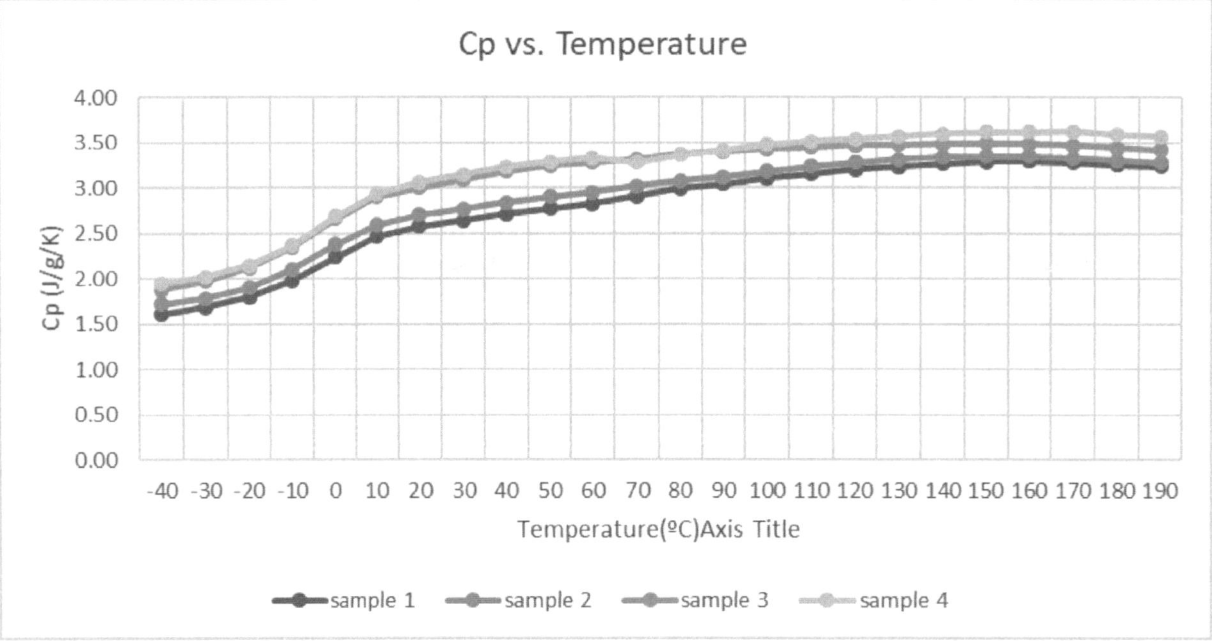

Figure 4. Specific Heat vs. Temperature

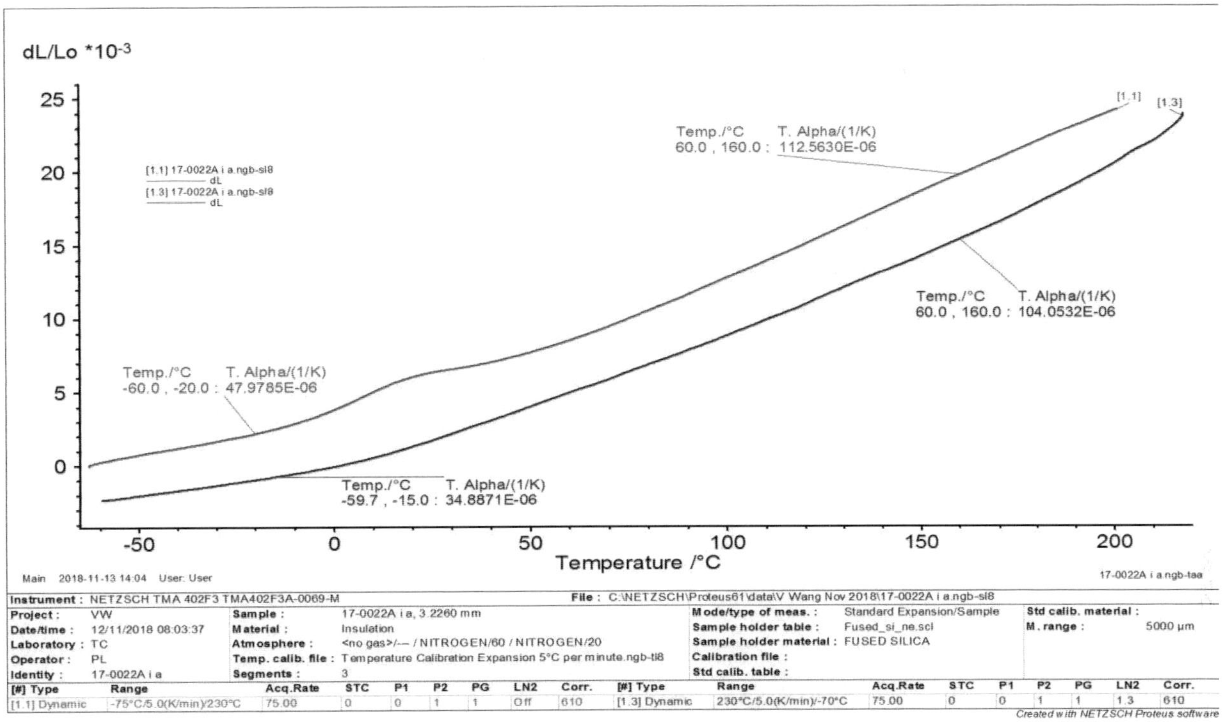

Figure 5. Coefficient of Thermal Expansion Test

Phase 3 Scale-up testing and process definition

Batch Scale Up

Initial wet samples were manufactured in the laboratory which allowed for the small-scale laboratory experiments to be progressed.

During the development phase, the batch size was increased from laboratory small scale to full production run to allow for large volumes of material to be produced for use in application trials and qualification testing.

A total of five 35-gallon batches of material were manufactured. The physical and thermal properties of material sampled from different locations in the batch mix were measured. These results are tabulated below:

Material	Batch #1-Top	Batch #1-Middle	Batch #1-Bottom
Manufacture date	2/14/2017	2/14/2017	2/14/2017
Density (Part A), g/cc	0.6912	0.6928	0.7129
Density (Target), g/cc	0.6980	0.6980	0.6980
Difference, %	0.97	0.74	2.13

Table 2 Batch # 1 Results

Material	Batch #2-Top	Batch #2-Middle	Batch #2-Bottom
Manufacture date	2/23/2017	2/23/2017	2/23/2017
Density (Part A), g/cc	0.6759	0.6805	0.6875
Density (Target), g/cc	0.6980	0.6980	0.6980
Difference, %	3.16	2.5	1.50

Table 3 Batch # 2 Results

Material	Batch #3-Top	Batch #3-Middle	Batch #3-Bottom
Manufacture date	3/23/2017	3/23/2017	3/23/2017
Density (Part A), g/cc	0.6912	0.6928	0.7129
Density (Target), g/cc	0.6980	0.6980	0.6980
Difference, %	0.97	0.74	2.13

Table 4 Batch # 3 Results

Material	Batch #4-Top	Batch #4-Middle	Batch #4-Bottom
Manufacture date	4/13/2017	4/13/2017	4/13/2017
Density (Part A), g/cc	0.6759	0.6805	0.6875
Density (Target), g/cc	0.6980	0.6980	0.6980
Difference, %	3.16	2.5	1.50

Table 5 Batch # 4 Results

Material	Batch #5-Top	Batch #5-Middle	Batch #5-Bottom
Manufacture date	5/11/2017	5/11/2017	5/11/2017
Density (Part A), g/cc	0.7022	0.7065	0.7079
Density (Target), g/cc	0.6980	0.6980	0.6980
Difference, %	0.6	1.21	1.41

Table 6 Batch # 5 Results

During the manufacture scale-up program some modifications were made to the procedure to optimize the process and ensure consistency in batch manufacture.

Field Trial

Several field trials have been conducted with different applicators during the development stages of the material

Application Process Block Diagram

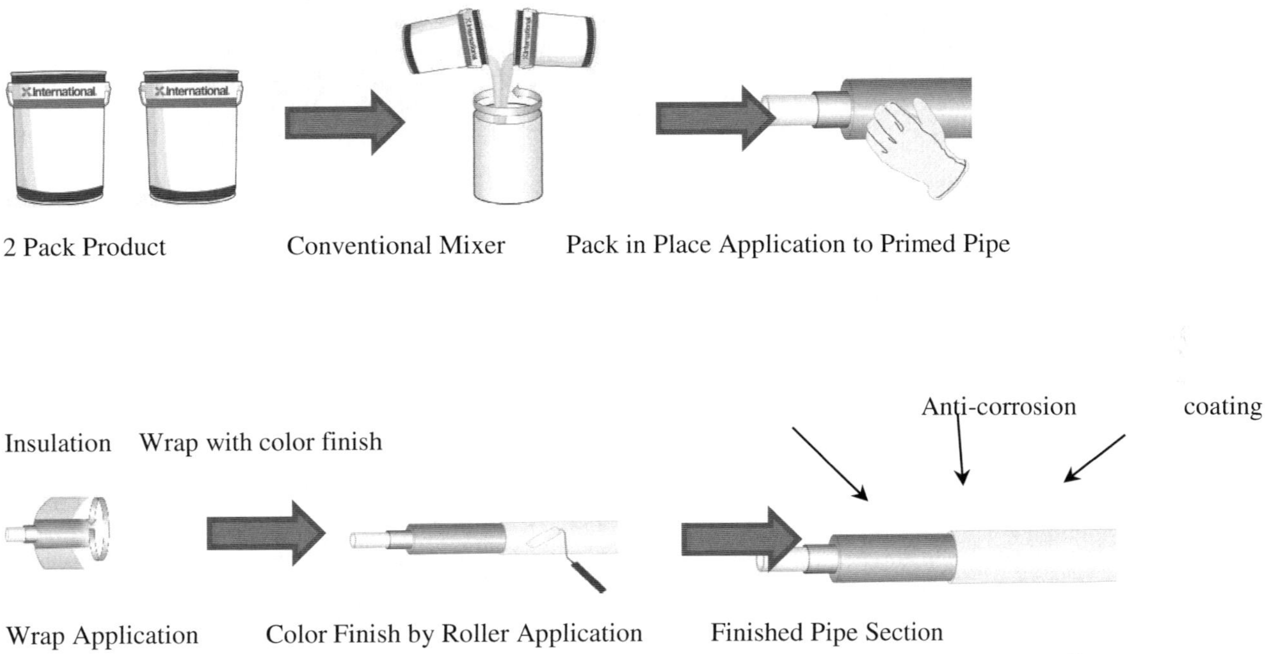

The goal was to coat mock structures under real field conditions and evaluate material performance after the it was fully cured. The material performed well during the application trial and met our expectations except for the curing time which was slower than our initial design. To reduce the process time even further, the possibility of speeding up the curing time was investigated by changing the process conditions or by slight modification of the formulation. Applicator feedback was very positive.

Photo 1 –Mock up structure

Photo 2 –Mock up structure

Photo 3 –Mock up structure

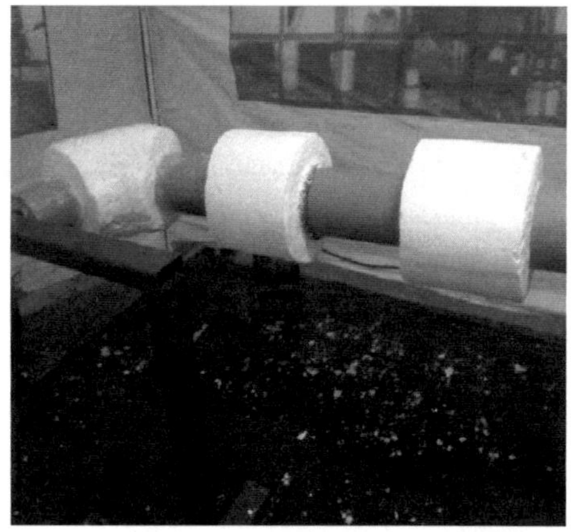

Photo 4 –Applying Insulation onto structure

Photo 5 –Finish with wrap

Photo 6 –Finish with wrap

Third-party Simulated Service Test (SST)

Part of Exxon GP qualification program is the Simulated Service Testing (SST); the purpose is to verify the mechanical and thermal performance of the insulation under 11-days stress test program.

A simulated service test of the insulation system was performed, by Southwest Research Institute (SWRI), over an 11-day period in accordance with ExxonMobil GP65-08-01 on samples prepared as per Figure 6 with 3" (75mm) insulation thickness.

The pipe section internal test maximum operating temperature (TmxOT) was 302°F (150°C) and the external, hydrostatic, sea water pressure was 4,400 psi (equivalent to ˜10,000 ft. water depth) at a temperature of 39°F (4°C).

During the 11-day test period, the test samples were cycled between TmxOT, minimum safe operating temperature (SOT) and hydrate formation temperature (HFT), where minimum SOT is 122°F (50°C) and HFT is 66°F (19°C). (See Figure 7)

The purpose of the simulated service testing is to primarily determine the time taken for cool down of the internal pipe bore to the specified temperatures and to verify the consistency of the cool down time after several periods of operating at maximum project temperatures. Post SST material property testing was performed by SWRI, in accordance with Exxon GP65-08-01, and allows an evaluation of the insulation material after operation at both temperature and hydrostatic pressure. (See Figure 8)

Before exposure testing, in accordance with Exxon GP65-08-01, a series of tests were also be performed by SWRI on material removed from the short pipe section that was produced during the application periods. (see Figure 9)

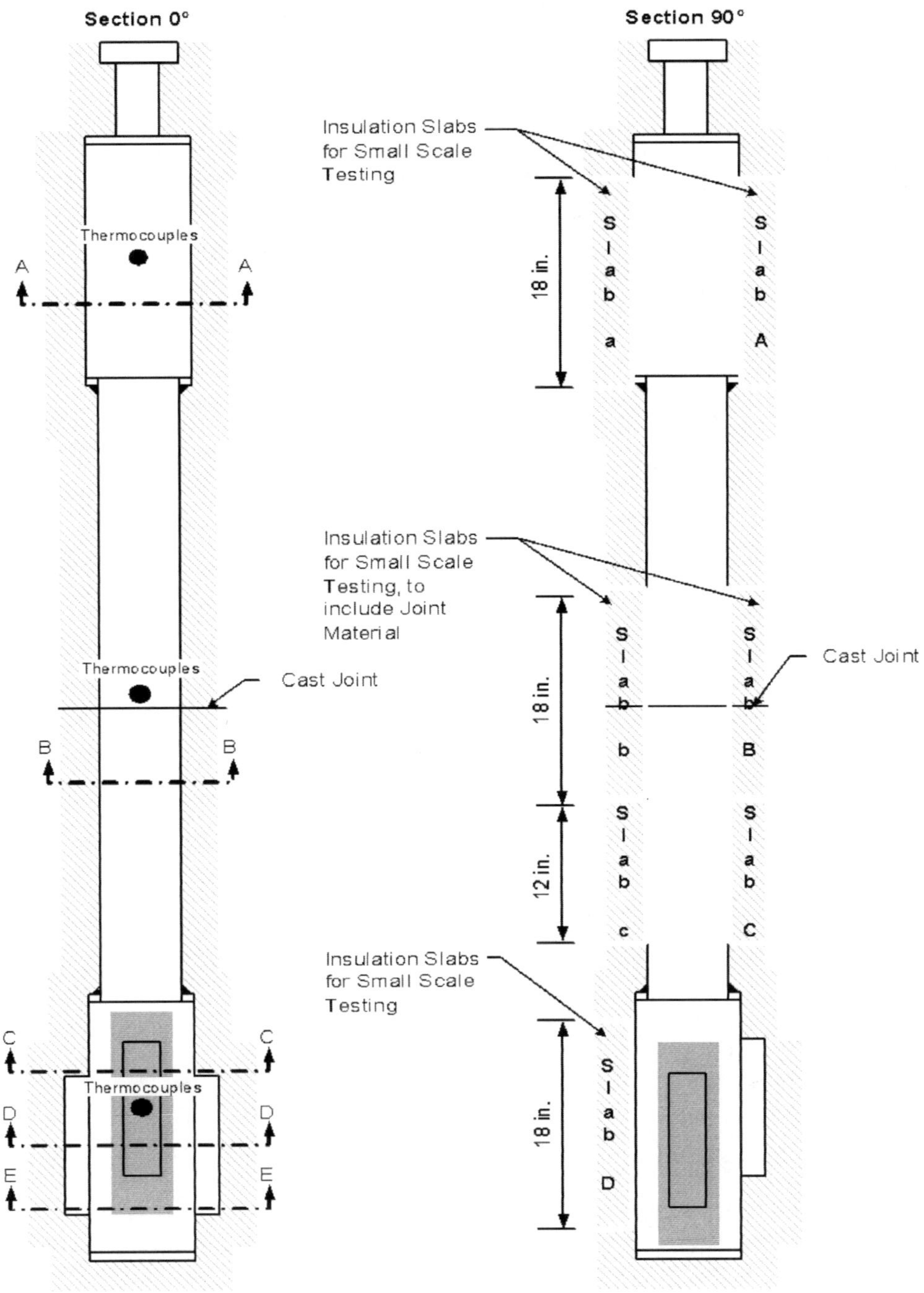

Figure 6 Simulated Service Testing

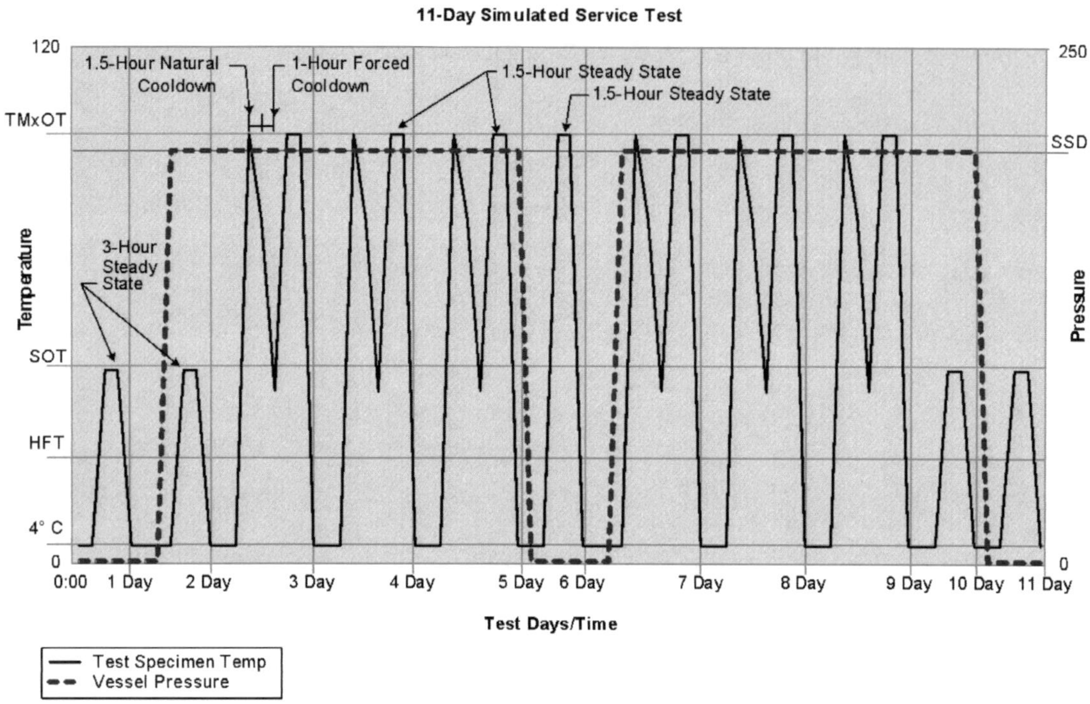

Figure 7. SST cycling test

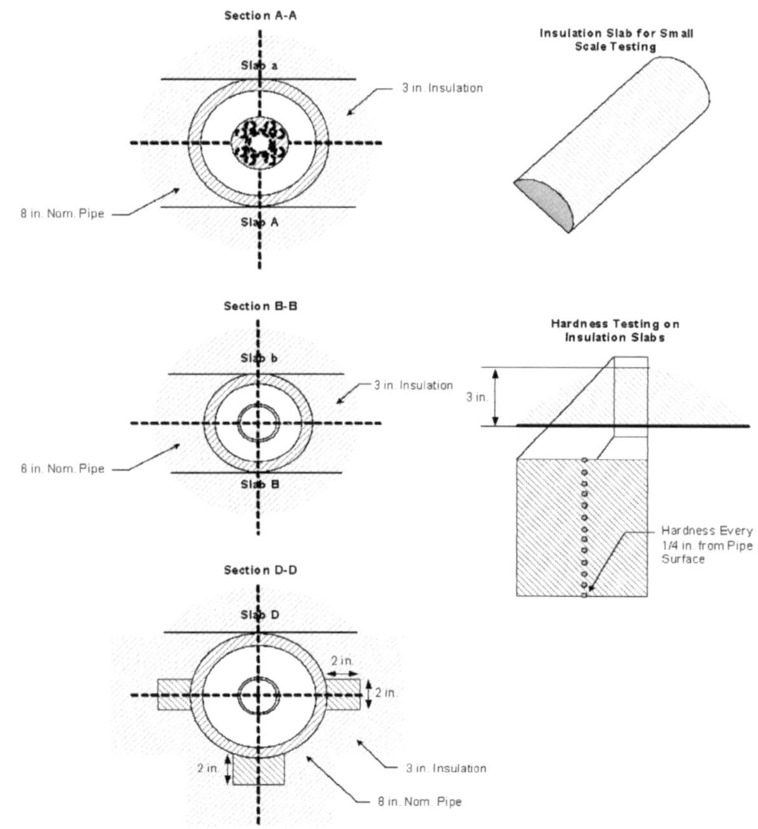

Figure 8. Post SST Test Sample Details

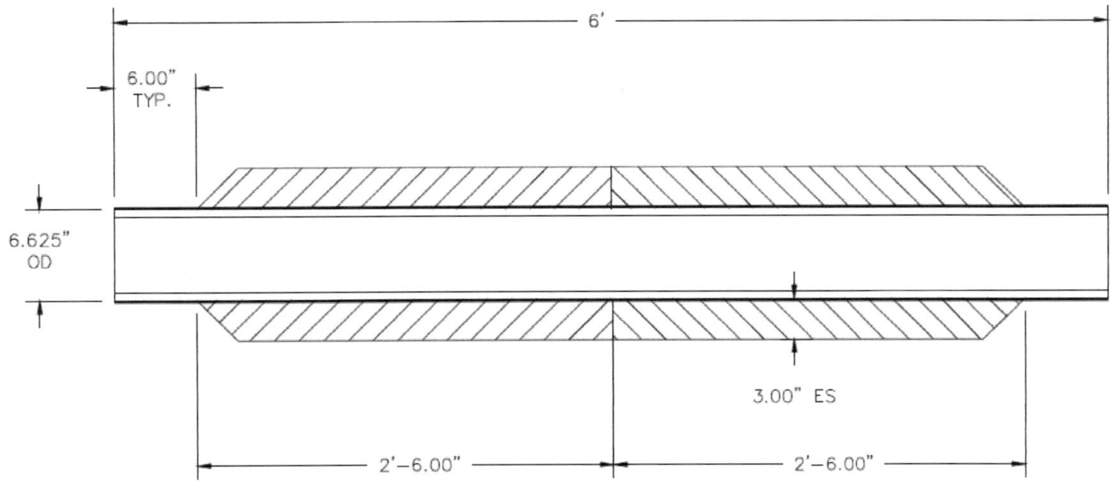

Figure 9. Short Pipe Section 3" Insulation Thickness Details

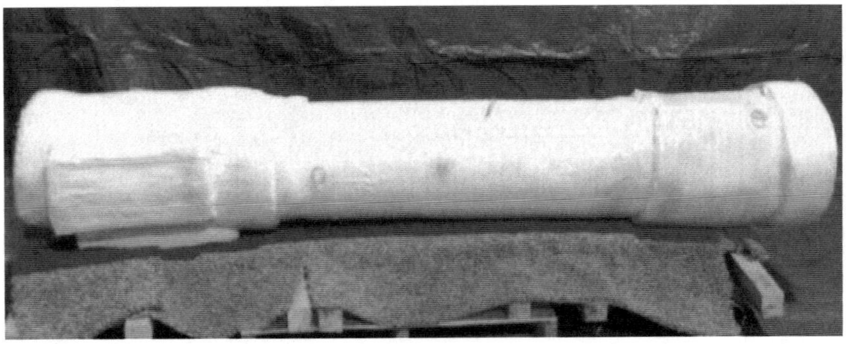

Photo 7. Finish Pipe for SST Test

Photo 8 – Preparing the pipe for testing

Photo 9 – Pressure vessel for SST

Photo 10 – Connecting the pipe with vessel

Photo 11 – Lowering the pipe into vessel

Post Simulated Service Testing

After completing the required cycling tests; the sample pipe was removed from the pressure vessel and visually inspected for any defects such as cracks, disbonding, separation, etc. After careful inspection no physical damages were observed. The initial cool-down time data indicated that insulation material can provide an additional 1-2 hours compared to available data on other commercial products.

Sectioning of the sample and additional tests outlined in the Exxon GP spec were performed.

Photo 12 – Removal of pipe after the test completed

Photo 13 – Inspecting the pipe after the test

Photo 14 –Dimension check after test

Photo 15 –Sectioning the pipe for post SST tests

Long-Term Aging Test

This test attempts to quantify the long-term effects on the insulation material's thermal and mechanical properties in the worst-case scenario (hot/wet) and to prove the mechanical stability of the insulation material as it is exposed to its maximum design service conditions.

Testing started in mid of February,2019 and is expected to complete the one-year aging in February 2020.

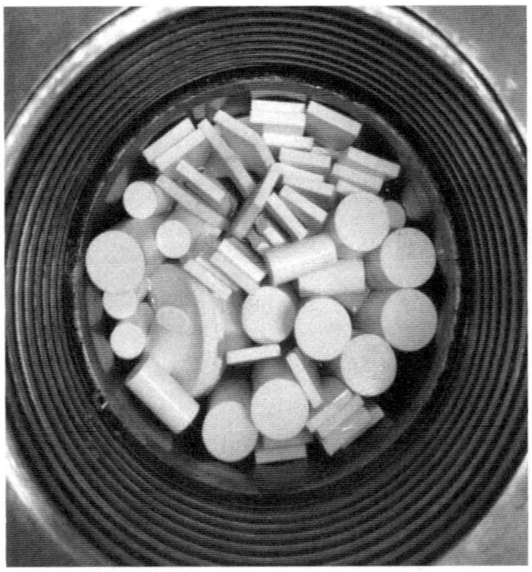

Photo 15 Lon-Term Aging Test

Photo 16 Lon-Term Aging Test Setup

A review of preliminary data for up to 180 days exposure showed that most of the properties were starting to stabilize or only minimum changes which is a very positive indication.

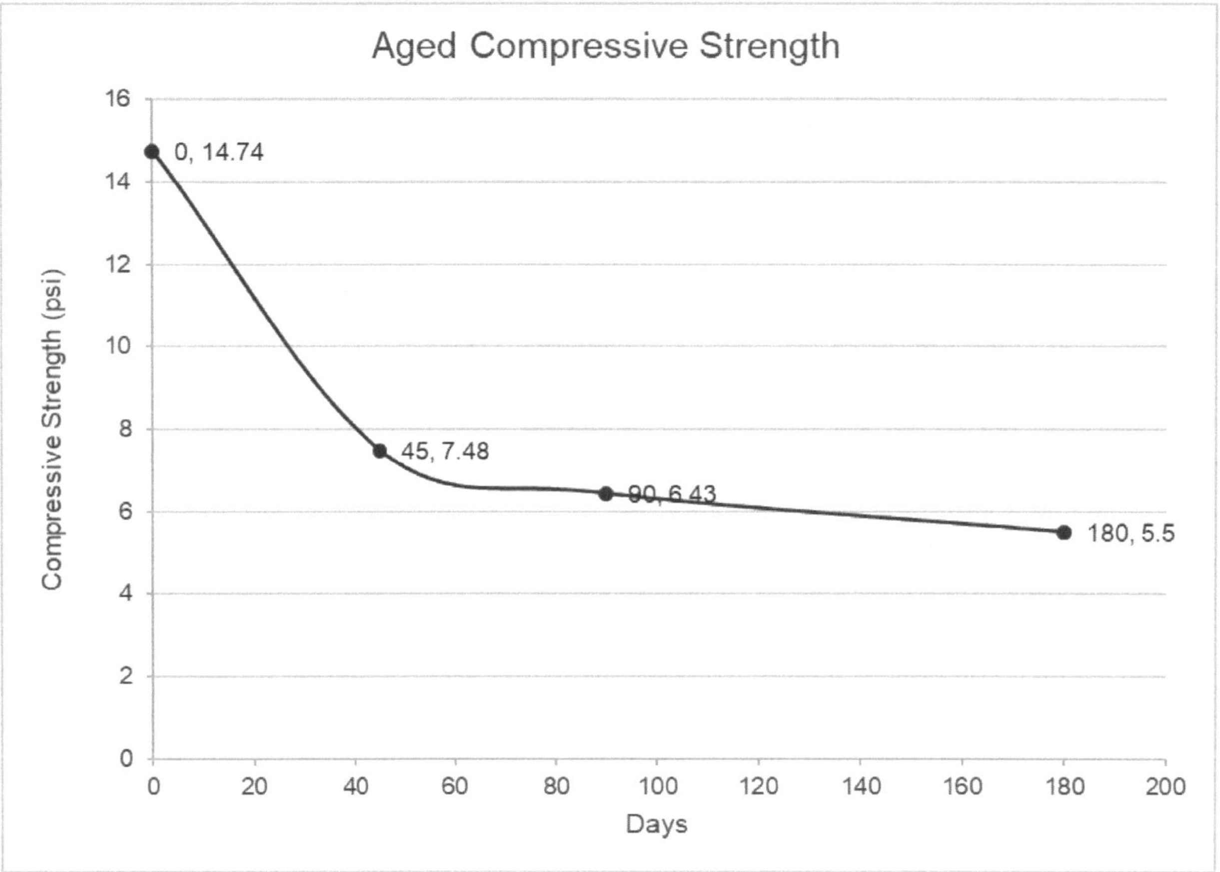

Figure 10 Aged Compressive Strength

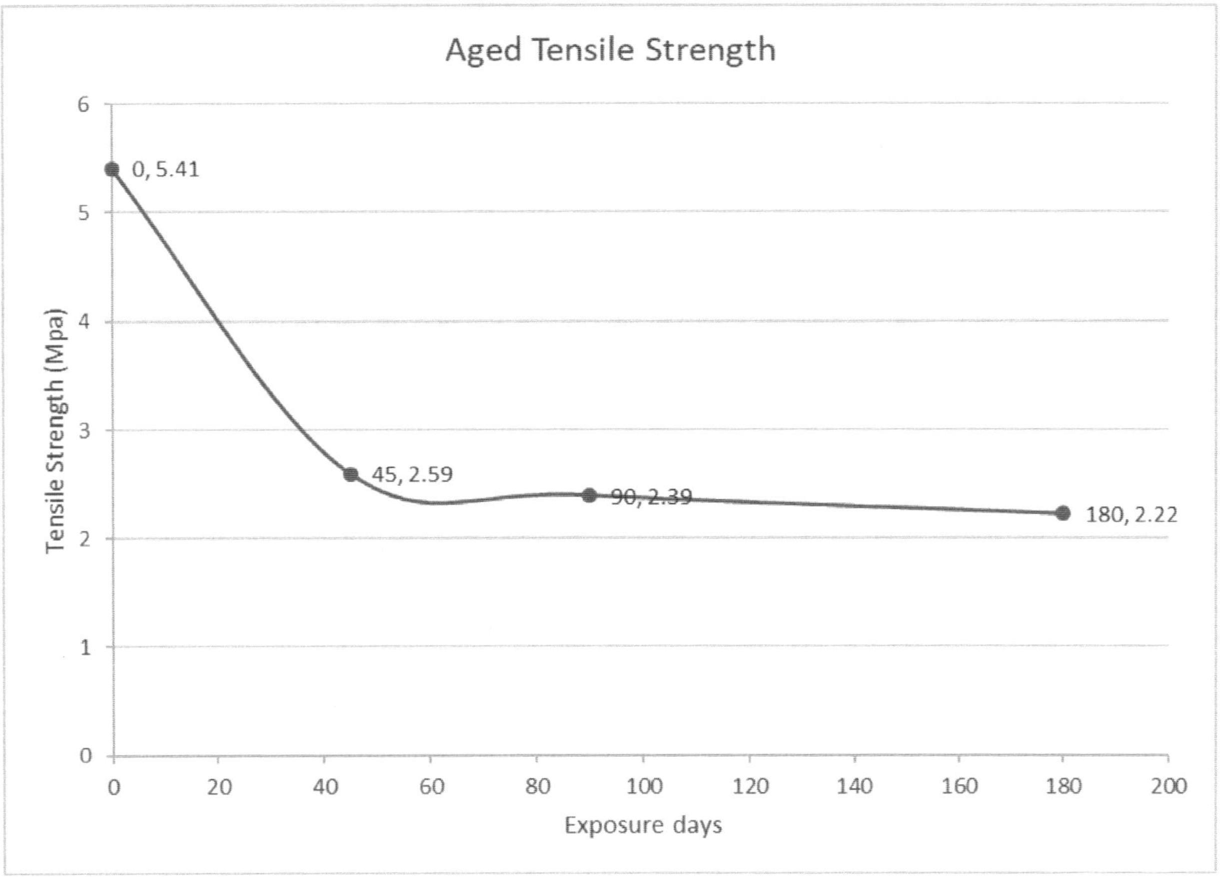

Figure 11 Aged Tensile Strength

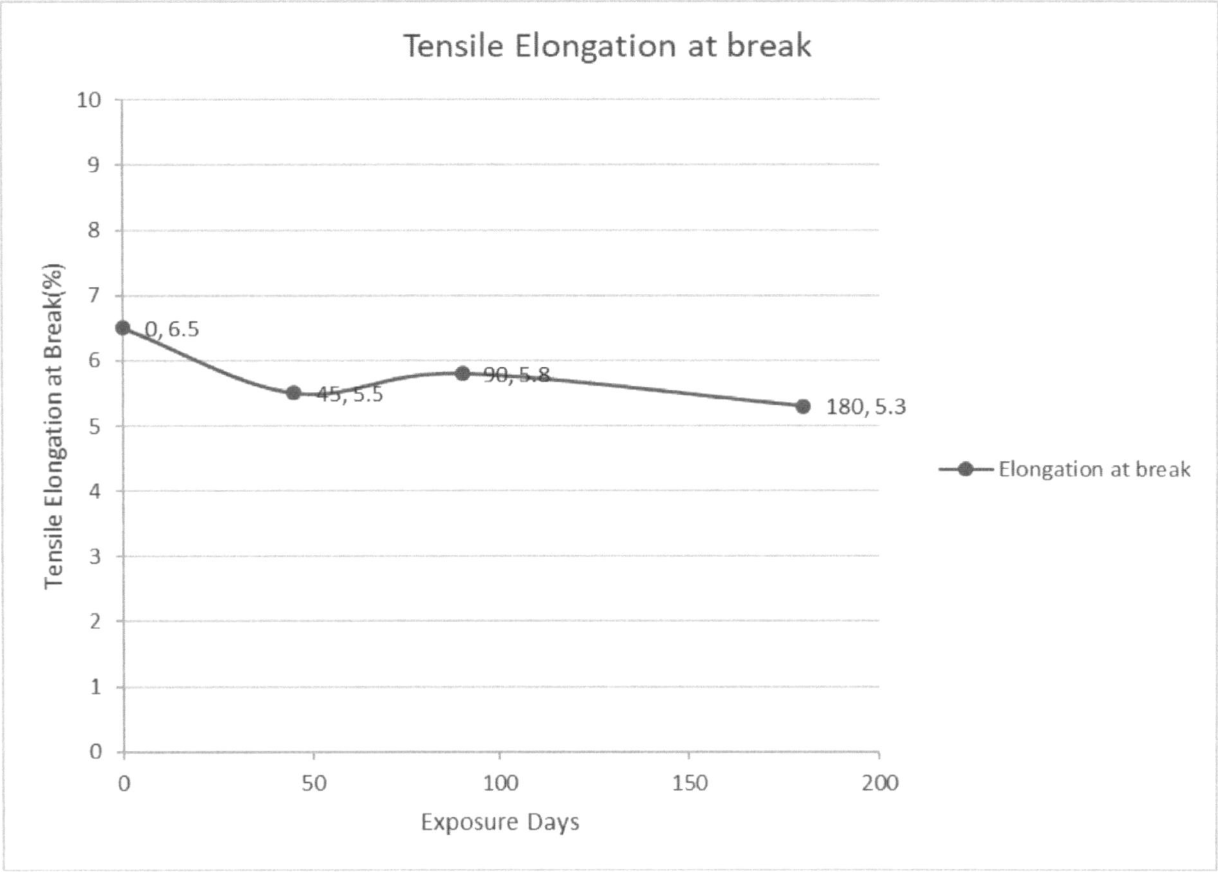

Figure 12 Aged Tensile Elongation at Break

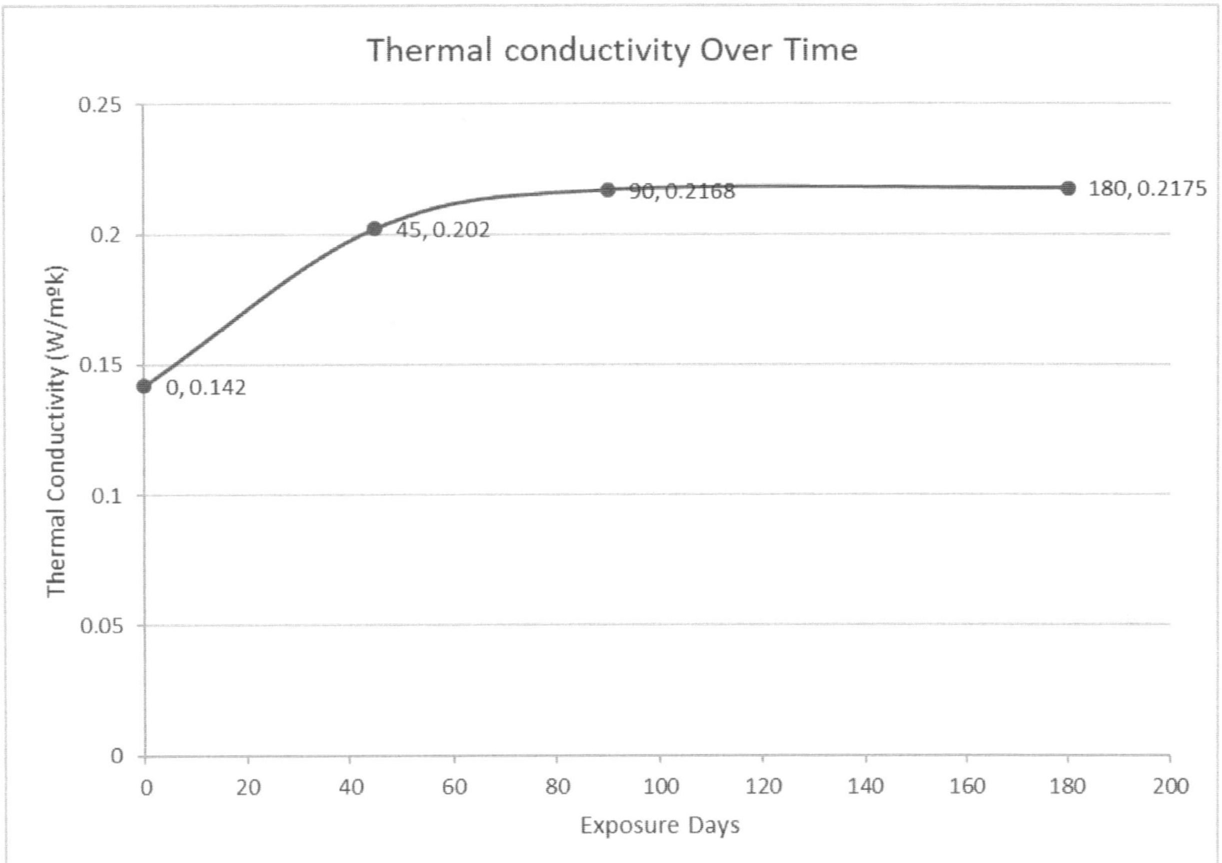

Figure 13 Thermal Conductivity over Time

CONCLUSION

- The insulation material met all the necessary technical requirements from the project brief.
- The insulation material is formulated with special epoxy resin with Nano technology to address rigidity issue for the traditional epoxy resin and the results were very positive compared with the traditional epoxy-based material.
- Maximum achievable application thickness in one application is 1.5to 3 inches depending on the location and staged application approach of building 3 inches layers with short intervals in between.
- Very low exotherm means no volume restriction and less installation time. Provides a cost-effective solution for customer.
- The insulation material provides systems with much better constructability compared to the current market material with superior thermal resistance, more rapid thickness build, faster handling and better mechanical resistance.
- Lower thermal conductivity can provide better cool down time and less material and cost saving.
- Field application has been progressed through trials in the UK and USA and detailed application procedures have been established.

REFERENCES

1. "Handbook of Epoxy Resin" Lee & Neville ,1982
2. ASM Handbook, Volume 21 Composite pp.3 – 17
3. ExxonMobil GP-65-08-01
4. ISO 12736

SNAME
THE INTERNATIONAL COMMUNITY
FOR MARITIME AND OCEAN PROFESSIONALS

AUTOMATED SITE-SPECIFIC ASSESSMENT OF JACK-UP PLATFORMS USING THE CALYPSO COMPUTER PROGRAM

Maas Hoogeveen
Calypso development, Standard Offshore Consultants and Operators

Ronald Rotteveel
Calypso development, Rebels Lab

Proceedings of the 25th Offshore Symposium, February 27th 2020, Houston, Texas
Texas Section of the Society of Naval Architects and Marine Engineers

ABSTRACT

There is a growing demand for Site-Specific Assessments of jack-ups, due to a trend of including an SSA in tender proposals and of optimization of the variable deck load and required preload. In addition, and as a result, there is a growing interest among rig owners to perform SSAs in-house. The Calypso computer program was developed to fulfill this demand.

Calypso enables automated analysis of jack-ups. The core of the program is a full implementation of the ISO 19905-1 standard of 2016. The graphical user interface enables any skilled offshore engineer to perform a Site-Specific Assessment. The jack-up engineering expert may interface deeply with the program through the Application Programming Interface (API). This allows performing customized calculations and inspection of all steps of the calculation process.

This paper summarizes the implementation details of the program. In addition, the graphical user interface is presented. Furthermore, a novel method for simulation of the dynamic response of a jack-up platform using frequency domain methods is introduced.

Keywords: Jack-up analysis, SSA, push-over analysis, time-domain response simulation

Figure 1 : Visualization of a jack-up in Calypso

INTRODUCTION

There is a growing demand from the industry, from drilling contractors, from wind-turbine installation contractors and from field operators to make Site-Specific Assessments of jack-ups an integral part of jack-up operations. SSAs could be used for determination of suitability for a prospect, for support of a tender, during the stay of the platform at the field to support variations such as in variable deck load. Currently, the SSA is typically used only for obtaining insurance and to comply with regulations. This is mainly due to the fact that rig owners have to rely on third parties to perform SSAs, which results in relatively long lead times and relatively high cost.

The Site-Specific Assessment of a jack-up is a complex undertaking. Expertise is required in structural analysis, structural dynamics and hydrodynamics. Furthermore, geotechnical skills are required as well as jack-up specific expertise such as for the modeling of the leg-to-hull interface. This makes performing SSAs the sole domain of jack-up engineering experts.

Calypso implements all this expertise in a computer program, accessible through a graphical user interface. It enables any skilled engineer, not necessarily a jack-up engineering expert, to perform a Site-Specific Assessment for any site using a predefined parametric model of the platform. A typical SSA can be performed in a matter of hours.

The core of the program is a full implementation of the ISO 19905-1 standard of 2016 [1] and of a number of referred documents. The purpose and benefit of this implementation of the ISO standard as a computer program is that the knowledge contained in the standard is made available to a large group of people. Currently, performing Site-Specific Assessments (SSAs) is the domain of jack-up engineering experts. Calypso opens up this domain to any skilled offshore engineer, tender specialist or estimator.

Calypso has an open, glass-box (as opposed to black-box) structure. This means that the jack-up engineering expert is able to interface with the program at a deep level. He can inspect the model and even use specific parts of the program to perform detailed analysis, such as push-over analysis and modal analysis.

The Calypso program was first presented in Hoogeveen [2] at the London jack-up conference of 2019. The focus of that paper is the inner workings of the program, some of the theoretical background, three implementation details and a validation case. Since the publication of that paper, a graphical user interface was developed as well as a novel method of simulation of dynamic and quasi-static platform response using a frequency domain approach. These are the key topics of this paper, along with a summary of the inner workings of the program.

SUMMARY OF THE IMPLEMENTATION

The ISO standard [1] explains how calculation models should be created based on the general arrangement and structural details of a jack-up platform, how the loading on these models should be calculated and how the suitability of the platform should be assessed based on the results. The Calypso program explicitly incorporates a distinction (see Figure 2) between:

- *the parametric platform model,*
- *the site data* and
- *the site-specific platform model.*

The *parametric platform model* is a never-changing description of the structural details of the platform. This comprises the hull shape, the legs, the footings and the leg-to-hull interface. The *site data* contains all information that is site-specific, such as metocean data and soil data. The *site-specific model* is automatically generated by the program based on the *parametric model* and the *site data* and contains all models that are required to assess the structural response and suitability of the platform for the site. More details about the implementation can be found in the next section and in Hoogeveen (2019) [2].

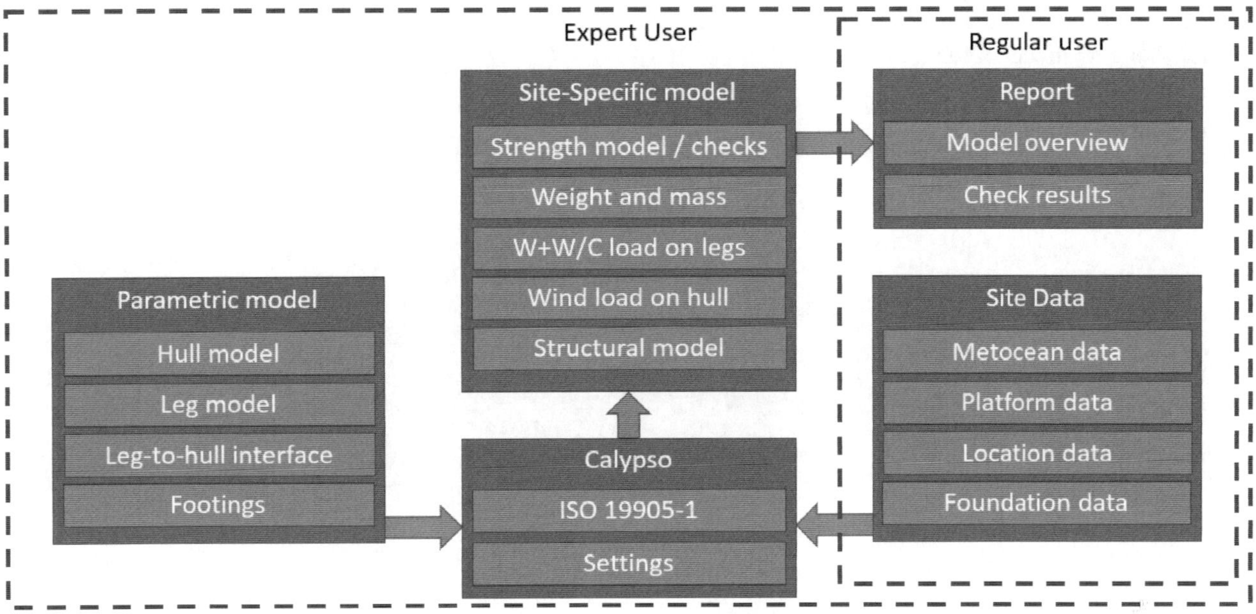

Figure 2 : Schematic overview of Calypso

IMPLEMENTATION DETAILS

The Calypso program is built up modularly and is made up of four groups of objects, that follow the distinction as presented in the previous section, see Figure 2. The expert user can interact directly with these objects through the well-documented Application Programming Interface (API), as presented in the following section.

The first group is called *the parametric model* and consists of all objects that make up the parametric model. This comprises the hull, the legs, the footings and the leg-to-hull interface. All details regarding the platform are stored in these objects. The required information can typically be found in detailed drawings of the platform. The level of detail that can be entered is high. For instance, the number of different sections that make up a leg is unlimited. The objects in the parametric model come with functionality to build the objects of the fourth group, *the site-specific model.*

The *parametric model* is to be created by an expert engineer. This is done once in the lifetime of a jack-up, or after the platform is improved such as with add-on footings (e.g. Jansen et al. [3]). Calypso could be used to determine the business case for such an improvement.

The second group of objects is called the *site data* and comprises all information regarding the site. These objects are the metocean data, platform data, location data and foundation data. The platform data is a set of parameters that pertain to the jack-up and is site-specific, such as variable deck load and the eccentricity box. The location data reflects parameters such as water depth. The foundation data can be one or multiple layers of soil. The penetration can be calculated by the program or given as a parameter if a penetration analysis was performed. These objects are mainly used to store data and come with limited functionality.

The *site data* can be entered by any skilled offshore engineer and does not require jack-up engineering expertise. The required data can typically be collected from a metocean report and a soil investigation report or a leg penetration analysis report if available.

The third group of objects is the calculating and reporting engine of *Calypso*. It takes the *parametric model* and the *site data* and generates the *site-specific model*, which is the fourth group of objects. After the site-specific model is created, the calculations are performed and a report is generated.

The fourth group is the *site-specific model* and contains a set of calculation models. These models are automatically generated based on the *parametric model* and the *site data.* This group consists of:

- The wind, wave and current load model of the legs
- The wind load model of the hull
- The structural (bar stool) model, containing nodes connected by structural elements (visualized in Figure 3)
- The weight/mass model
- The strength model

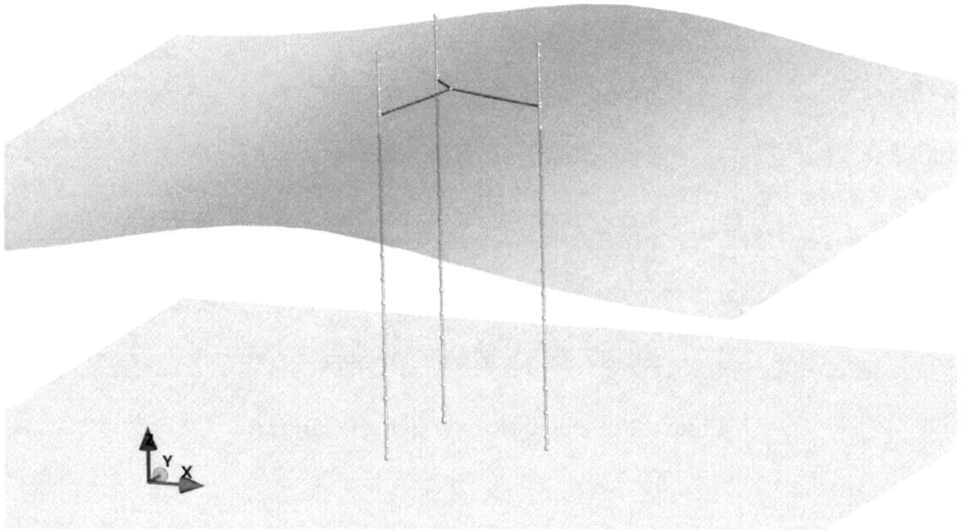

Figure 3 : Visualization of the structural model

THE USER INTERFACE

Calypso is made available as Software as a Service (SaaS). Users can login to a secured private area where they can access a browser-based interface and all of the user's jack-up models, site data and previous analyses. Typical calculations are run directly on the server, whereas batch runs involving time domain simulation can be run on a cluster for accelerated processing time. The program lay-out allows for parallelization and scalability, so that even large batches of calculations could be run swiftly. This section introduces different ways for a user to interact with Calypso as well as two examples of usage of the expert-user interface.

THE GRAPHICAL USER INTERFACE

The graphical user interface is intended to quickly perform a Site-Specific Assessment (SSA) according to ISO 19905-1 without requiring in-depth knowledge of jack-up engineering. It allows for selection of a pre-defined parametric jack-up model and for detailed specification of site data. All site data can be entered in a browser-based webform or as a spreadsheet. The main screens of the interface are presented in Appendix A.

A single calculation takes about 20 seconds to run on a single-core and generation of graphical output takes another 20 seconds if requested. After the calculation is done, the results are reported in a browser-based document, as a PDF file and in a database. A (blurred-out) example of the report of a recently performed SSA is included in the last figure in appendix A. The first section of the report summarizes the site data. The following sections summarize the results in an increasing level of detail.

THE EXPERT USER INTERFACE

The expert user interface allows the user to interface with Calypso directly through the Application Programming Interface (API). This requires jack-up engineering expertise as well as a basic understanding of the usage of an API. The API allows full control over all of the objects in Calypso. Some possible uses are listed in the following sections.

DIRECT USE OF THE STRUCTURAL MODEL

The structural model of Calypso is purpose-built and is mainly based upon Ghali et al. [4]. It is a miniature Finite Element Model, where the structure is made up of nodes that have a mass and are connected by elements. Each node has 6 Degrees of Freedom (DOFs). Nodes can be connected to other nodes through a stiffness matrix or by rigid elements. A stiffness matrix can be defined by a beam element, a spring element or direct input by the user. Rigid elements define one or more of the DOFs of one node (the slave node) as a function of the displacement of another node (the master node).

In typical use of Calypso, the structural model is automatically generated by the program. It can be visualized, see Figure 3, but other than that the regular user does not interface with it. The expert user may take the automatically generated structural model and add or modify any of the elements. For any node, the user may define the load and solve for its displacement, or define the displacement and solve for its load.

PUSH-OVER ANALYSIS

Calypso is suitable for a push-over analysis as described by HSE [5]. This type of analysis involves plastic deformation of structural components in a quasi-static approach. The environmental loading can be scaled up until failure is reached, where failure is defined as the collapse of the structure. Even severe plastic deformation could be deemed acceptable for an accidental case.

In typical use of Calypso, i.e. for an SSA, the foundation is modeled according to the ISO yield interaction envelope. This inherently non-linear model is included through an iterative engine. The stiffness matrix is updated every iteration until the solution has converged sufficiently. Any plastic element can be added by the expert-user in a similar fashion. A simple beam element can be extended by a yield function which basically caps the total loading. Similarly, the leg-to-hull interface can have a load-dependent stiffness to account for gaps and increasing contact area for increasing deformation.

DYNAMIC SIMULATION OF THE PLATFORM RESPONSE IN THE FREQUENCY DOMAIN

The ISO standard provides two methods to use time-domain simulation in an SSA. The first method is the deterministic two-stage approach. This approach is the default method of Calypso. Stage one is a time domain simulation of the structural response of a platform to irregular waves. The response of a dynamic model is compared to that of a quasi-static model in order to determine the Dynamic Amplification Factor (DAF). This DAF is then used in a quasi-static, non-linear design wave approach calculation, which makes up stage 2. This approach is pervasive in the ISO standard. For instance, the ISO yield interaction foundation model is suitable for a quasi-static design wave calculation only and should not be used in a dynamic simulation. ISO recommends, for dynamic simulations, to model the foundation as a rotational spring of constant stiffness.

The second method to use time-domain simulation according to the ISO standard is the stochastic one-stage approach. In this method, the SSA is based directly on the output of the time domain simulation, for instance a three-hour maximum. This method is mentioned in the ISO standard, but not detailed enough to be suitable for automated implementation. For instance, if a linear wave model is used for irregular wave simulation, such as implemented in most commercially available software, the wave forces may be severely underestimated as compared to the Stokes fifth order design wave approach.

The following describes the method of implementation of time domain simulation in Calypso, which the authors believe to be novel. The wave force on a jack-up platform is non-harmonic. The irregular wave elevation and kinematics are best described by second order Stokes theory, which is non-linear. The drag component of the wave force, which is governing for truss leg jack-ups, is quadratic with velocity and thus is non-linear as well. Therefore, the wave force on each node must be calculated in a time domain simulation. The result of this simulation is a force signal on each of the nodes.

The platform response due to dynamic excitation can be modeled as a *linear* process. In the case that the deterministic two-step approach is used, the purpose of the time domain simulation is to calculate the DAF. Therefore, interest goes out to overall platform behavior only, not to component loads. It is shown by Hoogeveen [2] that the Calypso implementation of the P-delta effect, which involves a negative contribution to the stiffness matrix, yields the same results as an iterative method for overall platform behavior, such as deck excursion, overturning moment and base shear.

A linear spring is typically used to model the foundation in dynamic simulations. All other components of the jack-up, including the leg-to-hull interface, are also modeled to behave linearly.

Based on the above, it can be stated that the jack-up can be modeled as behaving completely linear for dynamic simulations that are used to calculate the DAF. In matrix notation it can be stated that:

$$\boldsymbol{m\ddot{u}(t) + c\dot{u}(t) + ku(t) = F(t)} \tag{1}$$

Where $\boldsymbol{F(t)}$ is a time trace of force on every node in every DOF, which was found through time-domain simulation of wave kinematics. This signal can be transformed to the frequency domain using the Discrete Fourier Transform (DFT), so that the force on every node is expressed as a set of complex amplitudes. Now equation **(1)** can be restated as:

$$\boldsymbol{m\ddot{u}(\omega) + c\dot{u}(\omega) + ku(\omega) = F(\omega)} \tag{2}$$

Or, using $\boldsymbol{\ddot{u}(\omega) = -\omega^2 u}$ and $\boldsymbol{\dot{u}(\omega) = i\omega u}$,

$$\boldsymbol{(k + ci\omega - \omega^2 m)\, u(\omega) = F(\omega)} \tag{3}$$

Equation **(3)** can be used to calculate the transfer function between complex nodal force excitation amplitudes and complex nodal response amplitudes. Using that relation, the dynamic platform response due to the wave force excitation can be calculated. Similarly, the quasi-static response can be calculated by taking only the stiffness into account and not damping and inertia. Using the Inverse Fourier Transform, the complex amplitudes of response can be converted back to the time domain. Further analysis, such as determination of the maximum response or most probable maximum response, can be done based on this signal.

This approach is efficient because the platform response is not integrated by a numerical solver but is being solved analytically. Moreover, more insight is provided to the (expert) user because the transfer functions are made available. These can aid in-depth research, such as on reinforcement / cancellation.

CONCLUSIONS

The Calypso computer program makes available all expertise contained in ISO 19905-1:2016 [1] to engineers who are not necessarily jack-up engineering experts. It allows for automated performance of Site-Specific Assessments (SSAs). Performing SSAs become so accessible and low-cost that SSAs may be performed in any phase of a project:

- Before the tender phase, to determine the suitability of a platform for a specific site.
- During the tender phase, to support the tender and instill trust.
- Before the execution phase, to perform the SSA and optimize the required preload or the allowable variable deck load.
- During the execution phase, if operation outside of the limits stated in the SSA is required. This could be due to unforeseen penetrations, imminent weather, operational decision making, etc. Calypso is fast and due to automation has an inherently high level of quality, making it a suitable tool for support of operations.
- For determination of the business case of a platform in a more general sense, as opposed to the design cases.

Because of the glass-box structure, Calypso allows those who are experts in the domain of jack-up engineering to perform an in-depth review of the performed calculations. Moreover, through the use of the well-documented API, the expert user has direct access to all components of Calypso.

REFERENCES

[1] ISO 19905-1:2016, Petroleum and natural gas industries – Site-specific assessment of mobile offshore units – Part 1: Jack-ups

[2] M. Hoogeveen, *Coding the Code: Applying ISO 19905-1:2016 as a software package for site-specific assessments,* 17th International Conference: The Jack-up platform, 2019

[3] J. Jansen, H. Hofstede, M. Hoogeveen, *Improvement of Jack-up Operating Capability Using Add-on Spudcans,* 21st Offshore Symposium, 2016

[4] A. Ghali, A.M. Neville, T.G. Brown, *Structural Analysis, A unified classical and matrix approach,* Fifth Edition, 2003

[5] HSE Research Report 177, *Review of structural modeling of deep water jack up structures,* 2004

APPENDIX A – SCREENS OF THE GRAPHICAL USER INTERFACE OF CALYPSO

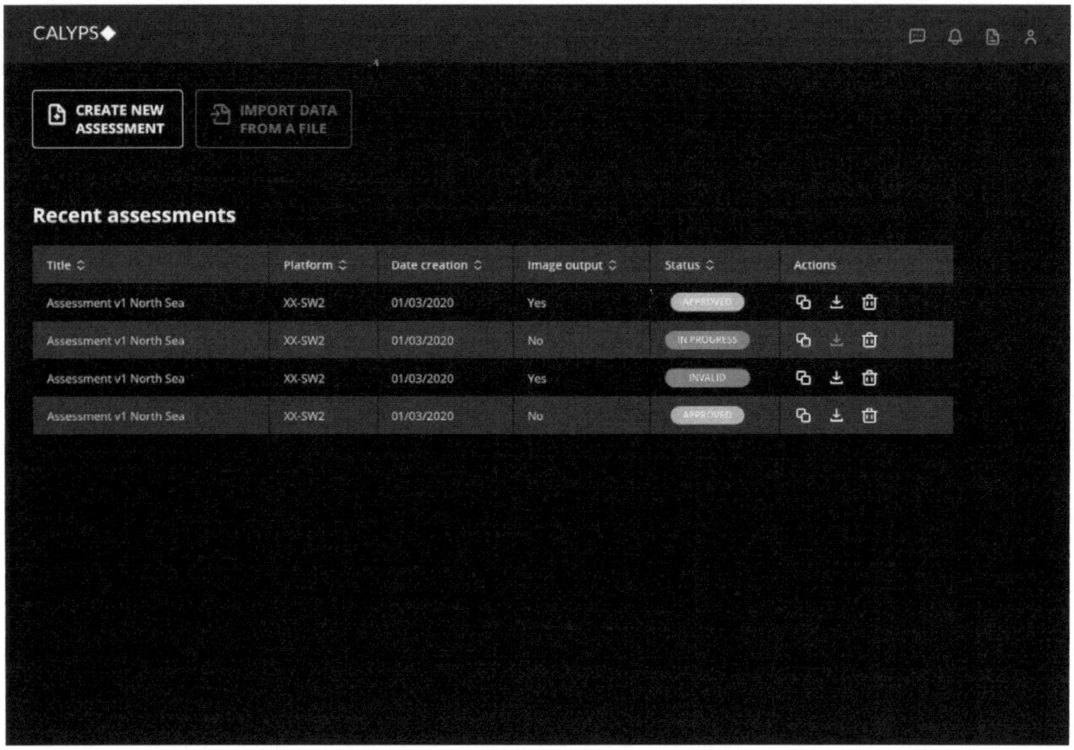

Figure 4 - Calypso Graphical User Interface – Dashboard

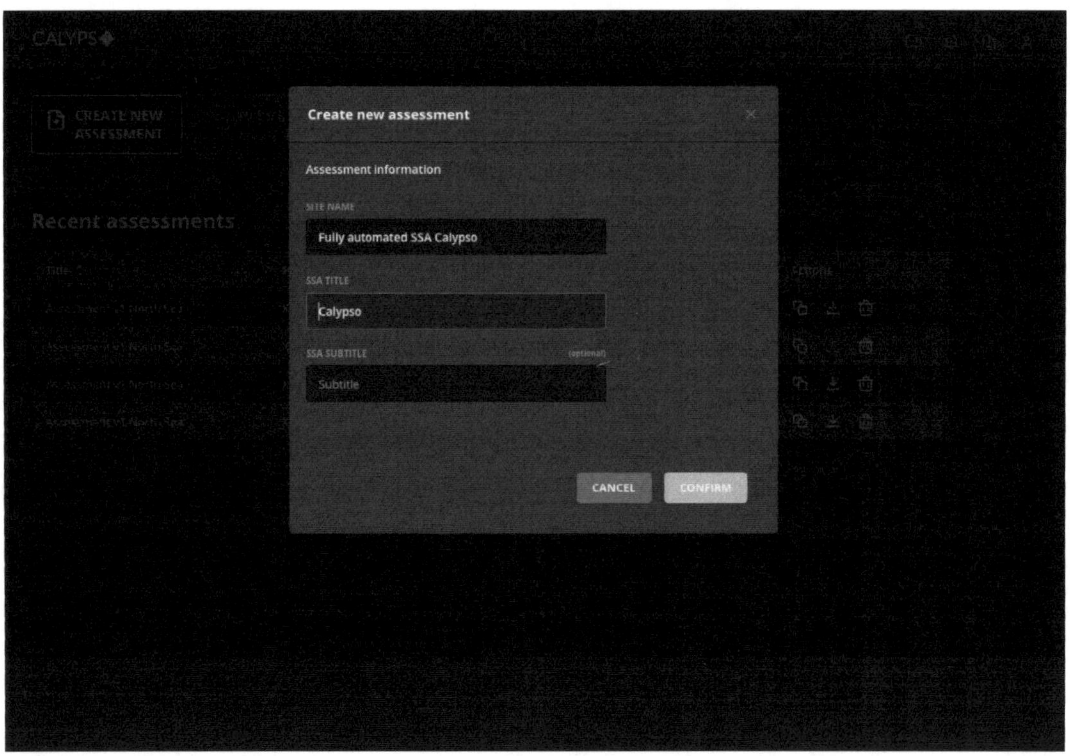

Figure 5 - Calypso Graphical User Interface - Create new SSA

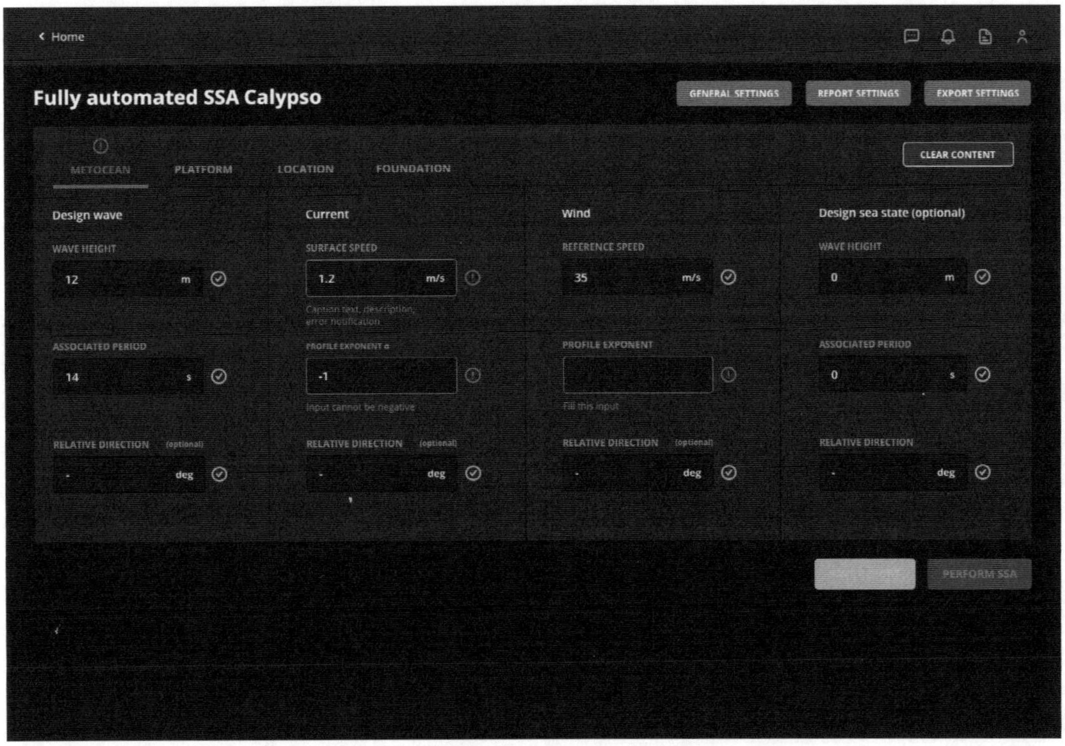

Figure 6 - - Calypso Graphical User Interface - Metocean tab

Figure 7 - Calypso Graphical User Interface - Platform tab

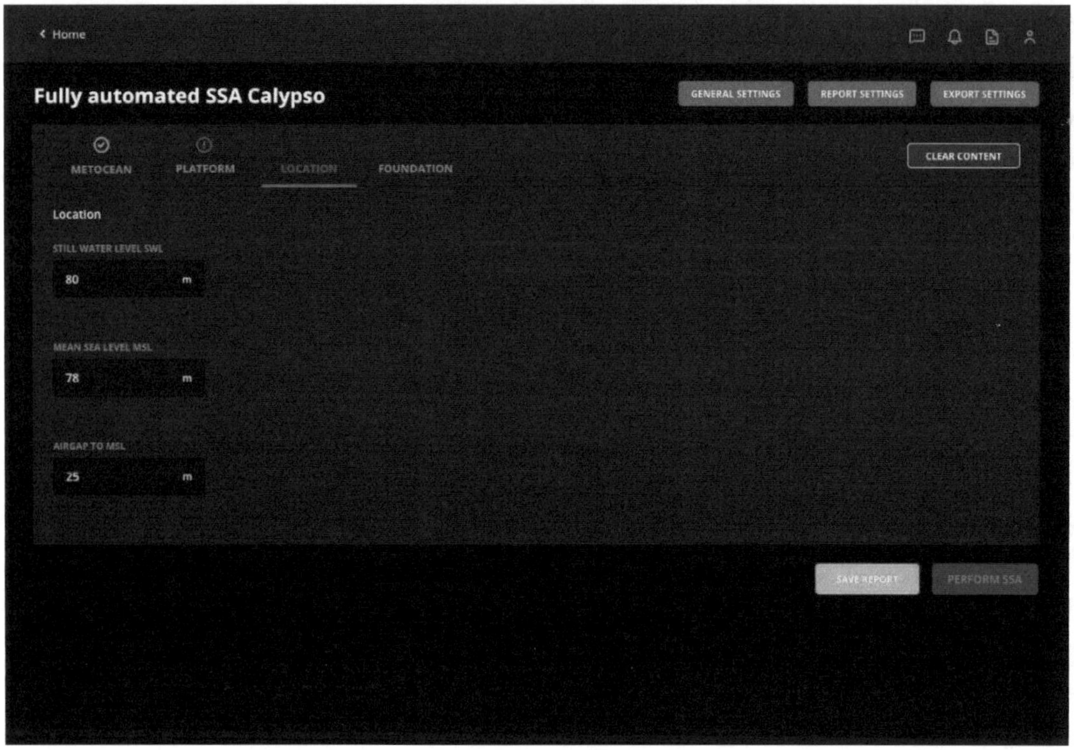

Figure 8 - Calypso Graphical User Interface - Location tab

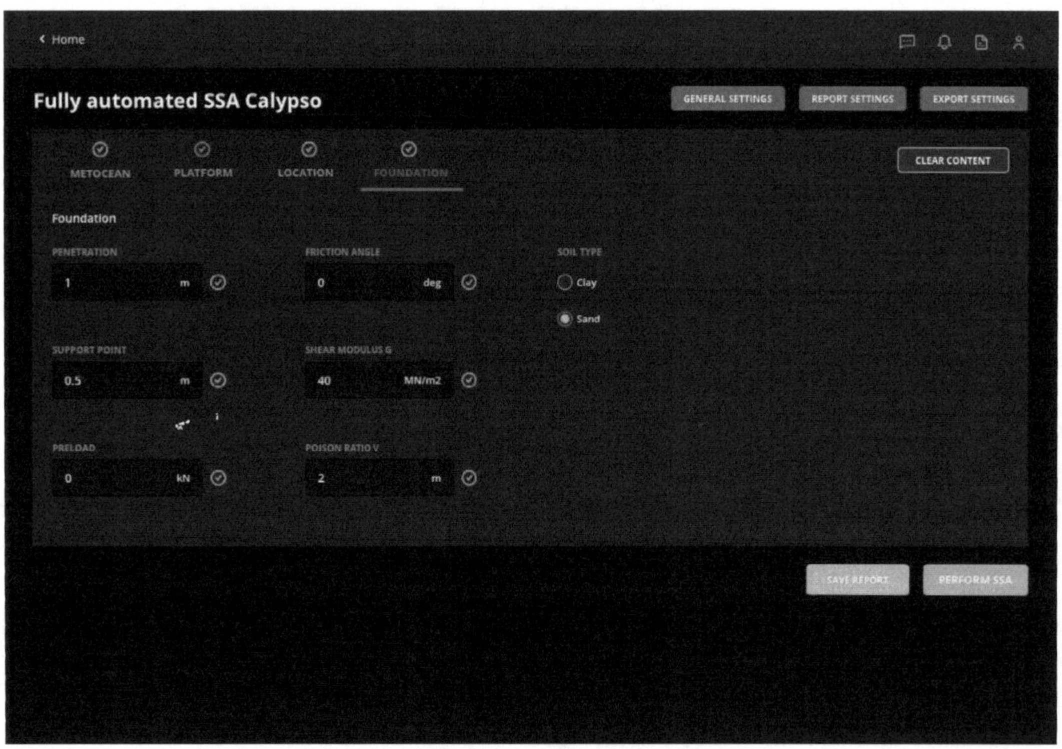

Figure 9 - Calypso Graphical User Interface - Foundation tab

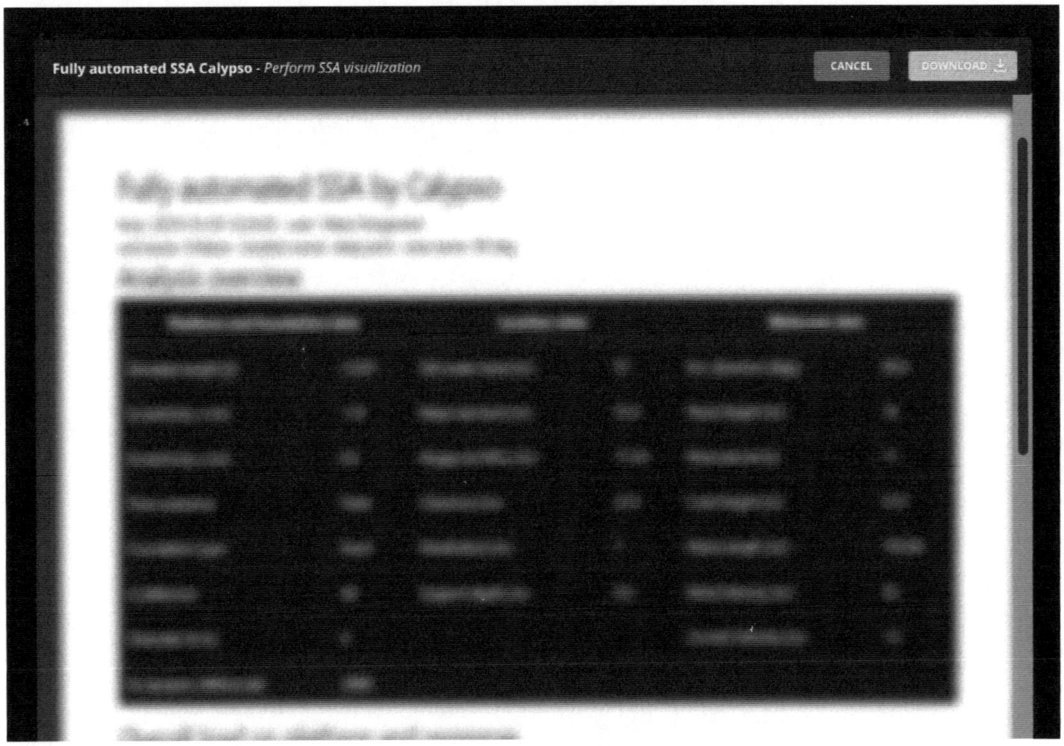

Figure 10 - Calypso Graphical User Interface - Example of report

AUTHOR INDEX

Society of Naval Architects and Marine Engineers (SNAME) Texas Section
99 Canal Center Plaza, Suite 310
Alexandria, VA 22314
USA

ISBN 978-1-7138-0813-8

9 781713 808138

CURRAN ASSOCIATES INC.
proceedings.com